COMMON LOGIC

Jon Bosak

PRENTICE HALL, Englewood Cliffs, New Jersey 07632

Library of Congress Cataloging-in-Publication Data

Bosak, Jon.
    Common logic / Jon Bosak.
        p.   cm.
    ISBN 0-13-152331-7
    1. Logic design.   2. Digital integrated circuits—Design and
construction.   I. Title.
TK7868.L6B67   1990   89-36469
621.381'5—dc20   CIP

Editorial/production supervision
   and interior design: MARY P. ROTTINO
Cover design: BRUCE KENSELAAR
Manufacturing buyer: RAY SINTEL

© 1990 by Prentice-Hall, Inc.
A Division of Simon & Schuster
Englewood Cliffs, New Jersey 07632

The publisher offers discounts on this book when ordered
in bulk quantities. For more information, write:

Special Sales/College Marketing
College Technical and Reference Division
Prentice-Hall
Englewood Cliffs, New Jersey 07632

Printed in the United States of America

10 9 8 7 6 5 4 3 2 1

ISBN 0-13-152331-7

PRENTICE-HALL INTERNATIONAL (UK) LIMITED, *London*
PRENTICE-HALL OF AUSTRALIA PTY. LIMITED, *Sydney*
PRENTICE-HALL CANADA INC., *Toronto*
PRENTICE-HALL HISPANOAMERICANA, S.A., *Mexico*
PRENTICE-HALL OF INDIA PRIVATE LIMITED, *New Delhi*
PRENTICE-HALL OF JAPAN, INC., *Tokyo*
SIMON & SCHUSTER ASIA PTE. LTD., *Singapore*
EDITORA PRENTICE-HALL DO BRASIL, LTDA., *Rio de Janeiro*

for Robert Bosak
who wondered how we would update it

Contents

Introduction

Common Logic is a new information access and decision-making tool for the digital electronics engineer. In one package it gives the designer the essential information, both functional and electrical, needed to make specific choices among hundreds of the most commonly used commercial integrated circuits. This information has been carefully arranged for maximum speed of access and ease of comparison between different types and different versions. At the same time, *Common Logic* provides a concise view of digital logic devices for nonengineers engaged in work associated with electronic components, such as sales and inventory control.

THE EVOLUTION OF "COMMON LOGIC"

A few years ago this book could not have been written; the digital IC world was just too complex. Three very different technologies—TTL, ECL, and CMOS—divided the realm of digital applications into three specialized camps oriented toward different priorities.

For applications that demanded high speed above all other considerations, there was ECL (Emitter-Coupled Logic), which achieved extremely high switching speeds at the cost of enormously increased power requirements. For applications demanding very low current draw, on the other hand, there were MOS (Metal-Oxide Semiconductor) circuits, in particular CMOS (Complementary MOS) devices. These had such low power requirements that they could be run off batteries or even solar cells—but only at the price of a severe loss in speed. And for the majority of everyday logic applications, there were the various forms of

TTL (Transistor-Transistor Logic), which gave engineers an easy-to-use, straight-forward technology offering a reasonably broad range of choices in speed, power consumption, and cost without the high speed of ECL or the low current requirements of CMOS.

In theory, almost any digital logic requirement could be met by choosing the appropriate IC from among these three families. In reality, however, choice was far more restricted, because the differences in electrical requirements between the three technologies often made it impractical to mix TTL, CMOS, and ECL on the same circuit board. A decision to use any CMOS or ECL devices in place of the usual TTL tended to result in a decision to implement whole subsystems, or even the entire design, in CMOS or ECL. Once this decision had been made, it was easier from both an engineering and an inventory-control standpoint to build all products of a given application type with components of the same logic family, regardless of whether the unique characteristics of that family were appropriate to the demands placed on each component within a system.

Given the difficulty of learning and completely mastering any logic family, the outcome of this was to fragment and specialize digital engineering itself. Some designers specialized in high-speed ECL and some in low-power CMOS, while the majority stayed with middle-of-the-road TTL and made the best of its limitations.

But in the late 1970s and early 1980s, technological advances began to make possible a crucial simplification. Realizing that the incompatibility of CMOS and ECL components would forever prevent them from penetrating the TTL-dominated mainstream of digital logic applications, integrated circuit manufacturers developed new logic families that largely achieve the benefits of CMOS and ECL in forms that are compatible with TTL voltage levels.

Put simply, these new components offer the mainstream TTL designer the best features of CMOS and ECL circuits (and in fact often *are* CMOS or ECL circuits) in components that otherwise look and act like the corresponding TTL circuits. The result is that *the designer need no longer look beyond TTL-compatible devices* for the overwhelming majority of digital logic applications. It now seems likely that the clear advantages, both conceptual and financial, of using a single set of compatible circuits will eventually restrict non-TTL-compatible logic families to just a few very specialized applications.

The realization upon which this book is based is that TTL-compatible ICs have become the logic family of choice for almost everyone. In recognition of this fact we have coined the phrase "common logic" to designate *all* of those digital logic circuits that are directly compatible with TTL components, regardless of the underlying technology used in manufacturing them.

A NEW KIND OF DATA BOOK

Until now, reference materials aimed at helping the designer choose among digital logic components have reflected the basic incompatibility between major logic families. Manufacturers' data books, the primary sources of information, are usually compiled by assembling a number of individually published data sheets on one logic family in one binding. This not only creates a great deal of redundancy, but it also results in an organization of data that makes meaningful comparisons between families very difficult. The independent sources have not been able to provide a significantly better alternative. On the one hand, there are comprehensive listings the size of a metropolitan telephone directory, slow to use and dominated by manufacturer-prepared descriptions. On the other hand, there are independently produced functional overviews that explain the types of components without ever giving enough information to enable the user to actually choose one particular IC over another.

Due to the recent technological simplification noted above, however, it is now possible to provide a directory of the most widely used commercial digital ICs in a reasonable space by restricting the entries to TTL-compatible "common logic" circuits. In this edition the simplification has been carried even further, by including only those components in the group known as "the 7400 series" (or "54/74 series"), which includes the most commonly used logic and interface types. Since this is the most widely supported of all IC series, the majority of types are available from more than one manufacturer, ensuring a reliable source of supply.

FOCUSING ON THE ESSENTIALS

In addition to limiting this book to the series that contains the most useful components, several other measures have been taken to reduce bulk and speed access to needed information.

First, there are almost no references to chip manufacturing itself. In the past, discussions of integrated circuits have typically devoted a great deal of space to detailing the physical structures of these circuits and the specific manufacturing techniques used to achieve them. But such discussions become less relevant to the practicing engineer as these technologies become more numerous. In the mature phase we are entering now, concern about the underlying details of IC implementation will increasingly become limited to a small group of experts in IC manufacture. What the rest of us will need is data on the external behavior of these circuits, and that is what this book provides. The focus throughout is on logical functions, which stay the same regardless of changes in the underlying technology.

Second, redundant technical data have been drastically reduced by eliminating little-used information and collecting all the remaining data that apply to whole families into an introductory General Information section. This has made it possible to organize the important data relating to each type into a single table of key parameters. Worst-case data for each version of that type are arranged side by side for immediate comparison between different implementations.

Third, the side-by-side arrangement of performance data has been complemented by a feature unique in digital logic documentation—a complete source profile for each type. This chart shows at a glance all the available implementations of each type and the manufacturing source or sources for each implementation. This not only speeds up the search for suppliers of a particular circuit but also serves as reference to manufacturer's data; it is a kind of index to all the other sources of information on a given type.

Finally, access to specific components has been substantially improved by arranging the circuits by function rather than by the arbitrary order of the part numbers. Functional types appear in a logical order, and similar circuits are always located near one another. An extensive, detailed functional index serves as a key to the whole book and provides the shortest path for getting from a general idea of functional requirements to comparative performance data on specific types. This advanced reference mechanism is supplemented by a complete numerical index.

In sum, the streamlined arrangement adopted for *Common Logic* highlights the critical data needed to actually choose a component—its manufacturing sources and its logical, electrical, and timing characteristics—in a format that speeds access and promotes easy comparisons between logic families and related circuit types.

Though this book is designed primarily for working designers, technicians, component engineers, industrial purchasing agents, and quality control personnel, the logical order of presentation and the simplified brief discussions that introduce each related group of components make it an excellent learning tool for engineering students and a unique all-purpose reference for anyone needing practical information on basic digital integrated circuits.

A NOTE ON THE FIRST EDITION

Eventually *Common Logic* will include information on all 7400-series digital ICs. Because of the rapid introduction of new types and new logic families, however, it has been necessary to issue the first edition of this work in a form that covers only the two most important and widely used categories. In addition to the introductory section entitled General Information, the text includes Part 1: Basic Circuits (gates, buffers/drivers, latches, flip-flops, and one-shots) and Part 2: Parallel Circuits (bus drivers, bus latches, registers, bus registers, and bus transceivers). These two broad classifications include all the simple one-bit devices used for high-speed control and "hardware programmed" applications plus all the parallel I/O circuits used for storage and interface applications.

Issuing the first edition in this form was the only way we could prevent the first compilations from becoming obsolete before the entire field of logic components could be properly surveyed. As a result of this strategy, we were able to include all "standard TTL," LS, HC, HCT, ALS, S, AS, F, AC, and ACT versions of these fundamental devices for which data were publicly available when their listings were compiled, giving a view of the two most important categories that is not duplicated by any other reference work.

Future releases of *Common Logic* will cover the more specialized 7400-series circuits such as multiplexers, decoders, counters, and shift registers. For clarity and efficiency, the numerical index lists all known commercially available 7400-series devices and shows which ones are discussed in this edition.

How to Use This Book

For the most efficient use of this work you should have some understanding of how it is organized and how it was intended to be employed.

The organizational framework adopted for this book is that of logical function. The great mass of TTL-compatible logic devices has been broken up into small groups, each group consisting of devices that share a similarity of logical operation. An indication of this logical similarity is that all of the devices in each group can be summarized by just a small set of function tables and simplified functional diagrams.

The text for each of these functional groups typically begins with a nontechnical descriptive passage summarizing the shared logical function of the group. This is followed by a series of capsule descriptions of all the commercially available circuits in the category under discussion. In the case of single-source components, key parameters are given directly under the capsule description for that component. For components with multiple sources, key electrical and switching parameters are given in a subsequent set of data sheets, which are set up to allow direct comparison of both performance and availability among all the currently available versions of a given type. Between the capsule descriptions and the data pages are the function tables and simplified functional diagrams that show the essential external behavior of these devices.

THE DATA PAGES

From a purely informational standpoint, the core of this book lies in the condensed data sheets, which in one place give a summary of performance characteristics and availability for each type supported by more than one manufacturer. (For example, see the data sheet for type 04 inverters on page 71.)

At the top of each data sheet is a chart or *source profile* that graphically displays the production status of each version for each manufacturer that produces some form of that type. A solid box indicates full production, crosshatching indicates preliminary production (parameters are still subject to change), and the letter D indicates that the particular implementation does not yet exist in reality but is still in the design stage.

Below the source profile is a table of *key parameters* that lists the most important performance specifications for each version together with footnotes that detail the differences (if any) among implementations of each version produced by different manufacturers. Where a wide range of values appears in the specifications for a given parameter in versions produced by different manufacturers, as often occurs in HC and HCT components, a representative value is chosen from among the set of actual values, which are always given in a footnote below.

Each column in the source profile and in the table of key parameters is headed by the brief name of a 7400-series technology family. (These families are discussed in the General Information section.) In every case except that of the original TTL versions, the heading name is the same code used in the 7400-series part number (LS, HC, ALS, F, etc.), with "TTL" signifying "standard" TTL, which is indicated in part numbers by the absence of a letter code. When a code is followed by a solidus or "slash" and then another letter, it indicates that the part number for components in that technology normally has the specified letter as a suffix. Thus, for instance, the heading "ALS/A" on the data sheet for the 04 (hex inverters) means that these components normally have the part number "74ALS04A" rather than "74ALS04." Exceptions, if any, are noted just below the source profile. These suffixes or *change control symbols* are used by manufacturers to signal minor changes in the electrical characteristics of a given device, usually minor improvements.

The technology families are displayed across the top of each table in the following order:

TTL   LS   HC   HCT   ALS   S   AS   F   AC   ACT

This arrangement does not follow any particular logical sequence but was designed to put the categories most frequently compared in adjacent or at least neighboring columns while preserving certain groupings that would make the order easy to remember. The arrangement is roughly in order of increasing speed going from left to right. The actual order of the implementations of a typical gate (the 00 quad NAND gates) in terms of worst-case propagation delays is

TTL   HCT   HC   LS   ALS   ACT   AC   F   S   AS

For a typical set of bus drivers (the 240) the order is

[TTL]   LS   HCT   HC   ACT   AC   ALS   F   S   AS

For a typical flip-flop (the 74), in terms of its worst-case maximum clock frequency, the order is

$$TTL \quad HCT \quad HC \quad LS \quad ALS \quad S \quad F \quad AS \quad ACT \quad AC$$

The order used in this book is also roughly the order in which the families were introduced and their order in terms of speed/power ratio. The CMOS-based HC and HCT families have been placed between the TTL-based families with which they are most frequently compared, LS and ALS, while the grouping of F, AC, and ACT at the right end reflects not only their relative newness but also the fact that all three were introduced by the same manufacturer. An introduction to these logic families will be found at the beginning of the General Information section.

TWO FORMS OF ACCESS

The key to using the resources of this book most efficiently is to understand that it was designed to be equally accessible to two very different groups of readers.

The first group consists of those who are not electrical engineers (EEs) and who may have very little background in electronics, but who nevertheless need to know something about integrated logic circuits in general and on occasion need to know quite a bit about certain ICs in particular. This group includes salespersons, purchasing agents, quality assurance personnel, technical writers, and managers, as well as engineers in fields outside of electronics who need to become familiar with the electronic control circuits that are now found in many mechanical devices.

Persons without a strong background in electronics can gain a comprehensive overview of simple integrated logic circuits by reading this book straight through, as if it were a typical work of nonfiction. The reader who takes this approach will find that the deliberately nontechnical language of the descriptive material allows for easy comprehension and that the logical sequencing of types from simpler to more complex allows each new category to build on a solid foundation of understanding. For most purposes the non-EE can ignore the key parameters given for single-sourced types and can turn to the data pages briefly for an overall glance at the status of each device in terms of sources and availability, referring to the comparative performance specifications only when the occasion demands it.

The second type of reader for which this book is intended is, of course, the designer, presumed to be an electrical engineer. A special mechanism has been set up to bypass the descriptive material and provide fast access to the comparative data in which this reader will be most interested. This mechanism is the *Functional Index,* which operates like a set of pointers in a computer program. The structure of the index replaces the descriptive text for most purposes of the designer or other experienced reader, with just enough information given for each type to allow the reader to focus in on a small set of possible candidates very quickly.

The items in the Functional Index are given in the same order in which they appear in the text. On the right are two columns of page numbers. The column headed "Descr. Page" (description page) gives a direct reference to the capsule description of each type. In the case of single-sourced types (indicated by "§" after the type number), availability and key parameters are given in the capsule description, and there is no further page reference. In the case of the more widely supported types, however, the source profile and table of key parameters will be found on the data sheet whose number is given in the far right-hand column under the heading "Data Page."

In practice, designers who already have some familiarity with commercial logic circuits will do their browsing in the Functional Index, turning directly to the indicated data pages for further information. The function table and diagram for each type can be found by turning backward a few pages to the beginning of that particular set of data sheets. The same technique can be used when the type number is known, by referring to the numerical index to find the appropriate data page.

For most ordinary purposes, therefore, general users should refer to the table of contents to find the beginning of a particular category and then read *forward* from the indicated page as far as necessary to gain the level of information required. Designers and experienced users should refer to the functional index or numerical index to find the appropriate data sheet on a given type and then work their way *backward* in the text as far as necessary to gain a more generalized view.

NOTE ON "SINGLE-SOURCED" TYPES

An important practical distinction is made throughout this book between types available from only one source and types available from more than one source. Though care has been taken to provide key parameters for nearly all types, the single-sourced types (indicated by "§") are given a definite second-class status in keeping with the usual practice of most commercial designers. Excluding the single-sourced designs from the main data sections increases speed of access to the more widely supported designs and makes them easier to compare.

Since single-sourced status is so strongly differentiated here, it must be understood that *single-sourced status is only relative in this edition*. The term "single-sourced" means that *as of this writing* only one manufacturer among the manufacturers surveyed for this edition was found for the device in an implementation given a 7400-series type number.

Subsequent editions of *Common Logic* will provide more comprehensive information on second-sourcing between 7400-series components and functionally equivalent types from other series than was possible to provide in this preliminary offering. At present, the "single-sourced" designation should not be taken literally but just as a differentiator between the more widely supported types and their less widely supported brethren. Purchase decisions should never be made without first consulting with suppliers to determine the actual availability of possible candidates labeled "single-sourced" in these listings.

Functional Index of Circuits Included in This Book

For notes on using this index see the preceding "How to Use This Book" section. Status symbols sometimes attached to type numbers are:

§   single-sourced
[ ]   still in the design stage (not commercially available at this writing)
%   numbering conflict (% flags the aberrant part)

The abbreviations "OC" and "OD" stand for "open-collector" and "open-drain."

I. BASIC CIRCUITS

	Type Number	Descr. Page	Data Page
1. GATES			
Inverters (NOT gates)			
>>> SEE ALSO Chapter 4 (parallel buffers/drivers)			
for 3-state versions			
Standard inputs			
Hex, 4-20 mA outputs	04	69	71

	Type Number	Descr. Page	Data Page
OR			
2-input OR			
Quad, 4-20 mA outputs	32	117	119
Quad, 24-48 mA outputs	1032	117	119
Hex, 6-48 mA outputs	832	117	120
2-input OR with Schmitt triggers			
Quad, [4] mA outputs	[7032]	117	
3-input OR			
Triple, 4 mA outputs	4075	117	120
4-input OR			
Dual, 4 mA outputs	4072	117	121
OR/NOR			
8-input OR/NOR			
4 mA output	4078	121	122
Exclusive NOR			
2-input exclusive NOR			
Quad, 4 mA outputs	7266	124	126
Quad, 4 mA outputs	266%	124	126
Quad, 8-20 mA outputs	810§	124	
Quad, 4-8 mA OC/OD outputs	266	124	127
Quad, 8-20 mA OC outputs	811	124	127
Exclusive OR			
2-input exclusive OR			
Quad, 4-20 mA outputs	86	128	130
Quad, 4-8 mA outputs	386	128	131
Quad, 8-20 mA OC outputs	136	128	131
Exclusive OR/NOR			
4-input exclusive OR/NOR			
Quad, 20 mA outputs	135	132	133
AND-OR-invert			
2-2-input and 3-3-input AND-OR-inverts			
4-20 mA outputs	51	133	137
AND-OR-invert			
8mA output	55	133	137
2-2-2-2-input or 2-2-3-3-input AND-OR-invert			
8-16 mA output	54	133	138
2-2-3-4-input AND-OR-invert			
20 mA output	64	134	138
20 mA OC output	65	134	139
AND-OR			
2-2-input and 3-3-input AND-ORs			
4 mA outputs	58	139	140
Miscellaneous			
Complementary output gates			
Inverter/noninverters and 2-input AND/NANDs			
Quad (2+2), 16 mA outputs	265§	141	
4-input AND/NAND			
Triple, [48] mA outputs	[800]	141	

	Type Number	Descr. Page	Data Page
4-input OR/NOR			
Triple, [48] mA outputs	[802]	141	
Current-sensing interface gates			
Hex, 8 mA outputs	63§	141	

2. LATCHES AND FLIP-FLOPS

	Type Number	Descr. Page	Data Page
Quad $\overline{S}$-$\overline{R}$ (set-reset) latches			
Noninverting 4-16 mA outputs	279	145	147
Quad transparent latches (enabled in pairs)			
Complementary 4-16 mA outputs	75	148	150
Complementary 4-8 mA outputs	375	148	151
Noninverting 4-8 mA outputs	77	148	152
Dual D flip-flops			
>>> SEE ALSO Chapter 6 (registers)			
>>> SEE ALSO type 7074§ multifunction circuits at the end of this chapter			
Dual D-flip-flops with individual clocks and asynchronous presets and clears			
Complementary 4-24 mA outputs	74	153	155
Dual J-K flip-flops			
Individual clocks and asynchronous clears			
Complementary 4-16 mA outputs	73	158	162
Complementary 4-16 mA outputs	107	158	163
Individual clocks and asynchronous presets			
Complementary 4-20 mA outputs	113	158	164
Individual clocks and asynchronous presets and clears			
Complementary 4-16 mA outputs	76	158	165
Complementary 4-20 mA outputs	112	158	166
Complementary 16 mA outputs	111§	158	
Individual asynchronous presets, common clock, and common asynchronous clear			
Complementary 4-8 mA outputs	78	159	167
Complementary 4-20 mA outputs	114	159	167
Dual J-$\overline{K}$ flip-flops			
Individual clocks and asynchronous presets and clears			
Complementary 4-24 mA outputs	109	168	170
Quad J-$\overline{K}$ flip-flops			
Individual clocks and common asynchronous preset and clear			
Noninverting 16 mA outputs	276§	168	
Common clock and common asynchronous clear			
Noninverting 16 mA outputs	376§	168	
Single J-$\overline{K}$ flip-flops with gated inputs and asynchronous preset and clear			
AND-gated J1, J2, J3 inputs and AND-gated K1, K2, K3 inputs			
Complementary 16 mA outputs	72	171	173

5. PARALLEL LATCHES

6. REGISTERS

	Type Number	Descr. Page	Data Page
8-bit registers with synchronous clear			
Noninverting			
24-48 mA 3-state outputs	575	312	314
Inverting			
24-48 mA 3-state outputs	577	312	314
8-bit registers with clock enable, asynchronous clear, and triple AND-gated output controls			
Noninverting			
48 mA 3-state outputs	825	315	318
Inverting			
48 mA 3-state outputs	826	315	319
8-bit registers with readback			
Noninverting			
24 mA 3-state readback outputs	794§	319	
8-bit diagnostic registers			
Noninverting			
32 mA 3-state outputs	818	320	321
8-bit multifunction registers with synchronous true and complement load, preset, and clear			
Inverting and noninverting			
24 mA 3-state outputs	380§	321	
8/8-bit multiplexed registers			
Noninverting			
6-24 mA 3-state outputs	604	322	325
24 mA 3-state outputs	606	323	325
24 mA OC outputs	605	323	326
24 mA OC outputs	607	323	326
8/8-bit two-stage pipelined registers			
Noninverting			
32 mA 3-state outputs	548§	327	
9-bit registers with clock enable and asynchronous clear			
Noninverting			
24-48 mA 3-state outputs	823	328	331
Inverting			
48 mA 3-state outputs	824	328	332
10-bit registers			
Noninverting			
24-48 mA 3-state outputs	821	333	335
Inverting			
48 mA 3-state outputs	822	333	336

8. BUS TRANSCEIVERS

	Type Number	Descr. Page	Data Page
4-line bidirectional bus transceivers with dual complementary output enables			
Inverting			
6-64 mA 3-state outputs	242	339	341
16-24 mA 3-state outputs	1242	339	342

	Type Number	Descr. Page	Data Page
Noninverting			
32 mA 3-state outputs	546§	377	
8-line bidirectional bus transceivers with dual edge-triggered registers and register status flags			
Inverting			
24/64 mA 3-state outputs	551§	377	
Noninverting			
24/64 mA 3-state outputs	550§	377	
8-line bidirectional bus transceivers with dual edge-triggered registers, register status flags, and parity generator/checker			
Noninverting			
24/64 mA 3-state outputs	552§	378	
8-line bidirectional bus transceivers with dual edge-triggered registers and multiplexed real-time or stored-data transmission			
Dual complementary output enables			
Inverting			
6-48 mA 3-state outputs	651	380	383
24-48 mA 3-state and OC outputs	653	380	384
Noninverting			
6-48 mA 3-state outputs	652	380	385
24-48 mA 3-state and OC outputs	654	380	386
Single output-enable and direction-control inputs			
Inverting			
6-48 mA 3-state outputs	648	380	387
24-48 mA OC outputs	649	381	388
Noninverting			
6-48 mA 3-state outputs	646	381	389
24-48 mA OC outputs	647	381	390
8-bit universal transceiver port controllers (transceivers with internal parallel/serial-in, parallel/serial-out shift-right register)			
Noninverting			
48 mA 3-state outputs	852§	390	
48 mA 3-state outputs	856§	391	
48 mA 3-state outputs	877§	391	

Numerical Index of 7400-Series Logic Circuits

The following index lists all 7400-series logic types known to be in production at the time of this writing, provides each type with a short descriptive label, and, if it is listed in this edition, gives the page number of its capsule description and its data sheet, if any. Thus, a reference to this index will give at a glance the functional category of virtually any 7400-series IC and show whether the circuit is covered in this edition.

This index includes several new types, almost all of them single-sourced, whose data became available too late for inclusion in the text. It also lists some older types that are not included in this book for other reasons; for example, expandable gates, now considered obsolete. The terminology adopted for circuits not included in this edition is provisional and subject to change in future editions.

Abbreviations used in this index:

ack.	acknowledge	compl.	complementary
addr.	address	cont.	control(s)
asynch.	asynchronous	demux.	demultiplexer(s)
bidir.	bidirectional	EDAC	error detection and
CE	clock enable		correction circuit
com.	common	exp.	expandable (obsolete)

FF	flip-flop	OE	output enable
gen.	generator	par.	parallel
ind.	individual/	reg.	register(s)/registered
	independent	R	resistor(s)
inp.	input	ser.	serial
inv.	inverting	synch.	synchronous
mult.	multiple/multiuser	term.	termination
noninv.	noninverting	trans.	transceiver(s)
OC	open-collector	tridir.	tridirectional
OD	open-drain	w/	with

§   Single-sourced (among manufacturers surveyed for this edition)

[ ]   Still in the design stage (at the time of this writing)

		Descr. Page	Data Page
00	Quad 2-input NAND gates	81	85
01	Quad 2-input NAND gates with OC outputs	82	88
02	Quad 2-input NOR gates	108	110
03	Quad 2-input NAND gates with OC or OD outputs	82	87
04	Hex inverters	69	71
05	Hex inverters with OC or OD outputs	69	72
06	Hex inverting drivers with high-voltage OC outputs	69	73
07	Hex drivers with high-voltage OC or OD outputs	76	79
08	Quad 2-input AND gates	101	103
09	Quad 2-input AND gates with OC or OD outputs	102	104
10	Triple 3-input NAND gates	91	93
11	Triple 3-input AND gates	105	106
12	Triple 3-input NAND gates with OC outputs	91	94
13	Dual 4-input Schmitt-trigger NAND gates	95	98
14	Hex inverters with Schmitt-trigger inputs	69	74
15	Triple 3-input AND gates with OC outputs	105	107
16	Hex inverting drivers with high-voltage OC outputs	69	73
17	Hex drivers with high-voltage OC outputs	76	79
18	Dual 4-input Schmitt-trigger NAND gates	95	98
19	Hex inverters with Schmitt-trigger inputs	69	74
20	Dual 4-input NAND gates	94	96
21	Dual 4-input AND gates	105	107
22	Dual 4-input NAND gates with OC outputs	94	97
23§	Expandable dual 4-input NOR gates with strobe		
24	Quad 2-input Schmitt-trigger NAND gates	83	91

		Descr. Page	Data Page
25	Dual 4-input NOR gates with independent strobes	113	116
26	Quad 2-input NAND gates with OC outputs	82	88
27	Triple 2-input NOR gates	113	115
28	Quad 2-input NOR drivers	108	110
30	8-input NAND gates	98	100
31§	Delay elements (inverters, noninverters, NAND buffers)		
32	Quad 2-input OR gates	117	119
33	Quad 2-input NOR drivers with OC outputs	108	113
34	Hex noninverters	76	78
35§	Hex noninverters with OC outputs	76	
36§	Quad 2-input NOR gates	108	
37	Quad 2-input NAND drivers	81	86
38	Quad 2-input NAND drivers with OC outputs	82	89
39	Quad 2-input NAND drivers with OC outputs	82	90
40	Dual 4-input NAND drivers	94	96
41§	4-line to 10-line BCD decoders/Nixie™ drivers		
42	4-line to 10-line BCD decoders/drivers		
43§	4-line to 10-line excess-3 decoders/drivers		
44§	4-line to 10-line Gray code decoders/drivers		
45	4-line to 10-line BCD decoders/drivers with OC outputs		
46	BCD to 7-segment decoders/drivers		
47	BCD to 7-segment decoders/drivers		
48	BCD to 7-segment decoders/drivers		
49	BCD to 7-segment decoders/drivers		
50	Expandable dual 2-wide 2-input AND-OR-invert gates		
51	2-2-input and 3-3-input AND-OR-invert gates	133	137
52§	Expandable 2-2-2-3-input AND-OR gates (H only)		
53	Expandable 2-2-2-2-input or (H) 2-2-2-3-input AOI gates		
54	2-2-2-2-input or 2-2-3-3-input AND-OR-invert gates	133	138
55	4-4-input AND-OR-invert gates	133	137
56§	50-to-1 frequency dividers		
57§	60-to-1 frequency dividers		
58	2-2-input and 3-3-input AND-OR gates	139	140
60§	Dual 4-input expanders		
61§	Triple 3-input expanders (H only)		
62§	3-2-2-3-input AND-OR expanders (H only)		
63§	Hex current-sensing interface gates	141	
64	2-2-3-4-input AND-OR-invert gates	134	138
65	2-2-3-4-input AND-OR-invert gates with OC outputs	134	139
68§	Dual 4-bit BCD counters with ind. clears		
69§	Dual 4-bit binary counters with ind. clears		

		Descr. Page	Data Page
521	8-bit identity comparators		
522	8-bit identity comparators with OC output		
524§	8-bit registered comparators		
525§	Programmable counters		
[526]	Fuse-programmable identity comparators		
[527]	Fuse-programmable identity comparators		
[528]	Fuse-programmable identity comparators		
531§	8-bit 3-state latches	253	
532§	8-bit 3-state registers	303	
533	8-bit 3-state inverting latches	253	257
534	8-bit 3-state inverting registers	304	308
535§	8-bit 3-state inverting latches	253	
536§	8-bit 3-state inverting registers	304	
537§	4-line to 10-line 3-state decoders		
538§	3-line to 8-line 3-state decoders		
539§	Dual 2-line to 4-line 3-state decoders		
540	8-line 3-state inverting drivers with NOR-gated enables	232	236
541	8-line 3-state drivers with NOR-gated enables	233	238
543	8-line 3-state latched bus transceivers	374	376
544	8-line 3-state latched inverting bus transceivers	374	375
545	8-line 3-state bus transceivers	354	367
546§	8-line 3-state registered bus transceivers	377	
547§	8-line 3-state latched bus transceivers	374	
547%§	3-line to 8-line decoders with addr. latch and OC ack.		
548§	8/8-bit 3-state pipelined registers	327	
548%§	3-line to 8-line decoders with OC acknowledge		
549§	8/8-bit 3-state pipelined latches	266	
550§	8-line 3-state registered bus trans. with status flags	377	
551§	8-line 3-state reg. inv. bus trans. with status flags	377	
552§	8-line 3-state reg. bus trans. w/ status flags and parity	378	
556§	16 × 16 3-state flow-through multiplier slices		
557§	8 × 8 3-state multipliers		
558§	8 × 8 3-state multipliers		
560§	4-bit 3-state synchronous BCD counters		
561§	4-bit 3-state binary counters		
563	8-bit 3-state inverting latches	253	258
564	8-bit 3-state inverting registers	304	309
566§	8-line 3-state registered inverting bus transceivers	376	
567§	8-line 3-state latched inverting bus transceivers	374	
568	4-bit 3-state synchronous up/down BCD counters		

		Descr. Page	Data Page
993§	9-bit 3-state inverting readback latches		
994§	10-bit 3-state readback latches		
995§	10-bit 3-state inverting readback latches		
996§	8-bit 3-state inverting/noninverting readback latches		
1000	Quad 2-input NAND drivers	81	86
1002	Quad 2-input NOR drivers	108	111
1003	Quad 2-input NAND drivers with OC outputs	82	89
1004	Hex inverting drivers	69	71
1005	Hex inverting drivers with OC outputs	69	72
1008	Quad 2-input AND drivers	101	103
1010	Triple 3-input NAND drivers	91	93
[1010%]	16 × 16 multipliers/accumulators		
1011	Triple 3-input AND drivers	105	106
[1016]	16 × 16 multipliers		
[1017]	16 × 16 multipliers with common clock		
1020	Dual 4-input NAND drivers	94	97
1032	Quad 2-input OR drivers	117	119
1034	Hex drivers	76	78
1035	Hex drivers with OC outputs	76	79
1036	Quad 2-input NOR drivers	108	111
1240	4+4-line 3-state inv. drivers with active-LOW enables	215	226
1241	4+4-line 3-state drivers with complementary enables	220	231
1242	4-line 3-state inverting bus transceivers	339	342
1243	4-line 3-state bus transceivers	340	344
1244	4+4-line 3-state drivers with active-LOW enables	217	228
1245	8-line 3-state bus transceivers (= 1645)		
1620§	8-line 3-state inverting bus transceivers	351	
1621§	8-line bus transceivers with OC outputs	352	
1622§	8-line inverting bus transceivers with OC outputs	351	
1623§	8-line 3-state bus transceivers	351	
1638§	8-line 3-state/OC inverting bus transceivers	353	
1639§	8-line 3-state/OC bus transceivers	355	
1640	8-line 3-state inverting bus transceivers	352	363
1641§	8-line bus transceivers with OC outputs	354	
1642§	8-line inverting bus transceivers with OC outputs	353	
1643§	8-line 3-state inverting/noninverting bus transceivers	355	
1644§	8-line inv./non-inv. bus transceivers with OC outputs	356	
1645	8-line 3-state bus transceivers	354	366
2240§	4+4-line 3-state inv. MOS drivers with active-LOW enables	215	

General Information

TECHNOLOGY FAMILIES

Various integrated circuit technologies are used to implement the functions available in 7400-series logic. Each technology presents certain economic and performance characteristics that give it some relative advantage over other technologies. Ignoring component cost, which is beyond the scope of this book, the key performance factors that distinguish these technologies are speed, power consumption, and the ability to drive output loads. The purpose of this section is to introduce each logic family in terms of these basic performance considerations.

Currently available 7400-series technologies fall into two broad categories, one based on TTL technology and one based on CMOS. While there is a great deal of external similarly between devices belonging to these two categories, the very different underlying structures of the circuits themselves give rise to some important performance differences. Consequently, it will be convenient to introduce these categories separately, using a historical ordering within each, before turning to a general comparison of performance characteristics.

The engineer or other knowledgeable user may be more immediately served by turning directly to the data sheets for individual types given in the main body of this book. For a concise overview of key performance differences between the different families, compare the data sheets for the type 00 NAND gates (page 85), the type 74 or type 109 flip-flops (pages 155 and 170), and the type 240 3-state bus drivers (page 225) together with the comparative data in the table entitled "Global Operating Conditions and Electrical Characteristics" below (page 60).

TTL Technologies

The name "TTL" is commonly used with two somewhat different meanings, depending on the context. In its most narrow interpretation, TTL means the original gold-doped process now usually called "standard TTL," a technology that is slowly becoming obsolete. In the wider and more frequently encountered sense, TTL refers to all the families that were later derived from this technology (H, L, LS, S, ALS, AS, and F), even though the later families often use diode-transistor logic (DTL). Despite the ambiguity involved in this practice, the name "TTL" for the whole array of TTL-compatible families is too entrenched to change now.

In *Common Logic* the meaning of the term "TTL" depends on the part of the book in which it appears. In the General Information section, TTL used by itself always means the whole category of logic families that evolved from the bipolar transistor logic of the original 7400-series devices and refers especially to the shared characteristics of that category, such as power supply voltage and logic levels. When the original TTL is meant it is called "standard TTL." In the main body of the text, however, the label TTL always means the original gold-doped process itself, as opposed to LS, ALS, S, AS, or F.

The evolution of the TTL families can be divided into three phases. In each phase major improvements were made in the balance between power consumption and speed of operation.

Phase 1. Phase 1 encompasses the original TTL (transistor-transistor-logic) family and its variants, introduced in the late 1960s. In these early ICs, the transistors making up each circuit were turned on by applying a base current sufficient for worst-case conditions. However, under ordinary conditions there was much more base current than necessary, which drove the transistor to saturation. Turning off the transistor required removal of the excess base charge, resulting in considerable delay. The gold doping commonly used to speed up this process decreased the current gain.

The following performance families were offered in this original phase:

> *74 ("Standard" TTL):* The original gold-doped bipolar logic family, with maximum gate propagation delays of about 15 ns and maximum LOW-level output capability of 16 mA (48 mA in buffers/drivers). This technology is still useful for simple gates driving heavy loads. Unlike all the later 7400-series families, standard TTL is not signified in part numbers by a one-to-three-letter code, but rather by no code at all. For example, the type 04 (hex inverters) when implemented in the original gold-doped TTL has the part number 7404, not 74TTL04.

74L (Low-power TTL): An old TTL-based family in which all the internal resistances of the corresponding standard TTL circuits are increased about 10 times, resulting in lower power requirements (about 1/10 the current draw of standard TTL) but also much slower gate propagation delays (max. 60 ns, or about 4 times standard TTL), and much lower drive capability (max. LOW output 2 mA or 1/8 standard TTL). The family is now considered obsolete, and the components still available are not listed in this book.

74H (High-speed TTL): Another old TTL-based family, in this case with internal resistances lowered from those of standard TTL to provide higher speed (max. gate propagation delays around 10 ns) and somewhat higher drive (20 mA vs. 16 mA) at the cost of much higher power requirements (current draw about twice that of standard TTL). Now obsolete and not listed in this book.

Phase 2. Phase 2 comprises Schottky technologies. In the early 1970s, two new, much improved TTL families, LS and S, replaced L, H, and to a large extent standard TTL itself. In the Schottky families, an added diode (the Schottky diode) with a low forward voltage drop is connected between the base and collector of each transistor. When the transistor is switched on and begins to conduct, excess current is diverted from base input to collector output, and the transistor never reaches saturation. Consequently, when the base current is interrupted, the transistor recovers much more quickly. Since gold doping is no longer needed, the transistor achieves higher current gain and turns on faster.

74LS (Low-power Schottky): Same speed as standard TTL (max. propagation delay about 15 ns), but at a fraction of the power (about 1/5 the current draw per gate) while still maintaining a practical drive rating (max. LOW output of 8 mA, or 24 mA in buffers/drivers). The lower power consumption of LS makes possible substantial increases in component density and reductions in power supply size and cost compared to standard TTL. Lower junction temperatures increase component reliability, and lower operating currents generate less switching noise. This winning combination of speed and low power consumption has made LS the workhorse of the industry, and LS components are still the most widely supported and least expensive of all the available technologies.

74S (Schottky): Draws about half again as much current as standard TTL, but provides somewhat greater drive (20 mA, or 60 mA in buffers/drivers) and is much faster (5 ns)—fast enough, in fact, to put extreme demands on power supplies and make S versions much trickier to design with than their LS counterparts.

Phase 3. Phase 3 comprises advanced Schottky technologies. Beginning around 1980, improvements in component miniaturization resulting from techniques developed for large-scale integration—in particular, a new fabrication process in which components are isolated by an oxide layer rather than a P-N junction—reduced the size of the individual gates and junctions within each IC, enabling higher switching frequencies without increasing current requirements. This development led to the introduction of three new families: an improved version of the low-power LS family called ALS and two improved versions of the high-speed S family, AS and F. These technologies represent the current state of the art in TTL-based commercial ICs.

74ALS (Advanced Low-power Schottky): Improved version of LS, featuring speeds two to three times faster and current requirements somewhat lower than LS, while maintaining the same drive capability and equal or better input loading. Principal sources are National Semiconductor and Texas Instruments. In new TTL designs, ALS circuits are used in paths where speed is not critical to minimize power consumption, while AS or F (see below) are used for high-speed functions.

74AS (Advanced Schottky): Improved version of S, featuring speeds and drive ratings comparable to S but with substantially reduced current draw and input loading (both less than half of S). Principal sources are National Semiconductor and Texas Instruments.

74F (short for FAST or Fairchild Advanced Schottky Technology): An alternate form of advanced Schottky with performance similar, but not identical, to that of AS. Principal sources are Fairchild, Motorola, and Signetics. Like the other high-speed families, S and AS, the F group switches at speeds that require a multi-layer board for good power and ground distribution.

HCMOS Technologies

CMOS (Complementary Metal-Oxide Semiconductor) components were developed in the 1970s for applications (chiefly in battery-operated systems) where the most important requirement was low power usage. The 4000 and 14000 series of CMOS devices from RCA, Motorola, and others provided a full selection of logic functions in circuits featuring very low power consumption, nearly zero input loading, wide power supply voltage range, and high noise immunity, but never approaching the TTL devices in speed and drive capability.

While manufacturers of Schottky ICs were developing the ALS, AS, and F families, CMOS suppliers were working to develop CMOS technology into forms that could be used along with, or instead of, TTL circuits. As with the TTL families, this development can be divided into three phases.

Phase 1: 7400-series pin-compatible CMOS. Logic functions from the 7400 series implemented in CMOS forms.

74C (metal-gate CMOS): A selection of 7400-series functions implemented in ordinary CMOS. Duplicating the pinout of the 7400 types and increasing the output current to 1.75 mA (almost equivalent to the output of the old 74L series) were steps toward interchangeability, but the retention of other 4000-series characteristics, such as higher voltage levels and low speed, made these components incompatible in practice with most Schottky-based systems. They are consequently not listed in this book.

74SC (Silicon-gate CMOS): This improved version of the 74C family was a precursor of HCT logic. It is not listed in this book.

Phase 2: High-speed CMOS. Improvements in CMOS technology—specifically, reductions in gate size and junction size—combined to greatly reduce the internal capacitances that had limited the speed of earlier CMOS families. This development finally made it possible to produce 7400-series logical types having the same speed as the corresponding LS versions but with the greatly reduced current requirements and input loading of CMOS.

74HC (High-speed silicon-gate CMOS). The first CMOS family to offer viable substitutes for some LS components in Schottky-based systems, featuring speeds about ten times faster than 74C and about five times faster than 74SC. HC output buffers are larger than the corresponding C and SC circuits to handle the output currents characteristic of TTL systems; the increased currents provided by HC outputs (which have about ten times the current-handling capacity of C outputs) also reduce signal line crosstalk, which can be a problem in high-speed systems.

Compared to the corresponding LS versions, HC circuits can achieve very substantial reductions in power without sacrificing operating speed or significant amounts of drive capability. HC gate propagation delays are about the same as LS delays, about twice as long as S delays, and two to three times longer than ALS delays. While HC parts are speed-compatible with LS-based systems, however, they are not direct replacements for TTL components because of differences in logic input thresholds. Briefly, HC output voltage levels can always drive TTL inputs, but TTL output levels can't always drive HC inputs. (See "Interfacing," below.) However, TTL outputs can easily be made to drive HC by adding pull-up resistors.

With the influx of HC versions of the traditional 7400-series functions came a number of types originally belonging to the 4000 series of CMOS circuits, adding a selection of functions previously unavailable in a TTL-compatible form. These parts are identified by a 4000-series type number instead of a 7400-series type number following the 74HC designation; for example, the 74HC4538 is the 74HC implementation of the CMOS type 4538, a precision one-shot. Thus, the introduction of HC has added a number of important low-power designs to the array of types available to the designer of TTL-compatible systems. Principal sources are Motorola, National Semiconductor, RCA, SGS, Signetics, Texas Instruments, and Toshiba.

74HCT (High-speed CMOS with TTL-compatible inputs). HC devices with input circuitry that translates TTL logic levels to HC logic levels. HCT types are power-saving drop-in replacements for the corresponding LS versions of the same types. In a large number of applications where high output current levels are not needed and systems are not running at maximum speed, HCT components can be used to replace ALS, AS, S, and F devices as well. The drop-in substitution of HCT for LS devices will reduce power consumption by approximately a factor of five.

For applications in which HCMOS is driven by TTL, the advisability of choosing HCT over HC usually depends on the number of channels that would otherwise need added pull-up resistors to bring the TTL outputs up to HC input levels. On this basis, some manufacturers have concluded that the HCT approach is most appropriate to the implementation of circuits such as bus interfaces, which would need a large number of external resistors if implemented in HC. They have focused their efforts on providing just the interface types in HCT versions. Other manufacturers, such as RCA and Signetics, are emphasizing the drop-in compatibility of HCT with LS and are undertaking to provide all their high-speed CMOS circuits in both HC and HCT implementations.

Because HCT components have TTL-level inputs and HC-level outputs, they are particularly appropriate when interfacing the many output lines of a TTL microprocessor to HC logic or when embedding a high-speed TTL device in an otherwise HC-dominated system. Principal sources are RCA and Signetics, with limited offerings from Motorola, National Semiconductor, SGS, Texas Instruments, and Toshiba.

HCU (High-speed CMOS with unbuffered outputs). HC circuits with no buffer stage at the outputs. This very small assortment of types (currently consisting of just the HCU04) is designed for use in linear or high-speed oscillator applications and is not covered in this book.

Phase 3: Advanced High-speed CMOS. In recent years, two improved categories of high-speed CMOS devices have been introduced that raise performance (speed and output drive) to levels comparable to the high-speed S, AS, and F families. These types, designated AC and ACT, provide the low-power alternative to AS and F that HC and HCT provide for LS and ALS circuits.

74AC (Advanced high-speed CMOS). A CMOS family similar to HC but with greatly improved speed and drive capabilities (maximum gate propagation delays are around 8.5 ns, and outputs can source or sink 24 mA instead of 4–6 mA for HC). Principal sources are Fairchild and RCA.

74ACT (Advanced high-speed CMOS with TTL-compatible inputs). Selected AC types with input circuitry designed to accept TTL logic levels. Drop-in replacements for TTL versions in most applications. Principal sources are Fairchild and RCA.

Note.   Corresponding to the term TTL, which in the General Information section means all currently available TTL-based families, the term HCMOS is used here to refer to all the high-speed CMOS families—HC, HCT, AC, and ACT.

TTL vs. HCMOS Technologies

A decision that often confronts the designer is whether to use TTL or HCMOS components. In applications where maximum speed is the most important priority, the fastest TTL families (AS and F) still have the edge. There is likely to be some change, however, as more functions become available in AC and ACT versions. Aside from speed, the main considerations are power consumption, power-supply voltage, noise immunity, input loading, output drive capability, and protection from "latch-up" and electrostatic damage.

Power Consumption

Basing designs primarily on HCMOS rather than TTL families can result in substantially reduced power consumption and therefore lower cost, because the reduced demand can be met with simpler, less expensive power supplies and can often eliminate the need for cooling fans.

Interestingly, the reduction in power is not necessarily found in individual components, but rather at the system level. Individual HCMOS circuits can actually draw more power than the corresponding TTL versions if the switching frequency is high enough.

HCMOS gates draw practically zero current when quiescent (that is, in a stable logic state), but do draw current when switching from one state to the other. The higher the switching frequency, the more current they need and the more power they consume.

At some point the power consumption curve for an HCMOS component will cross the curve for the same type implemented in the corresponding TTL family. For simple HC gates compared with the corresponding LS gates, this crossover point occurs somewhere in the neighborhood of 1 MHz, while for more complex types the crossover frequency falls somewhere in the range of 10 to 30 MHz, which is approaching the maximum frequency at which these devices can operate.

While individual HCMOS components can draw significant currents at the highest frequencies, however, HCMOS systems almost always use substantially less power than TTL systems because of the statistical fact that only about 30 percent of the components in a typical system switch at the maximum clock frequency of that system; the average component switches at a speed considerably less than the clock frequency.

This effect favors HCMOS circuits over their TTL counterparts, because HCMOS current demand continues to fall with lower frequency (halving the frequency halves the power consumption), while current demand in TTL technologies is essentially fixed at frequencies much lower than 1 MHz. For example,

National Semiconductor estimates that an arbitrary HC system composed of 200 gates, 150 counters, and 150 full adders, with 50 pF loads at all the outputs and a supply voltage of 5 V, will in the worst case draw about an order of magnitude less power than the same system under the same conditions implemented with LS components. The savings in power (and the power cross-over frequency) increase with increasing system complexity.

Calculating expected power consumption in HCMOS devices is discussed in a separate section below. Table G.1, adapted from one published by RCA, compares the power consumption of HC and LS versions of four basic logic types at different frequencies. Power consumption in AC/T devices is comparable to that of HC/T.

Supply Voltage

The TTL families and the HCMOS TTL look-alikes (HCT and ACT) operate from the standard TTL supply voltage of nominally +5 V, which can acceptably range 5 percent above or below this value, i.e. from 4.75 V to 5.25 V, in LS and S circuits or 10 percent above or below, i.e., 4.5 V to 5.5 V, in ALS, AS, and F circuits. HC and AC, on the other hand, can operate from supply voltages ranging from 2 V to 6 V. (Speed of operation is much slower at the lower voltages. When HC devices are used in linear applications a supply of at least 3 V is recommended.)

Of course, the wider supply voltage range of HCMOS is irrelevant when these circuits are used in mixed systems, because the TTL circuits will require the standard 5 V TTL supply. But this feature does make HCMOS components ideal for use in battery backup subsystems within TTL devices, or by themselves as the primary logic components of high-speed battery-operated portable systems. Such systems can be "put to sleep" (without losing data stored in latches and registers) by turning off the system clock.

Figure G.1 shows how battery backup can be provided for an HCMOS subsystem. The battery must provide a minimum of 2 V plus one diode drop. The level shifters (HC4049 and HC4050) prevent positive input currents into the battery-supplied subsystem due to input signals greater than one diode drop above V_{CC}. Concerning this plan, RCA notes:

> If the circuit design is such that input voltages can exceed V_{CC}, then external resistors should be included to limit input currents to 2 mA. External resistors may also

TABLE G.1

	LS versions			HC versions		
	0.1 MHz	1 MHz	10 MHz	0.1 MHz	1 MHz	10 MHz
Gate	5.5 mW	5.5 mW	20 mW	0.2 mW	2 mW	20 mW
Flip-flop	10 mW	10 mW	15 mW	0.15 mW	1.5 mW	15 mW
Bus transceiver	60 mW	60 mW	90 mW	0.25 mW	2.5 mW	25 mW
4-stage counter	95 mW	95 mW	120 mW	0.24 mW	2.4 mW	24 mW

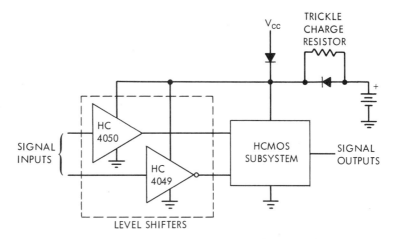

Figure G.1   Implementing an HCMOS Subsystem with battery backup.

be necessary in the output circuits to limit currents to 2 mA if the output can be pulled above V_{CC} or below GND. These currents are due to inherent V_{CC}/GND diodes that are present in all outputs, including 3-state outputs.*

Noise Margins

All digital logic circuits in commercial use implement a scheme of two-valued logic, that is, a logic in which there are two distinct values (usually named 1 and 0 or TRUE and FALSE) that are represented at the hardware level by two distinct electrical conditions. In the most commonly used scheme the two logical values are represented by two distinct voltage levels, which in the 7400 series and other TTL-compatible circuits are nominally 0 V or ground level (which usually represents FALSE or logic 0) and +5 V, that is, the V_{CC} or power supply level (which usually represents TRUE or logic 1). In this book the lower voltage level is called LOW and the higher voltage level is called HIGH.

In a perfect TTL system, there would be only 0 V and 5 V levels, with no in-between voltages to confuse the picture. In reality the LOW level is often a little above the nominal 0 V, and the HIGH level is often substantially below the nominal 5 V. In the TTL technologies, guaranteed maximum LOW output voltage over the commercial temperature range is 0.4 V to 0.5 V and guaranteed minimum HIGH voltage is 2.4 V to 2.7 V, depending on the family. (This is for standard outputs; at line driver outputs the guaranteed maximum HIGH output voltage can drop as low as 2.0 V under some conditions.)

Since a given component may "see" LOW levels ranging up to 0.5 V and HIGH levels down to 2.4 V in normal operation, its inputs must be designed to recognize these imperfect levels as representing the appropriate logic values. In

* RCA High-Speed CMOS Logic Integrated Circuits (1986), p. 27.

addition, some allowance must be made for extraneous noise (spurious voltage fluctuations) within the system.

To accommodate these real-world factors, all TTL components have inputs that recognize levels up to 0.8 V as LOW and levels down to 2.0 V as HIGH. This standard provides a LOW noise margin of 0.3 V to 0.4 V, which is the difference between the guaranteed maximum LOW output of 0.4 V to 0.5 V at the sending device and the maximum 0.8 V recognized as a LOW input by the receiving device, and it provides a similar HIGH noise margin of 0.4 V to 0.7 V, which is the difference between the guaranteed minimum HIGH output of 2.4 V to 2.7 V at the sending device and the minimum 2.0 V recognized as a HIGH input by the receiving device. As long as spurious voltage fluctuations are less than 0.4 V, therefore, they should have no effect on the operation of a TTL system.

Noise margins for HCMOS families are rather different because of their different input and output characteristics. At V_{CC} = 4.5 V (the standard supply voltage level for HCMOS specifications in this book), these circuits will recognize inputs down to 3.15 V as HIGH as up to 0.9 V (for SGS, Signetics, and Toshiba up to 1.35 V) as LOW. The 2.25 V (or 1.8 V) separation between these two input levels is substantially greater than the 1.2 V separation between the standard 0.8 V and 2.0 V worst-case levels of TTL devices.

More important, HCMOS circuits provide output voltages much closer to the theoretical ideal than do their TTL counterparts. CMOS families in general normally exhibit "rail-to-rail" output levels, that is, LOW outputs close to 0 V and HIGH outputs close to V_{CC}. Under ordinary circumstances, HCMOS outputs will be within a few millivolts of V_{CC} or ground.

The improved input and output characteristics of HCMOS devices substantially improve dc noise margins in all-HCMOS systems. At V_{CC} = 4.5 V, for example, HC devices driving a typical HC load of 0.02 mA (20 HC inputs) will put out a minimum HIGH level of 4.4 V and a maximum LOW level of 0.1 V. Taken with the standard HC input levels of 3.15 V and 0.9 V, this works out to a HIGH noise margin of 1.25 V and a LOW noise margin of 0.8 V (or 1.25 V for SGS, Signetics, and Toshiba). AC devices have the same input and output levels as the SGS, Signetics, and Toshiba HC devices, giving them the same 1.25 V HIGH and LOW margins.

Thus, an all-HC/AC system will have HIGH margins three to four times as large, and LOW noise margins two to three times as large, as the 0.5 V to 0.7 V HIGH margins and 0.3 V to 0.4 V LOW margins of TTL systems. The relatively higher built-in noise immunity of HCMOS systems gives them a real advantage over TTL systems in noisy environments.

Much of this advantage disappears, however, when HCMOS devices are used in TTL systems. For one thing, the voltage ratings of HCMOS outputs are considerably worse when providing the current needed to drive TTL inputs than when providing the few microamps needed for CMOS inputs. At their full output rating of 4 mA (6 mA for drivers) at V_{CC} = 4.5 V, HC/T outputs will provide a minimum HIGH level of 3.84 V (RCA 3.76 V, SGS and Toshiba 4.13 V) instead of

4.4 V, and a maximum LOW level of 0.33 V (RCA 0.37 V) instead of 0.1 V. Similarly, AC/T outputs provide minimum and maximum levels of 3.76 V and 0.37 V at their full output rating of 24 mA instead of the 4.4 V and 0.1 V they can provide when supplying just 20 microamps for HCMOS loads.

Thus, when driving a full array of LS inputs, for example, HCMOS outputs will have a HIGH noise margin of 1.76 V to 2.13 V (depending on the manufacturer) but a LOW margin of only 0.43 V to 0.47 V, which is no great improvement over the LS LOW margin. Some of the HCMOS input noise immunity is retained when TTL outputs are pulled up to drive HC or AC inputs, but even this is lost when HCT or ACT circuits are used, since these achieve complete TTL compatibility by adopting TTL input levels. In general, HCMOS devices used in a TTL system will be limited by the usual TTL-level noise margins.

The table of global operating conditions and electrical characteristics presented further on in this section contains voltage level data from which dc noise margins can be calculated for any combination of families.

Input Current and Fan-out

All TTL families use input structures that draw some current, though some families and some input types draw less than others. The amount of input current used by devices sharing the same input line puts a limit on fan-out, that is, on the number of devices that can be connected to the line. (See "Fan-out" below.) By contrast, HCMOS devices use virtually no current. The very small flow (less than a microamp) into these devices is due entirely to leakage currents. Consequently, a very large number of HCMOS devices (4000 or more) can in theory be driven by a single HCMOS output. The practical limiting factor for fan-out in HCMOS systems is not input current but input capacitance, which imposes no absolute limit but reduces speed to unacceptable levels if too many inputs are placed on a single line.

As pointed out above, HCT and ACT circuits have input voltage levels identical to those of LS. The input currents, however, are the same as for HC or AC. Consequently, the drop-in substitution of HCT or ACT for LS will not only greatly reduce power consumption but will also reduce input loading to nearly zero, practically eliminating the strict fan-out limitations encountered when driving LS.

Drive Capability

In general, TTL devices are larger and can put out more current than their HCMOS counterparts. At $V_{CC} = 4.5$ or 5 V, standard HC and HCT outputs can provide up to 4 mA at logic LOW or -4 mA at logic HIGH: for buffer/driver outputs these limits increase to 6 mA and -6 mA. AC and ACT outputs can provide 24 mA at LOW and -24 mA at HIGH, but they are now available in a limited selection of types. TTL circuits can typically provide from 8 mA to 64 mA of output current at LOW and up to -15 mA at HIGH, with some special-purpose drivers providing up to 200 mA.

For driving transmission lines or other high capacity loads, the TTL buffers/drivers are still the best choice. In the future, however, AC and ACT circuits, which can drive 50 Ω transmission lines over the commercial temperature range, will probably replace TTL drivers in many applications.

ESD and Latch-up Protection

The internal junctions of integrated circuits are extremely small and can be damaged by invisibly small discharges of static electricity entering the circuit during handling. This ESD (electrostatic discharge) problem becomes worse as internal components grow smaller. Since the major advances in IC technology over the last decade have come chiefly from reductions in size, the ESD threat has tended to become more acute as circuits have become more advanced, with HCMOS components the most vulnerable of all.

In recent HCMOS designs, however, added protection circuitry has largely eliminated the traditional advantage of TTL families in this regard, and some HCMOS versions are now better protected against ESD than their TTL counterparts. The best strategy, therefore, is to follow proper practices in handling all types of ICs. In particular, work stations, operators, and the parts themselves must be safely grounded at all times. (Ground circuits should include a series resistance of about 1 megohm to protect the operator in case of inadvertent contact with power current.) Detailed instructions for building and maintaining static-free work stations are included with the data published by nearly all manufacturers of HCMOS circuits.

Another problem chiefly associated with HCMOS families is latch-up. A latch-up condition occurs when outputs are subject to voltages much above $V_{CC} + 0.5$ V or much below -0.5 V. Enough current flowing through the output section under these conditions can cause the layers of the CMOS wafer to behave like an SCR (silicon-controlled rectifier). Once this has occurred, the output will stay latched on until power is removed or until uncontrolled supply current destroys the device. (In theory the inputs can also latch up if subjected to voltages above $V_{CC} +1.5$ V or below -1.5 V, but in practice the input protection circuits will usually fail before current levels sufficient to cause latch-up are reached.)

Here again, improvements in HCMOS circuit design have largely eliminated latch-up as a practical problem, but if voltage levels greater than indicated above are expected at the inputs or outputs of an HCMOS device, external protection diodes should be used to clamp the voltage. Alternatively, series resistors can be added to limit the current to a level less than the maximum rating, which is 20 mA for inputs, 25 mA for standard outputs, and 35 mA for bus-driver outputs.

HCMOS one-shots present a special latch-up problem because of large capacitors sometimes used to provide the timing constant. (See notes on this in Chapter 3, ''One-Shots.'')

INPUT TYPES

A number of standard input structures with names such as diffusion diode input, Schottky diode input, multiple-emitter input, diode cluster input, PNP input, and NPN input are used for various devices within the several 7400-series logic families. It is not unusual to find more than one kind used within the same family; in LS alone four different input structures are used for various types. The relative advantages and disadvantages of these input structures are matters of some technical interest but for the most part involve details beyond the scope of this book.

It should be noted, however, that PNP or NPN inputs can substantially reduce input loading in TTL devices, with input currents some 30 times lower than those required by the typical diode-type input structure in the logic LOW state. (In the logic HIGH state, all input types exhibit high impedance.) These transistorized inputs can make it possible for MOS devices to drive TTL inputs directly, without the added drivers that are often required to supply the current demand of conventional TTL inputs.

The presence of PNP or NPN inputs in certain types is indicated in this book in two different ways. In the one-bit gates, latches, flip-flops, and one-shots of Part I, these special inputs are rare; when they occur they are noted in the capsule description for that item. In the bus-oriented devices of Part II, however, this important modification occurs fairly often. For maximum utility and ease of comparison in Part II, the relevant parameters, the maximum input HIGH current I_{IH} (max.) and maximum input LOW current I_{IL} (max.) for each version, are given directly in the data table for each type. PNP and NPN inputs are found only in TTL families, since ordinary HCMOS inputs feature extremely high impedances to begin with.

In addition to all these normal input types, however, there is an entirely separate kind of input, the Schmitt trigger. Schmitt-trigger inputs have a much different transfer characteristic or response to changing input voltages than the one shown by conventional inputs.

In an ordinary inverter, for example, slowly raising the input from 0 V (LOW) will cause no change at first in the ordinarily HIGH output level, but when the input voltage reaches a certain level, the output will suddenly flip from HIGH to LOW. This threshold voltage, V_T, is different for different families, ranging from about 1 V to 1.5 V in TTL logic and centered on 50 percent of V_{CC} in HCMOS versions. If the input voltage is slowly lowered from V_{CC} (logic HIGH), the output will stay at low until the threshold voltage is reached, at which point the output will switch back to HIGH again. In other words, the input voltage versus output voltage curve looks much the same regardless of whether the input is going from LOW to HIGH or from HIGH to LOW (see Figure G.2).

In a Schmitt-trigger inverter, slowly raising the input voltage from 0 V will again have no effect on the output until a certain input level V_{T+} causes the output to suddenly change state. In this regard the Schmitt-trigger inverter acts much like an ordinary inverter of the same family, except that the change occurs at a

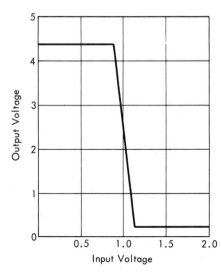

Figure G.2 Idealized transfer characteristic for ordinary LS inverter under typical conditions.

higher threshold voltage, typically about 1.6 V. Coming back the other way, however, decreasing the input voltage from V_{CC} will not cause the circuit to change back to its first state until the input reaches a substantially lower threshold level, V_{T-}, which is typically around 0.8 V. In other words, the Schmitt trigger's transfer curve and threshold voltage for inputs going from LOW to HIGH is not at all the same as its curve and threshold voltage for inputs going from HIGH to LOW (see Figure G.3).

The difference in response to inputs going in different directions is called hysteresis, and the distinctive shape formed by the upper and lower transfer

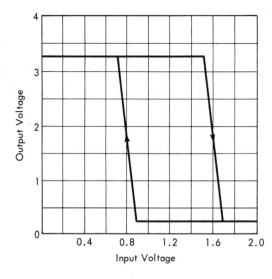

Figure G.3 Idealized transfer characteristic for Schmitt-trigger LS inverter under typical conditions.

curves is called a hysteresis loop. The amount of hysteresis is specified by the voltage difference between the two input thresholds V_{T+} and V_{T-}.

Hysteresis is an especially useful quality in circuits that must operate in electrically noisy environments. Consider the noisy, degraded input pulse shown in Figure G.4. Due to random voltage transients, the input voltage can recross the ordinary gate input threshold several times in the upward and downward course of a single pulse. In an ordinary gate this jiggling over the threshold voltage could well cause spurious logic changes at the gate's output.

In the Schmitt trigger, however, the same input will cause just one transition on the way up and one transition on the way down, because the separation between input thresholds is greater than the amplitude of the noise component in the incoming signal. The result is the regenerated square wave shown here. If the input signal is a regular train of such waves, the regenerated signal will have exactly the same frequency as the original square wave. Even a sine wave can be converted into a clean series of square waves this way. Schmitt triggers are also ideal for receiving very slow-changing signals, which would otherwise be vulnerable to system noise during their slow rise or fall past the ordinary input threshold.

In addition to such pulse-shaping applications, Schmitt triggers can be used as threshold detectors that ignore pulses less than a certain amplitude, as pulse-stretchers that can increase the width of input pulses without changing their frequency, and as the central components in elementary one-shots. (See Texas Instruments data on the 74LS14 for simple diagrams of these applications.)

The main disadvantage of Schmitt-trigger circuits is that they slow things down. For example, the 74LS04 (ordinary inverters) have a maximum propagation delay of 15 ns, while the 74LS14 (Schmitt-trigger inverters) have a maximum propagation delay of 22 ns, an almost 50 percent increase. This is the major reason that not all inputs are Schmitt triggers.

There are two levels of Schmitt-trigger capability in 7400-series circuits. True Schmitt triggers will exhibit a worst-case voltage difference of 0.4 V to 0.5 V (or a typical voltage difference of 0.8 V to 1.0 V), depending on the technology. This category is limited to the inverting Schmitt triggers 14 and 19, the 2-input NAND Schmitt triggers 132 and 24, and the 4-input NAND Schmitt triggers 13 and 18, the latter type in each pair being provided with PNP inputs to reduce input loading.

While not providing this level of noise rejection, a number of TTL bus trans-

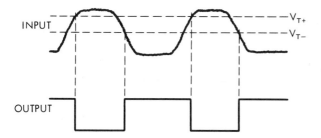

Figure G.4 Effect of Schmitt trigger on noisy input signal.

ceivers and other interface devices feature what in this book is called a "moderate" amount of Schmitt-trigger hysteresis, that is, a worst-case voltage difference of 0.2 V (or a typical voltage difference of 0.4 V) between the two input thresholds. Many clocked TTL devices have moderate levels of hysteresis at their clock inputs to decrease the likelihood of a noisy signal triggering the device.

HC and HCT versions of the 14 (Schmitt-trigger inverters) and the 132 (Schmitt-trigger NAND gates) are available from many sources; the 14 is also available in some AC and ACT versions. These HCMOS circuits provide the same amount of hysteresis as their TTL counterparts at comparable voltage levels. Generally, however, hysteresis is rarer in HCMOS versions than in the corresponding TTL versions. In particular, HCMOS bus drivers and transceivers do not have the moderate data input and clock input hysteresis often provided in TTL versions. For applications in which HCMOS devices must handle slowly changing clock inputs, one must either clean up the signal with an HC/T or AC/T Schmitt trigger (the 14 or 132) or build the device using the appropriate combination of the HC/T flip-flops 73, 74, 107, 109, or 112, which provide moderate hysteresis at their clock inputs.

Currently under development are HC implementations of a 2-input AND Schmitt trigger (the HC7001), a 2-input NOR Schmitt trigger (the HC7002), a 2-input OR Schmitt trigger (the HC7032), and a 2-input NAND Schmitt trigger with open-drain outputs (the HC7003). Check with suppliers for present availability.

OUTPUT TYPES

The function of a digital output is to provide one of two voltage levels, nominally 0 V and 5 V, to any device connected to it. In an ideal circuit, voltage levels would be the only consideration, but in reality circuits demand a certain amount of input current. This demand must be met by the outputs of the devices that are connected to those inputs. In general, the output stage must source current to a load to force its voltage to the HIGH logic level or sink current from the load to force its voltage to the LOW logic level.

HCMOS devices have high-impedance inputs that use very little current (around 1 microamp per input) and consequently put very little demand on the outputs of other devices. The inputs of TTL devices also require relatively little current (20 to 50 microamps per input) in the HIGH state. Unless provided with transistorized PNP or NPN inputs, however, the inputs of TTL devices see a LOW output from another device as a low-impedance path to ground and can require the sending device to sink a considerable amount of current, as much as 2 mA per input in some S designs. The ability of an output to meet this current demand at LOW is usually what determines how many devices of a given type can be driven successfully by that output, that is, it determines the sending device's fan-out.

In addition to determining fan-out, the supply of sink current also determines the speed at which an output load can be switched. Some output current beyond the amount needed to maintain a static load voltage must be available to charge or discharge the load capacitance.

There are three basic types of output structure used for 7400-series devices: totem-pole (push-pull), 3-state, and open-collector (or open-drain in HCMOS designs).

Totem-Pole (Push-Pull) Outputs. This is the usual output type for most applications requiring standard (non-driver) output currents. As with standard inputs, there are several variations on the basic output circuit with names such as the Darlington, the Darlington split, the Darlington with resistor to ground, and the transistor-diode; the differences between these are beyond the scope of this book. They are all variations on a single basic design in which two amplifying stages (in the simplest form, two transistors) are stacked above and below the output line as shown in Figure G.5—hence the name "totem-pole" to describe this configuration.

When the output is supposed to provide a logic HIGH, the pull-up driver is switched on and the pull-down driver is switched off by the device's control circuitry. The pull-down driver presents a high impedance, blocking the flow of current to ground, while the pull-up driver provides current as needed to pull the output load up to a logic HIGH voltage. When the output is supposed to provide a logic LOW, the upper driver is shut off, and the lower driver is turned on. This provides an open circuit to ground and pulls the output load down to a logic LOW voltage.

The drive capacity of totem-pole outputs classes them in one of two basic categories. In each logic family these outputs either have the standard drive capability associated with that logic family or are classified as buffer/driver outputs, which have current ratings 1.5 to 3 times higher than the standard outputs. For example, standard HC outputs can safely provide a maximum LOW current

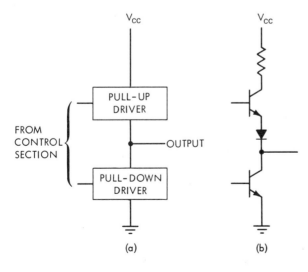

Figure G.5 (a) Simplified diagram of a totem-pole output. (b) Simplest TTL implementation (transistor-diode output).

of 4 mA and buffer/driver HC outputs a maximum LOW current of 6 mA, while standard LS outputs can provide 8 mA and buffer/driver LS outputs 24 mA of sink current. Thus, the terms "buffer" and "driver" are relative to the family; the standard outputs in some TTL families can have higher current ratings than the buffer/driver outputs in some HCMOS families.

It should be noted that HC outputs are identical to HCT outputs, and AC outputs are identical to ACT outputs; the input structures of the HCT and ACT versions differ from those of their HC and AC counterparts, but their outputs do not.

Three-State Outputs. A 3-state output is a totem-pole buffer/driver output in which both the pull-up driver and the pull-down driver can be shut off at the same time. This allows the output to achieve a third state in addition to the usual HIGH and LOW levels. When in the third state, these outputs present a high impedance to receiving devices, preventing any meaningful amount of current flow and thus effectively turning off the output signal. In the simplest terms, the 3-state feature can be thought of as an additional on/off switch placed in line between the output drivers and the output pin.

Three-state outputs are especially useful in devices in which outputs are connected to a common bus, or set of parallel transmission lines. The conflicting loads imposed by ordinary outputs would very quickly make a bus-oriented system unworkable if there were not some way of electronically disconnecting them when they are not designated to "speak" or send signals on the bus. For this reason, most of the interface circuits covered in Part II (bus drivers, bus latches, bus registers, and bus transceivers) are 3-state devices.

Typically, 3-state outputs are switched on or off the bus by a separate device input called the output enable (abbreviated OE in this book). In a few very complex devices the OE pin is eliminated, and the outputs are switched on or off the bus automatically in synchronization with the clock pulse. The enable and disable times—that is, the delays in switching 3-state devices on or off the bus—are always given here in the tables of key parameters for these types.

In designing systems where the outputs of a number of 3-state devices are tied together to a common bus, care must be taken to ensure that overlapping OE signals do not cause more than one device at a time to be taken out of the high-impedance state and put on line. Having two or more outputs active on the same line will not only cause logic errors but can actually cause physical damage by allowing destructive current flow between outputs that happen to be at opposite logic levels. Some 3-state devices are designed with output disable times shorter than enable times to minimize the possibility that such conflicts will occur when a single OE transition is used to switch one output on and another off at the same time.

When 3-state enables are activated by decoder outputs, the decoder should be disabled while its address is being changed to prevent decoding spikes from creating overlapping enable signals. Alternatively, the 3-state device can be protected by interposing an ordinary edge-triggered flip-flop like the 74 or a storage register (set of flip-flops) like the 174 or 273. Shift registers such as the 164 and 194 (not covered in this preliminary edition) can provide logical buffering for sequential enable signals.

Open-Collector or Open-Drain Outputs. Open-collector designs provide an alternate method for tying device outputs together. In an open-collector or open-drain output the pull-up driver of the totem-pole configuration is replaced by a single resistor, as shown in Figure G.6.

Here the resistor is an external component that must be added by the user. The output itself is just an open line to the collector of the pull-down transistor, or to the drain of the corresponding MOS transistor in HCMOS versions.

Open-collector outputs can be tied together or share a common bus without the overlap protection needed by 3-state devices. Of course, badly managed signal overlapping can still result in logic errors, but the devices are protected against physical damage, because there are no active pull-up drivers to supply large source currents. Also, the pull-up resistor used with open-collector outputs can pull the output to higher voltages than those usually achieved by TTL transistor pull-up drivers. The larger output voltage swing provides a larger HIGH-level noise margin, and in some special-purpose open-collector devices the higher voltage capability can be used to drive lamps or relays, which provide their own pull-up.

A final, very important advantage of open-collector outputs is that they can provide a free "wired-AND" (or negative-logic "wired-OR") function. When a number of open-collector outputs are tied together, the common output line will be LOW if any of the outputs is LOW. The pull-up resistor will be able to pull the line HIGH only if all paths to ground are cut off, that is, only if all the outputs are HIGH. This is the logic of AND, and the effect is to turn the common line into the equivalent of an extended AND gate the inputs of which are the open-collector outputs of the various devices connected to it. Such arrangements are very useful for applications such as common acknowledge lines, and open-collector circuits are often used for acknowledge and similar control outputs on complex devices that use totem-pole or 3-state circuits for their data outputs.

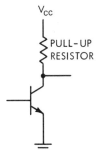

Figure G.6 TTL open-collector output (HCMOS open-drain versions are similar).

With all these advantages, open-collector outputs would no doubt find wider application if they were not much slower than standard 2-state and 3-state outputs. This significant disadvantage is another aspect of the same structural feature that gives them their other special characteristics, namely, the omission of an active pull-up driver. The passive pull-up resistor is necessarily slower in providing the current needed to raise the output to logic HIGH, and this limitation combines with a larger voltage swing in these devices to make the delay even longer. A basic open-collector gate will have maximum propagation delays 1.5 to 4 times as long (depending on the logic family) as those of the corresponding totem-pole version. Consequently, open-collector outputs are much less common in high-speed AS and F implementations than they are in the slower families.

The chief advantage of using open-collector devices for wired-AND applications is that it saves logic elements and therefore conserves power. This consideration is much less important in HCMOS systems, which in most cases can be served equally well by 3-state devices. Also, open-collector outputs have higher current ratings and a higher breakdown voltage than their open-drain counterparts. Consequently, open-drain HCMOS circuits are much less frequently encountered than open-collector TTL types. The only open-drain circuits in production at this writing are the HC/HCT03 (quad NAND gates) and the HC/HCT05 and AC/ACT05 (hex inverters), which are used primarily for level shifting; the HC09 (quad 2-input AND gates); and the HC266 (quad 2-input exclusive-NOR gates). Other open-drain circuits in the design stage at this writing include the HCT07 (hex buffers/drivers with high-voltage outputs) and the HC7003 (quad Schmitt-trigger NAND gates).

The value of the pull-up resistor R_L in wired-AND applications depends chiefly on the number of open-collector outputs acting as the inputs of the AND tie and the number of devices being driven by the AND tie. In practice the upper and lower limits are established, and then a suitable value is chosen from this range.

The minimum value that can be used for the pull-up resistor in a particular wired-AND circuit is the resistance sufficient to keep V_{CC} from pulling the wired-AND above the upper LOW-level voltage limit, even in the worst case when only one of the open-collector outputs is LOW. This minimum resistance, R_{MIN}, is given by the formula

$$R_{MIN} = \frac{V_{CC}(\text{max.}) - V_{OL}(\text{max.})}{I_{OL}(\text{max.}) - N\,[I_{IL}(\text{max.})]}$$

where

$V_{CC}(\text{max.})$ = maximum supply voltage

$V_{OL}(\text{max.})$ = maximum LOW voltage at open-collector outputs

$I_{OL}(\text{max.})$ = maximum LOW-level current at each open-collector output connected to the AND tie

N   = number of input loads being driven by the AND tie

I_{IL}(max.)   = maximum input LOW current of each input driven by AND tie (so the expression $N [I_{IL}(max.)]$ is the cumulative maximum LOW input current of all the devices driven by the AND tie).

This formula assumes that all the open-collector outputs belong to a single logic family, and all the driven inputs belong to a single logic family. In cases where families are mixed, the formula should be modified to ensure that the appropriate worst-case value appears for each term. For example, I_{OL}(max.) would be the maximum LOW current rating of the open-collector output with the least amount of LOW-level drive capability, and the term shown here as $N[I_{IL}(max.)]$ would be calculated by adding together the individual I_{IL}(max.) ratings of each input driven by the AND tie. The parameters I_{IL}(max.) and I_{OL}(max.) are given on the data page for each type listed in this book, while V_{CC}(max.) and V_{OL}(max.) are given for each logic family in the table entitled "Global Operating Conditions and Electrical Characteristics" that appears later in this chapter.

The maximum value for the pull-up resistor in a particular wired-AND circuit is the highest resistance that will still allow sufficient current flow from V_{CC} to supply all the HIGH level input and output leakages, thus maintaining a proper HIGH voltage level when all the connected open-collector outputs are HIGH. This maximum resistance is given by the formula

$$R_{MAX} = \frac{V_{CC}(min.) - V_{OH}(max.)}{M [I_{OH}(max.)] + N [I_{IH}(max.)]}$$

where

V_{CC}(min.)   = minimum supply voltage

V_{OH}(max.)   = maximum HIGH voltage at open-collector outputs

M   = number of connected open-collector outputs (inputs to the AND tie)

I_{OH}(max.)   = maximum HIGH leakage current of each open-collector output (so the expression $M [I_{OH}(max.)]$ is the cumulative maximum HIGH leakage current for all outputs driving the AND tie)

N   = number of input loads being driven by the AND tie

I_{IH}(max.)   = maximum HIGH current for each input driven by the AND tie (so the expression $N [I_{IH}(max.)]$ is the cumulative maximum HIGH current for all inputs driven by the AND tie).

As before, this formula will have to be modified in obvious ways to calculate the maximum pull-up resistance for circuits that mix logic families.

Once these limits have been determined, a suitable resistance can be chosen from the allowable range based on the desired trade-off between speed and power consumption. Resistor values closest to the minimum value will give the highest speeds, while values closest to the maximum will give the lowest power dissipation.

PARAMETER DEFINITIONS

Some of the information found in manufacturers' data has been excluded from *Common Logic* to simplify and speed up access to the most important items. The following set of definitions covers only the key parameters presented in this book.

1. Currents

I_{CC} (max.)   *Supply current (max.)*—The maximum supply current flowing into the V_{CC} terminal of a package under worst-case operating conditions. For the basic circuits of Part I this parameter is broken down into I_{CCH} (supply current with output HIGH) and I_{CCL} (supply current with output LOW). Because of their very different pattern of current consumption, maximum I_{CC} is not given for HCMOS circuits; instead, operating current is calculated from the quiescent supply current (see next) and the power dissipation capacitance, C_{pd} (see below). Refer to the section entitled "HCMOS Power Dissipation" for information on calculating HCMOS power consumption.

I_{CC} (quiesc.)   *Supply current (quiescent)*—The worst-case supply current for an HCMOS device when the device is quiescent (not switching). Note that I_{CC} (quiesc.) is given per device and must be multiplied by the number of devices in a package to calculate total quiescent current demand.

I_{IH} (max.)   *Input HIGH current (max.)*—The maximum current flowing into an input when V_{OH} (min.) is applied.

I_{IL} (max.)   *Input LOW current (max.)*—The maximum current flowing into an input when V_{OL} (max.) is applied. Note that this parameter is normally expressed as a negative quantity to indicate that the current is actually flowing away from the input.

I_{OH} (max.)   *Output HIGH current (max.)*—The maximum current that can flow into an output while still maintaining output voltage at or above V_{OH} (min.). Note that this parameter is normally expressed as a negative quantity to indicate that the current is actually flowing away from the output. In the case of open-collector outputs this parameter is a small positive quantity that gives the leakage current flowing into the output when V_{OH} (max.) is applied.

I_{OL} (max.) *Output LOW current (max.)*—The maximum current that can flow into an output while still maintaining output voltage at or below V_{OL} (max.).

2. Voltages

V_{CC} *Supply voltage*—The voltage applied to the power supply pin of an integrated circuit.

V_{IH} (min.) *Minimum input HIGH voltage*—The lowest voltage guaranteed to be recognized as a logic HIGH at a device input.

V_{IL} (max.) *Maximum input LOW voltage*—The highest voltage guaranteed to be recognized as a logic LOW at a device input.

V_{OH} (min.) *Minimum output HIGH voltage*—The lowest voltage guaranteed to be maintained by an output providing the maximum output current I_{OH} (max.) at the minimum supply voltage V_{CC} (min.). For purposes of comparison, V_{CC} (min.) for HCMOS circuits is considered to be 4.5 V, which is the worst-case voltage for the newer TTL families (ALS, AS, and F). It should be remembered that V_{CC} for HCMOS devices can actually range down to 2 V, and that V_{OH} (min.) is much lower at the lower end of that range.

V_{OL} (max.) *Maximum output LOW voltage*—The highest voltage that will occur at an output sinking the maximum output current I_{OL} (max.) at the minimum supply voltage V_{CC} (min.).

3. Switching Parameters

t_{PLH} (max.) *Maximum propagation delay, LOW to HIGH*—The worst-case delay between the arrival of an active voltage transition at an input and the occurrence of a corresponding LOW-to-HIGH voltage transition at the appropriate output.

t_{PHL} (max.) *Maximum propagation delay, HIGH to LOW*—The worst-case delay between the arrival of an active voltage transition at an input and the occurrence of a corresponding HIGH-to-LOW voltage transition at the appropriate output.

t_{pd} (max.) *Maximum propagation delay*—Either t_{PHL} (max.) or t_{PLH} (max.), whichever is the worst case. Used for the more complex devices to save space in the data charts.

t_{PZH} (max.) *Maximum output enable time, Z to HIGH*—The worst-case delay between the arrival of an enabling transition at a 3-state output enable and the corresponding transition of the enabled output from Z to HIGH.

t_{PZL} (max.)   *Maximum output enable time, Z to LOW*—The worst-case delay between the arrival of an enabling transition at a 3-state output enable and the corresponding transition of the enabled output from Z to LOW.

t_{en} (max.)   *Maximum output enable time*—Either t_{PZH} (max.) or t_{PZL} (max.), whichever is the worst case. Used for the more complex devices to save space in the data charts.

t_{PHZ} (max.)   *Maximum output disable time, HIGH to Z*—The worst-case delay between the arrival of a disabling transition at a 3-state output enable and the corresponding transition of the enabled output from HIGH to Z.

t_{PLZ} (max.)   *Maximum output disable time, LOW to Z*—The worst-case delay between the arrival of a disabling transition at a 3-state output enable and the corresponding transition of the enabled output from LOW to Z.

t_{dis} (max.)   *Maximum output disable time*—Either t_{PHZ} (max.) or t_{PLZ} (max.), whichever is the worst case. Used for the more complex devices to save space in the data charts.

t_W (min.)   *Minimum pulse width*—the smallest time between successive voltage transitions (measured from one V_S point to the next) in a clock pulse or other triggering signal that will guarantee recognition as a valid pulse.

t_{SU} (min.)   *Minimum setup time*—the amount of time before the arrival of a clock or trigger transition that a logic level has to be present at an input in order to be properly recognized by the device reading the input.

t_H (min.)   *Minimum hold time*—the amount of time after the arrival of a clock or trigger transition that a logic level has to be present at an input in order to be properly recognized by the device reading the input. A negative hold time means that the signal may be removed at the indicated time before the triggering edge arrives and still be recognized as valid data. (Negative hold times are sometimes seen in "typical" operating data but are extremely rare in the worst-case figures presented in this book.)

f_{MAX} (min.)   *Minimum maximum operating frequency*—the worst-case value for the fastest speed at which a device will operate properly given clock pulses meeting the pulse width, setup, and hold requirements; that is, the maximum frequency at which the device is guaranteed to operate under all allowable operating conditions.

TABLE G.2

Logic family	Switching voltage V_S	HIGH voltage V_{OH}
TTL	1.5 V	3.0 V
LS	1.3 V	3.0 V
HC	50% V_{CC}	V_{CC}
HCT	1.3 V	3.0 V
ALS	1.3 V	3.5 V
S	1.5 V	3.0 V
AS	1.3 V	3.5 V
F	1.5 V	3.0 V
AC	50% V_{CC}	V_{CC}
ACT	1.3 V	3.0 V

4. Reference Points

Most of the parameters defined above are measured between two voltage transitions. Since voltage transitions do not take place instantaneously, delays are measured from (or to) the moment that a signal reaches a certain switching voltage V_S that lies about halfway between typical LOW and HIGH level voltages for a given logic family. In HC and AC circuits, the usual LOW and HIGH levels are ground and V_{CC}, and V_S is defined as 50 percent of V_{CC} (2.5 V if V_{CC} = 5 V). In TTL circuits the LOW voltage level is again usually taken as ground (0.3 V for ALS and AS devices), but the HIGH voltage level V_{OH} is assumed to be rather lower than V_{CC}, and V_S is reduced accordingly to 1.3 V or 1.5 V.

This difference in reference points does introduce a slight incompatibility between the timing figures given for HC and AC circuits and those given for all the others (including HCT and ACT), but the difference just makes the HC or AC rating slightly more conservative. In any case it is too small (typically less than 1–2 ns) to be of much practical importance in conservatively designed systems.

The values of V_S and V_{OH} used in measuring switching parameters for each logic family are given in Table G.2.

It should also be noted that the ends of 3-state disable delays are measured not to the usual switching voltage V_S but only to a voltage about 10 percent of V_{CC} above V_{OL} or 10 percent of V_{CC} below V_{OH} due to the different shape of the disable curve.

GLOBAL OPERATING CONDITIONS AND ELECTRICAL CHARACTERISTICS

Table G.3 shows the basic voltage and current parameters that can be specified for each family as a whole. For ease of comparison, HCMOS circuits are specified at 4.5 V to compare with the normal worst-case performance of the newer TTL

TABLE G.3 7400 SERIES GLOBAL OPERATING CONDITIONS AND ELECTRICAL CHARACTERISTICS

Parameter		TTL	LS	HC	HCT	ALS	S	AS	F	AC	ACT	Units
V_{CC} Supply voltage	(min.)	4.75	4.75	2.0^e	2.0^e	4.5	4.75	4.5	4.75	2.0^e	2.0^e	V
	(max.)	5.25	5.25	6.0^e	6.0^e	5.5	5.25	5.5	5.25	6.0^e	6.0^e	V
V_{IH} Input HIGH voltage	(min.)	2.0	2.0	3.15	2.0	2.0	2.0	2.0	2.0	3.15	2.0	V
V_{IL} Input LOW voltage	(max.)	0.8	0.8	0.9^f	0.8	0.8	0.8	0.8	0.8	1.35	0.8	V
V_{OH} Output HIGH voltage	$(min.)^a$	2.4	2.7^g	3.84^h	3.84^h	$V_{CC} - 2^g$	2.7^g	$V_{CC} - 2^g$	2.7^g	3.76	3.76	V
V_{OL} Output LOW voltage	$(max.)^b$	0.4	0.5	0.33^i	0.33^i	0.5	0.5	0.5	0.5^j	0.37	0.37	V
I_{IH} Input HIGH current	$(max.)^c$	0.04	0.02	0.001	0.001	0.02	0.05	0.02	0.02	0.001	0.001	mA
I_{IL} Input LOW current	$(max.)^d$	−1.6	−0.4	−0.001	−0.001	−0.1	−2	−0.5	−0.6	−0.001	−0.001	mA
T_A Operating free-air temperature	(min.)	0	0	−40	−40	0	0	0	0	−40	−40	°C
	(max.)	70	70	85	85	70	70	70	70	85	85	°C

[a] Assuming I_{OH} = max. (see individual tables of key parameters)

[b] Assuming I_{OL} = max. (see individual tables of key parameters)

[c] Assuming $V_{IH} = V_{OH}$ (min.)

[d] Assuming $V_{IL} = V_{OL}$ (max.). Devices with PNP or NPN inputs will have much lower I_{IL} (max.). See individual type descriptions (Part I) or individual tables of key parameters (Part II)

[e] Operating voltage range in all-HCMOS systems. For ease in comparing with TTL-based types, all other HC, HCT, AC, and ACT specifications in this chart assume V_{CC} = 4.5 V

[f] SGS, Signetics, Toshiba 1.35 V

[g] Standard outputs; at line driver outputs V_{OH} (min.) ranges as low as 2.0 V

[h] RCA 3.76 V, SGS and Toshiba 4.13 V

[i] RCA 0.37 V

[j] Standard outputs; at line driver outputs V_{OL} (max.) ranges as high as 0.55 V

families. In all-HCMOS systems the supply voltage can be as low as 2 V or as high as 6 V, and at the extremes of this range HCMOS circuits will exhibit characteristics very different from those specified in this table.

INTERFACING

The distinguishing characteristic of all the logic families designated "common logic" in this book is that a component belonging to any of the families can be interfaced with a component belonging to any of the others with the addition of no more than a pull-up resistor, and in many cases with no additional circuitry at all. In the following paragraphs each possible combination is discussed briefly.

Any TTL Family to Any Other. All TTL families are completely input- and output-compatible; that is, any TTL component (standard TTL, LS, ALS, S, AS, or F) can be connected directly to any other without additional circuitry. This does not mean, of course, that all TTL versions of a given type are completely interchangeable. In fact, these compatible families differ in many performance characteristics (input loading, drive capability, speed, power consumption) that may affect a given implementation's ability to function properly in a particular application.

Any HCMOS Family to Any Other. All the HC, HCT, AC, and ACT circuits are completely input- and output-compatible. HCT and ACT inputs are designed to accept TTL-level outputs but will also work with HC/T or AC/T outputs if necessary, though in this case the normally higher HCMOS noise margins are reduced to TTL levels.

HCMOS to TTL. HC/T and AC/T outputs are completely compatible with TTL, LS, ALS, S, AS, and F inputs, and therefore any HCMOS device can drive TTL inputs directly. However, fan-out can vary a good deal depending on which TTL family is being driven and whether HC/T outputs (maximum 4–6 mA) or AC/T outputs (maximum 24 mA) are doing the driving.

TTL to HC or AC. TTL output levels are not generally compatible with HC and AC inputs; the minimum TTL output HIGH is 2.4 V to 2.7 V, while the minimum level recognized as logic HIGH at the inputs of an HC or AC device operating from the same power supply (at worst case 4.5 V) is 3.15 V. For complete compatibility, a pull-up resistor must be connected between each TTL output and V_{CC} to raise the output voltages to the HC/AC input range. The exact value of this resistor depends on a number of factors and is calculated by a process resembling the one used to calculate pull-up resistors for open-collector wired-AND circuits.

The pull-up resistor is not the ideal solution to this problem, because the time constant formed by the added resistor in combination with load capacitance and

stray capacitance will increase propagation delays and make them less predicta-
ble. Reducing the value of R to reduce propagation delay will increase power
consumption, contrary to the main reason for using HCMOS components in the
first place. When possible, a better solution is to use HCT or ACT versions,
which are specially designed to be driven by TTL outputs.

FAN-OUT

The fan-out, or number of devices that can be driven properly from a single
output, is determined chiefly by two factors: the drive capability of the output
(amount of current it can put out while maintaining acceptable HIGH and LOW
voltage levels) and the amount of current demanded by the driven inputs. Fan-out
in any particular case is determined by first adding up all the worst-case HIGH
input currents (I_{IH} (max.) for each input) and comparing the sum with the maximum HIGH current capability I_{OH} of the output, then adding up all the worst-case
LOW input currents (I_{IL} (max.) for each input) and comparing the sum with the
maximum LOW current capability I_{OL} of the output. Sometimes, as in all-ALS
circuits, it is the HIGH (source) currents that limit the fan-out; more often it is the
LOW (sink) currents.

   To simplify these calculations, it has been customary in all-TTL designs to
express fan-out in terms of unit loads, that is, the number of inputs of a given type
that can be driven by the same type of output. Thus, standard LS outputs are said
to have a fan-out of 20, because LS output ratings are 20 times larger than worst-
case LS input demands.

   The unit load concept was a handy shortcut as long as most systems were
designed around one logic family. As the number of TTL families increased,
however, this simplification became less and less helpful, because "unit loads" in
different logic families represent different actual input current requirements. The
technique lingered on in modified form, however, with fan-out expressed in standard LS unit loads to allow comparison between different logic families. For
example, standard F inputs are said to present a load of 0.5 LS unit loads at HIGH
and 0.375 LS unit loads at LOW. Such figures are no easier to remember or work
with than the current ratings themselves.

   With the coming of HCMOS devices, which require almost no input current,
the unit load concept has become virtually meaningless. In this book, all input
and output current ratings are given directly in milliamps. Output drive ratings
are given for each device in its table of key parameters. Input ratings are given in
two different ways. For the basic circuits of Part I, the standard input currents are
listed in the table, "Global Operating Conditions and Electrical Characteristics"
above. For the more variable parallel circuits in Part II, the input current ratings
for each type are given directly in its table of key parameters to allow comparisons
on a type-by-type basis.

HCMOS POWER DISSIPATION

As noted previously, power consumption in TTL families is practically constant over most of the operating frequency range, because current requirements are dominated by the static leakage component. The current requirements for TTL versions are given directly in the table of key parameters for each type.

In HCMOS families, on the other hand, the quiescent (non-switching) leakage currents are very small, and the actual current requirements are dominated by a switching component. This frequency-dependent component must be calculated to determine the actual current and power requirements of a given part under a particular set of conditions.

Besides the frequency, the switching component of power consumption depends on power supply voltage V_{CC}, load capacitance C_L, and internal capacitance within the device. This last factor is specified by manufacturers as the power dissipation capacitance, C_{pd}. The typical power requirements for a given device can be calculated from the equation

$$P_{TOTAL} = (C_L + C_{pd})\, V_{CC}^2\, f + V_{CC}\, I_{CC}(\text{quiesc.})$$

where

$$
\begin{aligned}
f \quad &= \text{switching frequency} \\
V_{CC} \quad &= \text{supply voltage} \\
C_L \quad &= \text{external load capacitance} \\
C_{pd} \quad &= \text{power dissipation capacitance} \\
I_{CC}(\text{quiesc.}) &= \text{quiescent device current (max.).}
\end{aligned}
$$

This formula gives the power requirement for each device within a package. To calculate power consumption for the package as a whole, this quantity may simply be multiplied by the number of devices in the package, or more sophisticated weighting may be employed based on estimates of how much time each device within the package will actually spend switching.

The factors f, V_{CC}, and C_L depend on the application; C_{pd} and $I_{CC}(\text{quiesc.})$ are given in the table of key parameters for each type. Unfortunately, the C_{pd} values provided by the industry are typical rather than worst-case figures, and C_{pd} itself varies widely among implementations provided by the different manufacturers. The single C_{pd} given in the table for each type is intended merely as a representative value; the actual values given by the manufacturers are provided in the footnote for each entry.

PART I
Basic Circuits

The simple circuits covered in this section are the basic "atoms" or building blocks of nearly all digital systems, even very complex devices that integrate thousands of such circuits on a single chip. In the series of components described here, these basic circuits are deliberately kept at a very low level of integration. This puts the individual circuits (or small groups of circuits) in separate packages, which can be interconnected in an infinite variety of ways to provide the designer with maximum flexibility in addressing the requirements of a particular application.

TEST CONDITIONS FOR BASIC CIRCUITS

Every effort has been made in this book to present data in a format that allows maximum ease of comparison among different types and among versions of the same type. It should be carefully noted, however, that components implemented in different technologies are tested under different test conditions; that buffers and drivers are often tested under different conditions than circuits with normal output ratings; and that circuits with open-collector outputs are often tested under different conditions than those with either 3-state or normal 2-state outputs. These differences in measuring technique can cause small but sometimes meaningful differences in the figures being compared.

Most important, the timing characteristics (for example, propagation delays) given in the data are somewhat dependent on differences in the output load capacitance C_L, the supply voltage V_{CC}, and the ambient temperature T_A. The performance of a device with small C_L or (within limits) a higher V_{CC} will always be somewhat better than it would be with a substantially higher C_L or a lower V_{CC}, and performance at 25°C is always better than performance over the rated temperature range. Normal-output components belonging to the older technologies— TTL, LS, and S—are usually rated at C_L =15 pF, V_{CC} = 5 V, and T_A = 25°C. Those components belonging to the newer technologies—HC, HCT, ALS, AS, F, AC, and ACT—are always rated at C_L = 50 pF and V_{CC} = 4.5 V (min.) over their entire operating temperature range. Older buffers/drivers stand somewhere between these two categories, with ratings usually given for C_L = 45 or 50 pF, but only at V_{CC} = 5 V (min. 4.75 V) and 25°C.

In short, the newer families are all rated more conservatively than the older ones. Consequently the figures for TTL, LS, and S components can be compared directly with each other but must be discounted to some extent when compared with corresponding parameters for the newer devices. The amount of this difference depends on the circuit, but it is always less for buffers/drivers than for ordinary gates.

A rough idea of the difference caused by a lower load capacitance in the older families can be gained by studying the data sheets published by National Semiconductor. These often give switching characteristics for both C_L = 15 pF and C_L = 50 pF, allowing a side-by-side illustration of the effect of load capacitance for many of the TTL, LS, and S versions of these basic gates. If a decision boils down to a choice between an HC or HCT version on one hand and a TTL or LS version on the other, it will often be useful to compare the corresponding data sheets for Motorola TTL/LS and HC/HCT, National TTL/LS and HC/HCT, and SGS LS and HC. In addition to HCMOS data at the standard C_L = 50 pF and V_{CC} = 4.5 V over the operating temperature range, each of these manufacturers usually gives maximum ratings for the same HCMOS components at C_L = 15 pF and V_{CC} = 5 V at 25°C to allow a more accurate comparison between its TTL or LS offering and its corresponding HC or HCT version.

For converting HCMOS performance specifications under standard HCMOS test conditions into figures that can be compared directly with LS, Signetics suggests the following rules of thumb:

> **Propagation delays and 3-state enable times.** Subtract 2.5 ns from the published HCMOS value to correct for the difference in C_L and then multiply by 0.9 to correct for the difference in V_{CC}.
>
> **3-state disable times.** Subtract 4 ns (for LS with 5 pF load) or 2 ns (for LS with 45 pF load) from the published HCMOS value and then multiply by 0.9.
>
> **Setup and hold times.** Multiply the published HCMOS value by 0.9 to correct for the difference in V_{CC}. (Not affected by C_L.)
>
> **Operating frequencies.** Multiply the published HCMOS value by 1.1.

TABLE 1.1

		Gates		Buffers/Drivers	
		Totem-pole	OC/OD	Totem-pole	OC/OD
TTL	25°C	15 pF	15 pF	45 pF or 50 pF	45 pF or 50 pF
	5 V	400 Ω	4 kΩ (t_{PLH})	133 Ω	133 Ω
			400 Ω (t_{PHL})		
LS	25°C	15 pF	15 pF	45 pF or 50 pF	45 pF or 50 pF
	5 V	2 kΩ	2 kΩ	667 Ω	667 Ω
HC	−40°C to 85°C	50 pF	50 pF	50 pF	
	4.5 V	——	1 kΩ	——	
HCT	−40°C to 85°C	50 pF	50 pF	50 pF	
	4.5 or 5 V	——	1 kΩ	——	
ALS	0°C to 70°C	50 pF	50 pF	50 pF	50 pF
	4.5 to 5.5 V	500 Ω	500 Ω	500 Ω	667 Ω or 680 Ω
S	25°C	15 pF	15 pF	45 or 50 pF	45 or 50 pF
	5 V	280 Ω	280 Ω	93 Ω	93 Ω
AS	0°C to 70°C	50 pF	50 pF	50 pF	50 pF
	4.5 to 5.5 V	500 Ω	2 kΩ	500 Ω	680 Ω
F	0°C to 70°C	50 pF	50 pF	50 pF	50 pF
	4.75 to 5.25 V	500 Ω	500 Ω	500 Ω	500 Ω
AC	−40°C to 85°C	50 pF			
	4.5 V	500 Ω			
ACT	−40°C to 85°C	50 pF			
	4.5 V	500 Ω			

Unless otherwise noted, the basic circuits in Part I are specified under the conditions summarized in Table 1.1. The figures on the left for each family are ambient operating temperature, T_A, and power supply voltage, V_{CC}; then, for each main grouping of output varieties, the load capacitance, C_L, and load resistance, R_L.

1

Gates

Gates are the most basic of the basic circuits. Flip-flops and latches, which together form the next most important category of basic circuits, are themselves made up of gates. Gates embody the one-bit logical operations that underlie all digital electronics. In this section they have been organized according to logical function as follows:

Inverters
Noninverters
NAND gates
AND gates
NOR gates
OR gates
OR/NOR gates
Exclusive-NOR gates
Exclusive-OR gates
Exclusive-OR/NOR gates
AND-OR-invert gates
AND-OR gates

Complementary output gates

Current-sensing interface gates

All of these gates are *combinational* circuits, that is, circuits in which output conditions are completely determined by present input conditions, with no reference to any previous or subsequent state. Such circuits have no "memory;" changes in input are followed almost instantaneously by changes in output. The two events are separated by just the small time lag or *propagation delay* introduced by the workings of the circuit itself. The delay in propagating an input change from LOW to HIGH is given in the data sheets as t_{PLH} and from HIGH to LOW as t_{PHL}.

Most of the logical functions listed above are implemented in several different forms. The major characteristics that distinguish these different implementations of the same function are:

Number of inputs.   Gates with different numbers of inputs are considered to occupy different functional subcategories. Gates with more inputs come after gates with fewer inputs.

Input type.   Gates with Schmitt-trigger inputs or high-voltage level-shifting inputs are separated from and placed later than gates with standard inputs within each functional category.

Output type.   Within each functional category and input type, gates are ordered according to output type: totem-pole, open-collector or open-drain (OC/OD), 3-state. (Almost all of the 3-state devices are listed in Part II.)

Output current.   Gates with output currents higher than normal for their technology are called "buffers" or "drivers," and the circuits called "drivers" tend to have a somewhat higher rating than those of the same technology called "buffers." But the differences in current-handling ability among the different technologies are so great that the terms "gate," "buffer," and "driver" have only relative meaning. In this book the single term "buffer/driver" or sometimes just "driver" is used to refer to any gate with a higher output-current rating than is normal for its technology. Within each category, buffers/drivers are placed after the ordinary gates that have the same logical function, input type, and output type.

INVERTERS

Inverters (NOT gates) perform the logical operation of *negation:* the logic value at the output, Y, is always the logical complement of the value at the input, A. (Logic 0 is the complement of 1, and 1 is the complement of 0.) Thus, if the voltage level at A is HIGH, the voltage level at Y will be LOW, and vice versa. From an electrical standpoint, the device is inverting the input signal; it is an inverting amplifier or "inverter."

All 7400-series inverters designed to be used as independent gates are provided in packages of six, usually as 14-pin DIPs (the 4049 is provided in a 16-pin package with two pins inactive). In addition to the inverters listed here, however, there are a number of inverting 3-state parallel buffers/drivers that are covered in Chapter 4.

04   *Hex inverters.* Standard components available in nearly all technologies. HCT versions can be used as level converters for interfacing TTL or NMOS to HC.

1004   *Hex inverting buffers/drivers.* High-current versions of the 04 with I_{OL} = 24 mA (ALS) or 48 mA (AS).

05   *Hex inverters with open-collector or open-drain outputs.* OC/OD versions of the 04 allow active-high wired-AND or active-low wired-OR function but require the addition of external pull-up resistors.

1005   *Hex inverting buffers/drivers with open-collector outputs.* High-current ALS versions of the 05 with I_{OL} = 24 mA.

06, 16   *Hex inverting buffers/drivers with high-voltage open-collector outputs.* High-current (I_{OL} = 40 mA) TTL versions of the 05 with the open-collector outputs rated at 15 V (16) or 30 V (06) for interfacing with high-level MOS circuits or for driving high-current loads such as lamps and relays.

14   *Hex inverters with Schmitt-trigger inputs.* Schmitt-trigger versions of the 04 providing a minimum of 0.4 V (ALS 0.5 V) input hysteresis.

19   *Hex inverters with Schmitt-trigger inputs.* LS versions of the 14 with PNP inputs to reduce system loading (max. I_{IL} = −0.05 mA instead of −0.4 mA for the LS14).

4049   *Hex inverters with high-voltage level-shifting inputs.* HC and HCT components with modified input structure that allows input levels as high as 15 V (SGS and Toshiba 8 V), enabling the inverters to function as high-voltage to low-voltage logic-level translators while operating from the low-voltage power supply. Typically used to interface 0–15 V CMOS to TTL systems with 5 V supplies or to battery-powered HC systems with supply voltages as low as 2 V. Can also function as ordinary inverters; useful in circuits with battery backup because input voltage which is not biased by V_{CC} can be applied.

FUNCTION TABLE

INPUT	OUTPUT
A	Y
L	H
H	L

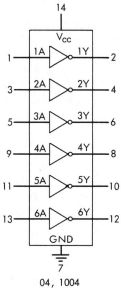

04, 1004

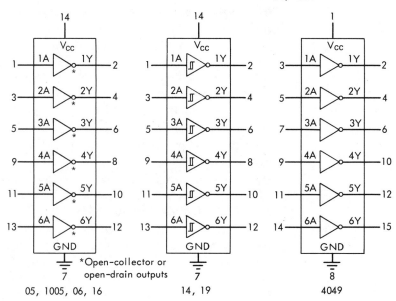

*Open-collector or open-drain outputs

05, 1005, 06, 16   14, 19   4049

70

04   Hex inverters

MANUFACTURERS	TTL	LS	HC	HCT	ALS/A	S	AS	F	AC	ACT
Fairchild										
Motorola				D						
National Semiconductor										
RCA										
SGS				D						
Signetics										
Texas Instruments										
Toshiba										

KEY PARAMETERS		TTL	LS	HC	HCT	ALS/A	S	AS	F	AC	ACT
I_{OH} (max.)	mA	-0.4[a]	-0.4	-4	-4	-0.4	-1	-2	-0.6[i]	-24	-24
I_{OL} (max.)	mA	16	8	4	4	8	20	20	20	24	24
I_{CCH} (max.)	mA	12	2.4			1.1	24	4.8	4.2		
I_{CCL} (max.)	mA	33	6.6			4.2	54	26.3	15.3		
I_{CC} (quiesc.)	mA			0.02[c]	0.02[f]					0.04	0.04
C_{pd} (typ.)	pF/gate			20[d]	20[g]					30[j]	30[i]
t_{PLH} (max.)	ns	22	15[b]	23[e]	25[h]	11	4.5	5	6.0	7.5[k]	8.8[m]
t_{PHL} (max.)	ns	15	15[b]	23[e]	25[h]	8	5	4	5.3	7[k]	8.8[m]

[a] Fairchild -0.8 mA
[b] Fairchild and National 10 ns
[c] SGS and Toshiba 0.01 mA
[d] RCA 21 pF, SGS and Toshiba 23 pF
[e] SGS 22 ns, National and TI 24 ns
[f] Toshiba 0.01 mA
[g] RCA 23 pF/gate
[h] Signetics 24 ns
[i] Motorola -1.0 mA
[j] Toshiba 18 pF/gate, RCA 105 pF/gate
[k] RCA 6 ns; Toshiba data not available
[l] RCA 115 pF/gate
[m] RCA; comparable Fairchild data not available

1004   Hex inverting buffers/drivers

MANUFACTURERS	TTL	LS	HC	HCT	ALS	S	AS	F	AC	ACT
National Semiconductor										
Texas Instruments										

Added change control symbol: National 74AS1004A

KEY PARAMETERS		TTL	LS	HC	HCT	ALS	S	AS	F	AC	ACT
I_{OH} (max.)	mA					-15		-48			
I_{OL} (max.)	mA					24		48			
I_{CCH} (max.)	mA					3		5			
I_{CCL} (max.)	mA					12		28			
t_{PLH} (max.)	ns					7		3.5[a]			
t_{PHL} (max.)	ns					6		3.5[a]			

[a] National 4 ns

05   Hex inverters with open-collector or open-drain outputs

MANUFACTURERS	TTL	LS	HC	HCT	ALS/A	S	AS	F	AC	ACT
Fairchild		■				■				
Motorola		■				■				
National Semiconductor	■	■		▨	■	■				
RCA									■	■
SGS		■								
Signetics	■	■				■				
Texas Instruments	■	■		■		■				

KEY PARAMETERS		TTL	LS	HC	HCT	ALS/A	S	AS	F	AC	ACT
I_{OH} (max.)	mA	0.25	0.1^{a}	0.005	n/a	0.1	0.25^{a}			-24	-24
I_{OL} (max.)	mA	16	8	4	4	8	20			24	24
I_{CCH} (max.)	mA	12	2.4			1.1	19.8				
I_{CCL} (max.)	mA	33	6.6			4.2	54				
I_{CC} (quiesc.)	mA			0.02	0.02					0.04	0.04
C_{pd} (typ.)	pF/gate			20	20					105	115
t_{PLH} (max.)	ns	55	32^{b}	29	25	54	7.5			d	f
t_{PHL} (max.)	ns	15	28^{c}	21	28	14	7			e	g

[a] Fairchild data not available
[b] National 20 ns
[c] National 15 ns
[d] t_{PLZ} (max.) = 7.5 ns
[e] t_{PZL} (max.) = 6 ns
[f] t_{PLZ} (max.) = 10.3 ns
[g] t_{PZL} (max.) = 8.8 ns

1005   Hex inverting buffers/drivers with open-collector outputs

MANUFACTURERS	TTL	LS	HC	HCT	ALS	S	AS	F	AC	ACT
National Semiconductor					■					
Texas Instruments					■					

KEY PARAMETERS		TTL	LS	HC	HCT	ALS	S	AS	F	AC	ACT
I_{OH} (max.)	mA					0.1					
I_{OL} (max.)	mA					24					
I_{CCH} (max.)	mA					3					
I_{CCL} (max.)	mA					12					
t_{PLH} (max.)	ns					30					
t_{PHL} (max.)	ns					10					

72

06   Hex inverting buffers/drivers with 30 V open-collector outputs

MANUFACTURERS	TTL	LS	HC	HCT	ALS	S	AS	F	AC	ACT
National Semiconductor	■									
Signetics										
Texas Instruments										

KEY PARAMETERS		TTL	LS	HC	HCT	ALS	S	AS	F	AC	ACT
I_{OH} (max.)	mA	0.25									
I_{OL} (max.)	mA	40									
I_{CCH} (max.)	mA	48[a]									
I_{CCL} (max.)	mA	51[b]									
t_{PLH} (max.)	ns	15[c]									
t_{PHL} (max.)	ns	23[c]									

[a] National 42 mA   [c] C_L = 15 pF, R_L = 110 Ωs
[b] National 38 mA

16   Hex inverting buffers/drivers with 15 V open-collector outputs

MANUFACTURERS	TTL	LS	HC	HCT	ALS	S	AS	F	AC	ACT
National Semiconductor	■									
Signetics										
Texas Instruments										

KEY PARAMETERS		TTL	LS	HC	HCT	ALS	S	AS	F	AC	ACT
I_{OH} (max.)	mA	0.25									
I_{OL} (max.)	mA	40									
I_{CCH} (max.)	mA	48[a]									
I_{CCL} (max.)	mA	51[b]									
t_{PLH} (max.)	ns	15[c]									
t_{PHL} (max.)	ns	23[c]									

[a] National 42 mA   [c] C_L = 15 pF, R_L = 110 Ωs
[b] National 38 mA

14   Hex inverters with Schmitt-trigger inputs

MANUFACTURERS	TTL	LS	HC	HCT	ALS	S	AS	F	AC	ACT
Fairchild								D		
Motorola										
National Semiconductor										
RCA									D	D
SGS										
Signetics										
Texas Instruments										
Toshiba										

KEY PARAMETERS		TTL	LS	HC	HCT	ALS	S	AS	F	AC	ACT
I_{OH} (max.)	mA	-0.8	-0.4	-4	-4	-0.4			-1	-24	-24
I_{OL} (max.)	mA	16	8	4	4	8			20	24	24
I_{CCH} (max.)	mA	36	16			12			22		
I_{CCL} (max.)	mA	60	21			12			32		
I_{CC} (quiesc.)	mA			0.02[a]	0.02					0.04	0.04
C_{pd} (typ.)	pF/gate			27[b]	20[e]						
t_{PLH} (max.)	ns	22	22	31[c]	39[f]	h			7.5	11	h
t_{PHL} (max.)	ns	22	22	31[c]	39[g]	h			8.0	9.5	h
Input Hyst. (min.)	V	0.4	0.4	0.4[d]	0.4	0.5			0.4	0.4	0.4

[a] SGS and Toshiba 0.01 mA

[b] Motorola, National; Signetics 7 pF/gate, RCA and TI 20 pF/gate, SGS 31 pF/gate

[c] Motorola, National, TI, Toshiba; SGS 34 ns, Signetics 39 ns

[d] Toshiba 0.36 V

[e] RCA; Signetics 8 pF/gate

[f] Signetics; RCA 34 ns

[g] Signetics; RCA 48 ns

[h] Maximum value not available

19   Hex inverters with Schmitt-trigger inputs

MANUFACTURERS	TTL	LS	HC	HCT	ALS	S	AS	F	AC	ACT
SGS										
Texas Instruments										

KEY PARAMETERS		TTL	LS	HC	HCT	ALS	S	AS	F	AC	ACT
I_{OH} (max.)	mA		-0.4								
I_{OL} (max.)	mA		8								
I_{CCH} (max.)	mA		18								
I_{CCL} (max.)	mA		30								
t_{PLH} (max.)	ns		20								
t_{PHL} (max.)	ns		30								
Input Hyst. (min.)	V		0.4								

4049	Hex inverters with high-voltage level-shifting inputs

MANUFACTURERS	TTL	LS	HC	HCT	ALS	S	AS	F	AC	ACT
Motorola			D							
National Semiconductor			■							
RCA			■							
SGS			■							
Signetics			■							
Texas Instruments			D							
Toshiba			■							

KEY PARAMETERS		TTL	LS	HC	HCT	ALS	S	AS	F	AC	ACT
I_{OH} (max.)	mA			-4^a							
I_{OL} (max.)	mA			4^a							
I_{CC} (quiesc.)	mA			0.02^b							
C_{pd} (typ.)	pF/gate			25^c							
t_{PLH} (max.)	ns			21^d							
t_{PHL} (max.)	ns			21^d							

[a] SGS and Toshiba have I_{OH} = -6 mA and I_{OL} = 6 mA   [d] RCA and Signetics; National 20 ns, SGS 24 ns, To-
[b] SGS and Toshiba 0.01 mA   shiba 25 ns
[c] Signetics 14 pF/gate, RCA 35 pF/gate

NONINVERTERS

Since the NOT operator (see inverters, above) is its own inverse, two NOT operations performed in sequence will leave the original logic value unchanged. If a pair (or any even number) of inverters are connected in series, the output of the resulting device will always have the same logic value as its input. Though it does not affect the logic, this arrangement does introduce certain electrical changes and causes a slight time lag between input and output due to the combined propagation delays of the inverters themselves. Commercial noninverters are inverters to which a further inverting stage has been added in order to produce a noninverted output. These circuits are typically used to provide buffering or level-translating functions without changing logic values.

In general, noninverting circuits tend to be more complex than simple inverters. Therefore, they operate more slowly and need more supply current than the inverting versions from which they are derived. In this book the secondary status of the basic noninverting components (noninverters, AND gates, OR gates, etc.) is indicated by their placement after the corresponding inverting versions (inverters, NAND gates, NOR gates, etc.)

All 7400-series noninverters designed to be used as independent gates are provided in packages of six, usually as 14-pin DIPs (the 4050 is provided in a 16-pin package with two pins inactive). In addition, there are many noninverting

3-state buffers/drivers intended to be used as bus drivers; these are listed in Chapter 4.

34   *Hex noninverters.* Standard HCT and AS noninverters. HCT versions can be used as level converters for interfacing TTL or NMOS to HC.

7007§   *Hex noninverters.* An HCT version of the 34. (Toshiba)

I_{OH} (max.)	-4 mA	t_{PLH} (max.)	33 ns
I_{OL} (max.)	4 mA	t_{PHL} (max.)	33 ns
I_{CC} (quiesc.)	0.01 mA		
C_{pd} (max.)	28 pF/gate		

1034   *Hex buffers/drivers.* High-current versions of the 34 with $I_{OL} = 24$ mA (ALS) or 48 mA (AS).

35§   *Hex noninverters with open-collector outputs.* ALS OC version of the 34 that allows "wired-AND" function but requires the addition of external pull-up resistors. (Texas Instruments)

I_{OH} (max.)	0.1 mA	t_{PLH} (max.)	*
I_{OL} (max.)	8 mA	t_{PHL} (max.)	†
I_{CCH} (max.)	1 mA		
I_{CCL} (max.)	3.5 mA		

* Max. value not available; typical is 25 ns

† Max. value not available; typical is 8 ns

1035   *Hex buffers/drivers with open-collector outputs.* High-current ALS versions of the 35§ with $I_{OL} = 24$ mA.

07, 17   *Hex buffers/drivers with high-voltage open-collector outputs.* High-current ($I_{OL} = 40$ mA) TTL versions of the 35§ with the open-collector outputs rated at 15 V (17) or 30 V (07) for interfacing with high-level MOS circuits or driving high-current loads such as lamps and relays.

4050   *Hex noninverters with high-voltage level-shifting inputs.* HC and HCT components with modified input structure that allows input levels as high as 15 V (SGS and Toshiba: 8 V), enabling the gates to function as high-voltage to low-voltage logic-level translators while operating from the low-voltage power supply. Typically used to interface 0–15 V CMOS to TTL systems with 5 V supplies or to battery-powered HC systems with supply voltages as low as 2 V. Can also function as ordinary noninverters (low-current buffers); useful in circuits with battery back-up because input voltage not biased by V_{CC} can be applied.

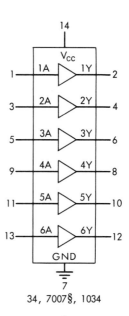

34, 7007§, 1034

FUNCTION TABLE

INPUT	OUTPUT
A	Y
L	L
H	H

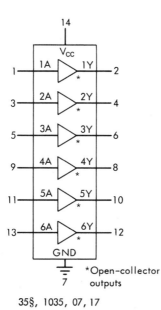

35§, 1035, 07, 17

*Open-collector outputs

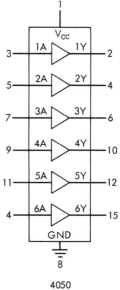

4050

34   Hex noninverters

MANUFACTURERS	TTL	LS	HC	HCT	ALS	S	AS	F	AC	ACT
National Semiconductor				▓			■			
Texas Instruments					■		■			

KEY PARAMETERS		TTL	LS	HC	HCT	ALS	S	AS	F	AC	ACT
I_{OH} (max.)	mA				-4	-0.4		-2			
I_{OL} (max.)	mA				4	8		20			
I_{CCH} (max.)	mA					5		12			
I_{CCL} (max.)	mA					8		34.6			
I_{CC} (quiesc.)	mA				0.02						
C_{pd} (typ.)	pF/gate				20						
t_{PLH} (max.)	ns				29	15		5.5			
t_{PHL} (max.)	ns				29	10		6			

1034   Hex buffers/drivers

MANUFACTURERS	TTL	LS	HC	HCT	ALS	S	AS/A	F	AC	ACT
National Semiconductor					■		■			
Texas Instruments					■		■			

KEY PARAMETERS		TTL	LS	HC	HCT	ALS	S	AS/A	F	AC	ACT
I_{OH} (max.)	mA					-15		-48			
I_{OL} (max.)	mA					24		48			
I_{CCH} (max.)	mA					6		15[a]			
I_{CCL} (max.)	mA					14		35[b]			
t_{PLH} (max.)	ns					8		6[c]			
t_{PHL} (max.)	ns					8		6[c]			

[a] National 14 mA   [c] National 5 ns
[b] National 33 mA

1035   Hex buffers/drivers with open-collector outputs

MANUFACTURERS	TTL	LS	HC	HCT	ALS	S	AS	F	AC	ACT
National Semiconductor					■					
Texas Instruments					■					

KEY PARAMETERS		TTL	LS	HC	HCT	ALS	S	AS	F	AC	ACT
I_{OH} (max.)	mA					0.1					
I_{OL} (max.)	mA					24					
I_{CCH} (max.)	mA					6					
I_{CCL} (max.)	mA					14					
t_{PLH} (max.)	ns					30					
t_{PHL} (max.)	ns					12					

07   Hex buffers/drivers with 30 V open-collector outputs

MANUFACTURERS	TTL	LS	HC	HCT	ALS	S	AS	F	AC	ACT
National Semiconductor	■									
SGS				D						
Signetics	■									
Texas Instruments	■									

KEY PARAMETERS		TTL	LS	HC	HCT	ALS	S	AS	F	AC	ACT
I_{OH} (max.)	mA	0.25			[-6]						
I_{OL} (max.)	mA	40			[6]						
I_{CCH} (max.)	mA	41									
I_{CCL} (max.)	mA	30									
t_{PLH} (max.)	ns	10^a			n/a						
t_{PHL} (max.)	ns	30^a			n/a						

$^a C_L = 15\,\text{pF}, \; R_L = 110\,\Omega$

17   Hex buffers/drivers with 15 V open-collector outputs

MANUFACTURERS	TTL	LS	HC	HCT	ALS	S	AS	F	AC	ACT
National Semiconductor	■									
Signetics	■									
Texas Instruments	■									

KEY PARAMETERS		TTL	LS	HC	HCT	ALS	S	AS	F	AC	ACT
I_{OH} (max.)	mA	0.25									
I_{OL} (max.)	mA	40									
I_{CCH} (max.)	mA	41									
I_{CCL} (max.)	mA	30									
t_{PLH} (max.)	ns	10^a									
t_{PHL} (max.)	ns	30^a									

[a] C_L = 15 pF, R_L = 110 Ω

4050   Hex noninverters with high-voltage level-shifting inputs

MANUFACTURERS	TTL	LS	HC	HCT	ALS	S	AS	F	AC	ACT
Motorola			D							
National Semiconductor			■							
RCA			■							
SGS			■							
Signetics			■							
Texas Instruments			D							
Toshiba			■							

KEY PARAMETERS		TTL	LS	HC	HCT	ALS	S	AS	F	AC	ACT
I_{OH} (max.)	mA			-4^a							
I_{OL} (max.)	mA			4^a							
I_{CC} (quiesc.)	mA			0.02^b							
C_{pd} (typ.)	pF/gate			25^c							
t_{PLH} (max.)	ns			21^d							
t_{PHL} (max.)	ns			21^d							

[a] SGS and Toshiba have I_{OH} = -6 mA and I_{OL} = 6 mA
[b] SGS and Toshiba 0.01 mA
[c] Signetics 14 pF/gate, RCA 35 pF/gate
[d] RCA and Signetics; National 20 ns, SGS 24 ns, Toshiba 25 ns

NAND GATES

A NAND gate is a device that normally provides a HIGH output except in the special case where all its inputs are HIGH, in which case its output is LOW. It is the inverting form of the AND gate, whose input is normally LOW except in the same special case where all its inputs are HIGH. In other words, the NAND gate signals the fact that its inputs have just all attained the HIGH state by a low-going transition at its output. Conversely, a high-going transition at the output signals that one or more of its inputs has just broken this uniformity by going LOW.

Internally, NAND gates are the simplest, and therefore fastest, of the common logic devices that have more than one input. They are also the most versatile; in theory, nearly any digital logic circuit can be built out of NAND gates. Also, any NAND gate can be made into an inverter by wiring its inputs together.

Common-logic NAND gates are presently available in implementations having 2, 3, 4, 8, 12, or 13 inputs. Having more inputs means, of course, that fewer of these gates will fit in a given package. If component density is not as critical a consideration as inventory control, the most versatile NAND gate to have on hand will be the one with the most inputs, because any NAND with too many inputs for a particular application can be cut down to the right size by tying the unused inputs to logic HIGH.

For convenience, 2-input NAND gates are listed below, and 3-input, 4-input, and 8/12/13-input NAND gates are listed in later sections.

2-Input NAND Gates

2-input NAND gates and NAND buffers/drivers with totem-pole outputs:

00	*Quad 2-input NAND gates.* Standard NAND gates, available in nearly all technologies. 14-pin DIP.
8003§	*Dual 2-input NAND gates.* ALS version of the 00 in an 8-pin package containing just two gates. (Texas Instruments)

I_{OH} (max.)	-0.4 mA	t_{PLH} (max.)	11 ns
I_{OL} (max.)	8 mA	t_{PHL} (max.)	8 ns
I_{CCH} (max.)	0.43 mA		
I_{CCL} (max.)	1.5 mA		

37	*Quad 2-input NAND buffers/drivers.* High-current versions of the 00 (max. I_{OL} of 24–64 mA).
1000	*Quad 2-input NAND buffers/drivers.* ALS and AS versions of the 37.
804	*Hex 2-input NAND buffers/drivers.* Basically the 1000 in a hex package with 20 pins. Max. I_{OH} increased to -15 mA in ALS

version, and −1 option raises ALS max. I_{OL} to 48 mA. HC version available.

3037§   *Quad 2-input NAND 30-ohm transmission-line buffers/drivers.* Very-high-current F versions of the 37. Dual V_{CC} and GND pins lower lead inductance to 3 nH when both pins are used. (Signetics; under development by Fairchild)

I_{OH} (max.)	−67 mA	t_{PLH} (max.)	6.5 ns
I_{OL} (max.)	160 mA	t_{PHL} (max.)	5.5 ns
I_{CCH} (max.)	6 mA		
I_{CCL} (max.)	40 mA		

2-input NAND gates and NAND buffers/drivers with open-collector or open-drain outputs:

03   *Quad 2-input NAND gates with open-collector or open-drain inputs.* Standard OC/OD versions of the 00.

01   *Quad 2-input NAND gates with open-collector outputs.* TTL, LS, and ALS implementations of the 03 with a variant pinout.

26   *Quad 2-input NAND gates with high-voltage open-collector outputs.* TTL and LS versions of the 03 designed to interface with 12 V systems. The open-collector outputs will withstand up to 15 V while these components remain powered by their standard 5 V power supply.

38   *Quad 2-input NAND buffers/drivers with open-collector outputs.* High-current versions of the 03 (max. I_{OL} to 64 mA in F implementation).

1003   *Quad 2-input NAND buffers/drivers with open-collector outputs.* An ALS high-current version of the 03 (max. I_{OL} = 24 mA).

39   *Quad 2-input NAND buffers/drivers with open-collector outputs.* Same as the 38 but with a variant pinout; high-current versions of the 01.

3038§   *Quad 2-input NAND 30-ohm buffers/drivers with open-collector outputs.* Very-high-current F versions of the 38. Dual V_{CC} and GND pins lower lead inductance to 3 nH when both pins are used. (Signetics; under development by Fairchild)

I_{OH} (max.)	0.25 mA	t_{PLH} (max.)	12.5 ns
I_{OL} (max.)	160 mA	t_{PHL} (max.)	5.5 ns
I_{CCH} (max.)	6 mA		
I_{CCL} (max.)	40 mA		

2-input NAND gates with Schmitt-trigger inputs:

132   *Quad 2-input Schmitt-trigger NAND gates. Schmitt-trigger versions of the 00 providing a minimum of 0.4 V (ALS versions 0.5 V, S versions 0.2 V) input hysteresis.*

24   *Quad 2-input Schmitt-trigger NAND gates.* LS version of the 132 with PNP inputs to reduce system loading (I_{IL} = −0.05 mA compared to −0.4 mA for the LS132).

[7003]   *Quad 2-input Schmitt-trigger NAND gates with open-drain outputs.* An HC circuit currently in the design stage. (Texas Instruments)

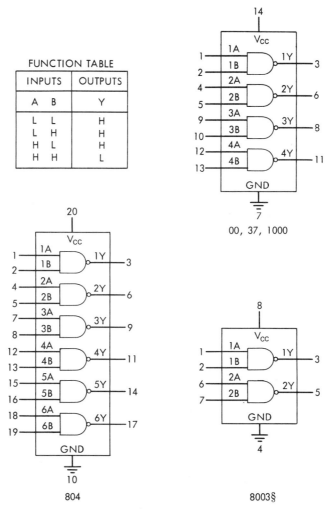

FUNCTION TABLE		
INPUTS		OUTPUTS
A	B	Y
L	L	H
L	H	H
H	L	H
H	H	L

00, 37, 1000

804

8003§

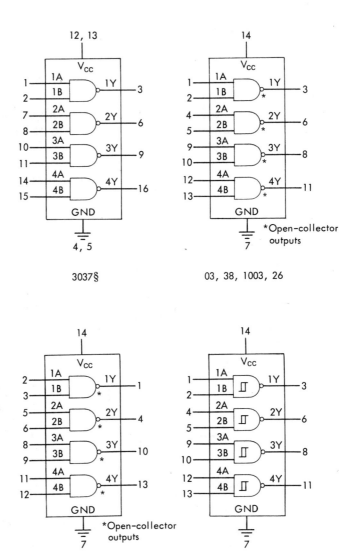

3037§

03, 38, 1003, 26

*Open-collector outputs

01, 39

*Open-collector outputs

132, 24

84

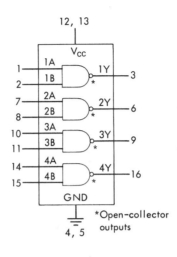

12, 13

V_{CC}

1A
1B
1Y
3

2A
2B
2Y
6

3A
3B
3Y
9

4A
4B
4Y
16

GND

*Open-collector outputs

4, 5

3038§

---

00   Quad 2-input NAND gates

MANUFACTURERS	TTL	LS	HC	HCT	ALS/A	S	AS	F	AC	ACT
Fairchild	■		■			■			■	
Motorola			░	D						
National Semiconductor	■			░						
RCA			■						░	░
SGS			░							
Signetics	■	■				■				
Texas Instruments	■	■	■							
Toshiba			■						░	

KEY PARAMETERS		TTL	LS	HC	HCT	ALS/A	S	AS	F	AC	ACT
I_{OH} (max.)	mA	-0.4[a]	-0.4	-4	-4	-0.4	-1	-2	-1	-24	-24
I_{OL} (max.)	mA	16	8	4	4	8	20	20	20	24	24
I_{CCH} (max.)	mA	8	1.6			0.85	16	3.2	2.8		
I_{CCL} (max.)	mA	22	4.4			3.0	36	17.4	10.2		
I_{CC} (quiesc.)	mA			0.02[c]	0.02					0.04	0.04
C_{pd} (typ.)	pF/gate			20[d]	20[f]					30[h]	30[j]
t_{PLH} (max.)	ns	22	15[b]	23[e]	25[g]	11	4.5	4.5	6.0	8.5[i]	9.5[k]
t_{PHL} (max.)	ns	15	15[b]	23[e]	25[g]	8	5	4	5.3	7[i]	8[k]

[a] Fairchild -0.8 mA
[b] Fairchild and National 10 ns
[c] SGS and Toshiba 0.01 mA
[d] Signetics and Toshiba 22 pF/gate
[e] SGS 22 ns
[f] Signetics 22 pF/gate

[g] National; Signetics 24 ns
[h] Toshiba 20 pF/gate, RCA 88 pF/gate
[i] RCA 6.2 ns; Toshiba data not available
[j] RCA 105 pF/gate
[k] RCA 11.8 ns

37   Quad 2-input NAND buffers/drivers

MANUFACTURERS	TTL	LS	HC	HCT	ALS/A	S	AS	F	AC	ACT
Fairchild								D		
Motorola	☐	■								
National Semiconductor	■	■			■					
SGS	☐	■								
Signetics	■	■			☐	■		■		
Texas Instruments	■	■			■					

KEY PARAMETERS		TTL	LS	HC	HCT	ALS/A	S	AS	F	AC	ACT
I_{OH} (max.)	mA	-1.2	-1.2			-2.6	-3		-15		
I_{OL} (max.)	mA	48	24			24	60		64		
I_{CCH} (max.)	mA	15.5	2			1.6	36		6		
I_{CCL} (max.)	mA	54	12			7.8	80		33		
t_{PLH} (max.)	ns	22	24[a]			8	6.5		6.5		
t_{PHL} (max.)	ns	15	24[a]			7	6.5		5		

[a] National 15 ns, Fairchild 20 ns

1000   Quad 2-input NAND buffers/drivers

MANUFACTURERS	TTL	LS	HC	HCT	ALS/A	S	AS/A	F	AC	ACT
National Semiconductor					■		■			
Texas Instruments					■		■			

KEY PARAMETERS		TTL	LS	HC	HCT	ALS/A	S	AS/A	F	AC	ACT
I_{OH} (max.)	mA					-2.6		-48			
I_{OL} (max.)	mA					24		48			
I_{CCH} (max.)	mA					1.6		3.5			
I_{CCL} (max.)	mA					7.8[a]		19			
t_{PLH} (max.)	ns					8		4			
t_{PHL} (max.)	ns					7		4			

[a] National 4.8 mA

804   Hex 2-input NAND buffers/drivers

MANUFACTURERS	TTL	LS	HC	HCT	ALS	S	AS/B	F	AC	ACT
National Semiconductor					■					
Texas Instruments			■		■	■	■			

Added change control symbol: TI 74ALS804A. ALS -1 version available from National

KEY PARAMETERS		TTL	LS	HC	HCT	ALS	S	AS/B	F	AC	ACT
I_{OH} (max.)	mA			-6		-15		-48			
I_{OL} (max.)	mA			6		24^a		48			
I_{CCH} (max.)	mA					2.5		5^d			
I_{CCL} (max.)	mA					12		27			
I_{CC} (quiesc.)	mA			0.08							
C_{pd} (typ.)	pF/gate			40							
t_{PLH} (max.)	ns			25		7^b		4^e			
t_{PHL} (max.)	ns			25		8^c		4^e			

^{a} -1 option has I_{OL} (max.) = 48 mA ^{d} National 4 mA
^{b} National 6 ns ^{e} National 3.5 ns
^{c} National 7 ns

03   Quad 2-input NAND gates with open-collector or open-drain outputs

MANUFACTURERS	TTL	LS	HC	HCT	ALS/A	S	AS	F	AC	ACT
Fairchild		■				■				
Motorola			▨							
National Semiconductor	■				■					
RCA			■							
SGS		D								
Signetics		■				■				
Texas Instruments	■				■					
Toshiba			■							

Variant change control symbol: TI 74ALS03B

KEY PARAMETERS		TTL	LS	HC	HCT	ALS/A	S	AS	F	AC	ACT
I_{OH} (max.)	mA	0.25	0.1	0.005	0.005	0.1	0.25				
I_{OL} (max.)	mA	16	8	4		8	20				
I_{CCH} (max.)	mA	8	1.6			0.85	13.2				
I_{CCL} (max.)	mA	22	4.4			3^g	36				
I_{CC} (quiesc.)	mA			0.02^c	0.02						
C_{pd} (typ.)	pF/gate			20^d	9^f						
t_{PLH} (max.)	ns	45	32^a	32^e	30	54	7.5				
t_{PHL} (max.)	ns	15	28^b	32^e	30	22	7				

^{a} National 20 ns, Fairchild 22 ns
^{b} National 15 ns, Fairchild 18 ns
^{c} Toshiba 0.01 mA
^{d} National, TI; Signetics 4 pF/gate, RCA 6.4 pF/gate, Toshiba 17 pF/gate. Motorola data not available

^{e} National; TI has t_{PLH} (max.) = 31 ns and t_{PHL} (max.) = 25 ns. Other manufacturers specify t_{PZL} (max.) = t_{PLZ} (max.) = 24 ns (Signetics) or 25 ns (RCA) or 31 ns (Toshiba) or 32 ns (Motorola)
^{f} RCA; Signetics 4 pF/gate
^{g} TI 4 mA

01   Quad 2-input NAND gates with open-collector outputs

MANUFACTURERS	TTL	LS	HC	HCT	ALS	S	AS	F	AC	ACT
Motorola		■								
National Semiconductor	■	■			■					
SGS	■	■								
Signetics		■								
Texas Instruments	■	■			■					

KEY PARAMETERS		TTL	LS	HC	HCT	ALS	S	AS	F	AC	ACT
I_{OH} (max.)	mA	0.25	0.1			0.1					
I_{OL} (max.)	mA	16	8			8					
I_{CCH} (max.)	mA	8	1.6			0.85					
I_{CCL} (max.)	mA	22	4.4			3					
t_{PLH} (max.)	ns	55[a]	32[b]			54					
t_{PHL} (max.)	ns	15	28[c]			28					

[a] National 45 ns
[b] National 20 ns
[c] National 15 ns

26   Quad 2-input NAND gates with high-voltage open-collector outputs

MANUFACTURERS	TTL	LS	HC	HCT	ALS	S	AS	F	AC	ACT
Fairchild		■								
Motorola		■								
National Semiconductor		■								
SGS	■	■								
Signetics	■	■								
Texas Instruments	■	■								

KEY PARAMETERS		TTL	LS	HC	HCT	ALS	S	AS	F	AC	ACT
I_{OH} (max.)	mA	0.05[a]	0.05[a]								
I_{OL} (max.)	mA	16	8								
I_{CCH} (max.)	mA	8	1.6								
I_{CCL} (max.)	mA	22	4.4								
t_{PLH} (max.)	ns	24	32[b]								
t_{PHL} (max.)	ns	17	28[c]								

[a] $V_O = 12$ V. At $V_O = 15$ V, $I_{OH} = 1$ mA
[b] Fairchild 22 ns
[c] Fairchild 18 ns

MANUFACTURERS	TTL	LS	HC	HCT	ALS/A	S	AS	F	AC	ACT
Fairchild		■	■					D		
Motorola		■			■					
National Semiconductor	■	■			■					
SGS	■	■								
Signetics	■	■				■		■		
Texas Instruments	■	■			■	■				

KEY PARAMETERS		TTL	LS	HC	HCT	ALS/A	S	AS	F	AC	ACT
I_{OH} (max.)	mA	0.25	0.25			0.1	0.25		0.25		
I_{OL} (max.)	mA	48	24			24	60		20		
I_{CCH} (max.)	mA	8.5	2			1.6	36		7		
I_{CCL} (max.)	mA	54	12			7.8	80		30		
t_{PLH} (max.)	ns	22	32[a]			33	10		13		
t_{PHL} (max.)	ns	18	28[a]			12	10		5.5		

[a] Fairchild and National 22 ns

MANUFACTURERS	TTL	LS	HC	HCT	ALS/A	S	AS	F	AC	ACT
National Semiconductor					■					
Texas Instruments					■					

KEY PARAMETERS		TTL	LS	HC	HCT	ALS/A	S	AS	F	AC	ACT
I_{OH} (max.)	mA					0.1					
I_{OL} (max.)	mA					24					
I_{CCH} (max.)	mA					1.6					
I_{CCL} (max.)	mA					7.8					
t_{PLH} (max.)	ns					33					
t_{PHL} (max.)	ns					12					

39   Quad 2-input NAND buffers/drivers with open-collector outputs

MANUFACTURERS	TTL	LS	HC	HCT	ALS	S	AS	F	AC	ACT
Signetics										
Texas Instruments	■									

KEY PARAMETERS		TTL	LS	HC	HCT	ALS	S	AS	F	AC	ACT
I_{OH} (max.)	mA	0.25									
I_{OL} (max.)	mA	48[a]									
I_{CCH} (max.)	mA	8.5[b]									
I_{CCL} (max.)	mA	54									
t_{PLH} (max.)	ns	22									
t_{PHL} (max.)	ns	18									

[a] TI 60 mA, or 80 mA if V_{CC} is maintained between 4.75 V and 5.25 V   [b] TI data not available

132   Quad 2-input Schmitt-trigger NAND gates

MANUFACTURERS	TTL	LS	HC	HCT	ALS	S	AS	F	AC	ACT
Fairchild		■				■		D		
Motorola			▓							
National Semiconductor			▓		▓					
RCA			■							
SGS		■	▓							
Signetics	■							■		
Texas Instruments	■		D			■				
Toshiba			■							

KEY PARAMETERS		TTL	LS	HC	HCT	ALS	S	AS	F	AC	ACT
I_{OH} (max.)	mA	-0.8	-0.4	-4	-4	-0.4	-1		-1		
I_{OL} (max.)	mA	16	8	4	4	8	20		20		
I_{CCH} (max.)	mA	24	11			8	44		12		
I_{CCL} (max.)	mA	40	14			8	68		19.5		
I_{CC} (quiesc.)	mA			0.02[b]	0.02[f]						
C_{pd} (typ.)	pF/gate			30[c]	30[f]						
t_{PLH} (max.)	ns	22	22[a]	31[d]	31[g]	n/a	10.5		8.5		
t_{PHL} (max.)	ns	22	22[a]	31[d]	31[g]	n/a	13		8.5		
Input Hyst. (min.)	V	0.4	0.4	0.4[e]	0.4[f]	0.5	0.2		0.4		

[a] Fairchild 20 ns
[b] SGS 0.01 mA
[c] RCA; Motorola 20 pF/gate, Signetics 24 pF/gate, SGS 34 pF/gate, Toshiba 35 pF/gate. National data not available
[d] Motorola and National 32 ns, SGS 33 ns
[e] Toshiba 0.36 V
[f] RCA; Signetics 20 pF/gate
[g] RCA; Signetics 41 ns

| **24** | **Quad 2-input Schmitt-trigger NAND gates** |

MANUFACTURERS	TTL	LS	HC	HCT	ALS	S	AS	F	AC	ACT
SGS		▒▒								
Texas Instruments		██								

KEY PARAMETERS		TTL	LS	HC	HCT	ALS	S	AS	F	AC	ACT
I_{OH} (max.)	mA		-0.4								
I_{OL} (max.)	mA		8								
I_{CCH} (max.)	mA		12								
I_{CCL} (max.)	mA		20								
t_{PLH} (max.)	ns		20								
t_{PHL} (max.)	ns		40								
Input Hyst. (min.)	V		0.4								

3-Input NAND Gates

All 7400-series 3-input NANDs are provided in groups of three housed in 14-pin DIPs.

10   *Triple 3-input NAND gates.* Standard 3-input NANDs, available in nearly all technologies.

1010   *Triple 3-input NAND buffers/drivers.* High-current ALS versions of the 10 with I_{OL} = 24 mA.

12   *Triple 3-input NAND gates with open-collector outputs.* OC versions of the 10.

FUNCTION TABLE

INPUTS			OUTPUT
A	B	C	Y
H	H	H	L
all other cases			H

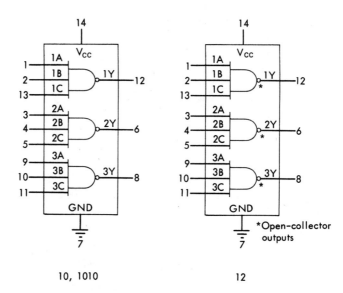

10, 1010

12

*Open-collector outputs

92

10   Triple 3-input NAND gates

MANUFACTURERS	TTL	LS	HC	HCT	ALS	S	AS	F	AC	ACT
Fairchild	■					■			■	
Motorola	■		▒							
National Semiconductor	■	■						■		
RCA			■		■				▒	▒
SGS	■		▒	■						
Signetics	■	■				■		■		
Texas Instruments	■	■								
Toshiba	■							D		

Added change control symbol: National 74ALS10A

KEY PARAMETERS		TTL	LS	HC	HCT	ALS	S	AS	F	AC	ACT
I_{OH} (max.)	mA	-0.4[a]	-0.4	-4	-4	-0.4	-1	-2	-1	-24	-24
I_{OL} (max.)	mA	16	8	4	4	8	20	20	20	24	24
I_{CCH} (max.)	mA	6	1.2			0.6	12	2.4	2.1		
I_{CCL} (max.)	mA	16.5	3.3			2.2	27	13	7.7		
I_{CC} (quiesc.)	mA			0.02[c]	0.02					0.04	0.04
C_{pd} (typ.)	pF/gate			24[d]	28[f]					25[h]	n/a
t_{PLH} (max.)	ns	22	15[b]	24[e]	30[g]	11	4.5	4.5	6	8[i]	12.1
t_{PHL} (max.)	ns	15	15[b]	24[e]	30[g]	18	5	4.5	5.3	6.5[i]	12.1

[a] Fairchild -0.8 mA
[b] National 10 ns
[c] SGS and Toshiba 0.01 mA
[d] RCA; Signetics 12 pF/gate, National 20 pF/gate, TI 25 pF/gate, SGS and Toshiba 30 pF/gate. Motorola data not available

[e] National and Toshiba 23 ns, RCA 25 ns.
[f] RCA; Signetics 14 pF/gate
[g] RCA; Signetics 29 ns
[h] Fairchild; RCA data not available
[i] RCA 10.9 ns

1010   Triple 3-input NAND buffers/drivers

MANUFACTURERS	TTL	LS	HC	HCT	ALS/A	S	AS	F	AC	ACT
National Semiconductor					■					
Texas Instruments					■					

NUMBERING CONFLICT: 74AC1010 and 74ACT1010 (Fairchild) are binary multipliers rather than NAND gates

KEY PARAMETERS		TTL	LS	HC	HCT	ALS/A	S	AS	F	AC	ACT
I_{OH} (max.)	mA					-2.6					
I_{OL} (max.)	mA					24					
I_{CCH} (max.)	mA					1.2					
I_{CCL} (max.)	mA					5.8					
t_{PLH} (max.)	ns					8					
t_{PHL} (max.)	ns					7					

12	**Triple 3-input NAND gates with open-collector outputs**

MANUFACTURERS	TTL	LS	HC	HCT	ALS/A	S	AS	F	AC	ACT
Motorola		■								
National Semiconductor		■			■					
SGS		■								
Texas Instruments	■	■			■					

KEY PARAMETERS		TTL	LS	HC	HCT	ALS/A	S	AS	F	AC	ACT
I_{OH} (max.)	mA	0.25	0.1			0.1					
I_{OL} (max.)	mA	16	8			8					
I_{CCH} (max.)	mA	6	1.4			0.6					
I_{CCL} (max.)	mA	16.5	3.3			2.2					
t_{PLH} (max.)	ns	45	32[a]			54					
t_{PHL} (max.)	ns	15	28[b]			18					

[a] National 20 ns
[b] National 15 ns

4-Input NAND Gates

All 7400-series 4-input NANDs are dual configurations supplied in 14-pin DIPs. In all except the 3040, two of the pins (3 and 11) are not connected internally.

20   *Dual 4-input NAND gates.* Standard 4-input NANDs, available in nearly all technologies.

40   *Dual 4-input NAND buffers/drivers.* Standard high-current versions of the 20 with max. I_{OL} of 24–64 mA.

1020   *Dual 4-input NAND buffers/drivers.* High-current (max. I_{OL} of 24 mA) ALS versions of the 20.

140   *Dual 4-input NAND 50-ohm buffers/drivers.* Very-high-current S versions of the 40 with max. I_{OH} = −40 mA and max. I_{OL} = 60 mA.

3040§   *Dual 4-input NAND 30-ohm transmission-line buffers/drivers.* Very-high-current F versions of the 40. Dual V_{CC} and GND pins lower lead inductance to 2 nH when both pins are used. (Signetics)

I_{OH} (max.)	−50 mA	t_{PHL} (max.)	7 ns
I_{OL} (max.)	160 mA	t_{PHL} (max.)	5.5 ns
I_{CCH} (max.)	2 mA		
I_{CCL} (max.)	17 mA		

22   *Dual 4-input NAND gates with open-collector outputs.* Standard OC versions of the 20.

13   *Dual 4-input Schmitt-trigger NAND gates.* Schmitt-trigger versions of the 20 providing a minimum of 0.4 V (ALS 0.5 V) input hysteresis.

18   *Dual 4-input Schmitt-trigger NAND gates.* LS versions of the 13 with PNP inputs to reduce system loading (max. $I_{IL} = -0.05$ mA instead of -0.4 mA for the LS13).

FUNCTION TABLE

INPUTS				OUTPUT
A	B	C	D	Y
H	H	H	H	L
all other cases				H

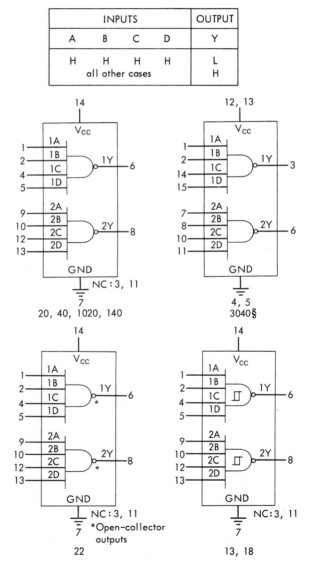

MANUFACTURERS	TTL	LS	HC	HCT	ALS/A	S	AS	F	AC	ACT
Fairchild		■	■			■		■	■	
Motorola		■	▒						■	
National Semiconductor	■			■	■					
RCA		■							▒	▒
SGS		■	▒							
Signetics	■	■				■	■	■		
Texas Instruments	■	■	■		■		■			
Toshiba			■						D	

KEY PARAMETERS		TTL	LS	HC	HCT	ALS/A	S	AS	F	AC	ACT
I_{OH} (max.)	mA	-0.4[a]	-0.4	-4	-4	-0.4	-1	-2	-1	-24	-24
I_{OL} (max.)	mA	16	8	4	4	8	20	20	20	24	24
I_{CCH} (max.)	mA	4	0.8			0.4	8	1.6	1.4		
I_{CCL} (max.)	mA	11	2.2			1.5	18	8.7	5.1		
I_{CC} (quiesc.)	mA			0.02[c]	0.02					0.04	0.04
C_{pd} (typ.)	pF/gate			25[d]	38[f]					40[g]	n/a
t_{PLH} (max.)	ns	22	15[b]	23[e]	35	11	4.5	5	6	8[h]	12.1
t_{PHL} (max.)	ns	15	15[b]	23[e]	35	10	5	4.5	5.3	7[h]	12.1

[a] Fairchild -0.8 mA
[b] National 10 ns
[c] SGS and Toshiba 0.01 mA
[d] TI; Motorola and National 20 pF/gate, Signetics 22 pF/gate, RCA 26 pF/gate, SGS and Toshiba 28 pF/gate

[e] RCA 25 ns, SGS 26 ns, TI 28 ns
[f] RCA; Signetics 17 pF/gate
[g] Fairchild; RCA data not available
[h] RCA 10.9 ns

MANUFACTURERS	TTL	LS	HC	HCT	ALS/A	S	AS	F	AC	ACT
Fairchild	■	■				■		D		
Motorola	■									
National Semiconductor	■	■			■	■				
SGS	■					■				
Signetics	■	■				■		▒		
Texas Instruments	■	■			■					

KEY PARAMETERS		TTL	LS	HC	HCT	ALS/A	S	AS	F	AC	ACT
I_{OH} (max.)	mA	-1.2	-1.2			-2.6	-3[b]		-15		
I_{OL} (max.)	mA	48	24			24	60		64		
I_{CCH} (max.)	mA	8	1			0.8	18		4		
I_{CCL} (max.)	mA	27	6			3.9	44		17		
t_{PLH} (max.)	ns	22	24[a]			8	6.5		7		
t_{PHL} (max.)	ns	15	24[a]			7	6.5		5.5		

[a] National 15 ns
[b] Fairchild -2.8 mA

1020   Dual 4-input NAND buffers/drivers

MANUFACTURERS	TTL	LS	HC	HCT	ALS/A	S	AS	F	AC	ACT
National Semiconductor					■					
Texas Instruments					■					

KEY PARAMETERS		TTL	LS	HC	HCT	ALS/A	S	AS	F	AC	ACT
I_{OH} (max.)	mA					-2.6					
I_{OL} (max.)	mA					24					
I_{CCH} (max.)	mA					0.8					
I_{CCL} (max.)	mA					3.9					
t_{PLH} (max.)	ns					8					
t_{PHL} (max.)	ns					7					

140   Dual 4-input NAND 50-ohm buffers/drivers

MANUFACTURERS	TTL	LS	HC	HCT	ALS	S	AS	F	AC	ACT
National Semiconductor						■				
Signetics						■				
Texas Instruments						■				

KEY PARAMETERS		TTL	LS	HC	HCT	ALS	S	AS	F	AC	ACT
I_{OH} (max.)	mA						-40				
I_{OL} (max.)	mA						60				
I_{CCH} (max.)	mA						18				
I_{CCL} (max.)	mA						44				
t_{PLH} (max.)	ns						6.5				
t_{PHL} (max.)	ns						6.5				

22   Dual 4-input NAND gates with open-collector outputs

MANUFACTURERS	TTL	LS	HC	HCT	ALS/B	S	AS	F	AC	ACT
Fairchild		■								
Motorola										
National Semiconductor		■			■					
SGS										
Texas Instruments	■	■			■					

KEY PARAMETERS		TTL	LS	HC	HCT	ALS/B	S	AS	F	AC	ACT
I_{OH} (max.)	mA	0.25	0.1			0.1	0.25				
I_{OL} (max.)	mA	16	8			8	20				
I_{CCH} (max.)	mA	4	0.8			0.4	6.6				
I_{CCL} (max.)	mA	11	2.2			1.5	18				
t_{PLH} (max.)	ns	45	32[a]			45	7.5				
t_{PHL} (max.)	ns	15	28[b]			18	7				

[a] National 20 ns, Fairchild 22 ns
[b] National 15 ns, Fairchild 18 ns

13 Dual 4-input Schmitt-trigger NAND gates

MANUFACTURERS	TTL	LS	HC	HCT	ALS	S	AS	F	AC	ACT
Fairchild		■						D		
Motorola										
National Semiconductor	■				▓					
SGS	■									
Signetics		■						■		
Texas Instruments		■								

KEY PARAMETERS		TTL	LS	HC	HCT	ALS	S	AS	F	AC	ACT
I_{OH} (max.)	mA	-0.8	-0.4			-0.4			-1		
I_{OL} (max.)	mA	16	8			8			20		
I_{CCH} (max.)	mA	23	6			4			8.5		
I_{CCL} (max.)	mA	32	7			4			10		
t_{PLH} (max.)	ns	27	22[a]			b			8		
t_{PHL} (max.)	ns	22	27[a]			b			13.5		
Input Hyst. (min.)	V	0.4	0.4			0.5			0.4		

[a] National and Motorola 22 ns

[b] Maximum values not available. Typically t_{PLH} = 8 ns and t_{PHL} = 13 ns

18 Dual 4-input Schmitt-trigger NAND gates

MANUFACTURERS	TTL	LS	HC	HCT	ALS	S	AS	F	AC	ACT
SGS		▓								
Texas Instruments		■								

KEY PARAMETERS		TTL	LS	HC	HCT	ALS	S	AS	F	AC	ACT
I_{OH} (max.)	mA		-0.4								
I_{OL} (max.)	mA		8								
I_{CCH} (max.)	mA		6								
I_{CCL} (max.)	mA		10								
t_{PLH} (max.)	ns		20								
t_{PHL} (max.)	ns		55								
Input Hyst. (min.)	V		0.4								

8-Input, 12-Input, and 13-Input NAND Gates

These multiple-input NANDs are useful in cases where a number of lines have to be monitored for the continuous existence of a given condition. All three types are packaged as single gates.

30   *8-input NAND gates.* Available in most technologies. 14-pin DIP with pins 1, 13, and 14 not internally connected.

133   *13-input NAND gates*. A standard device that provides the great-
est number of gate inputs possible in a 16-pin package. Available in
most technologies.

134   *12-input NAND gates with 3-state output*. These S components
use one of the 16 pins for an output control instead of a NAND
input, enabling a high-impedance "Z" state that can be used to
switch the resulting 12-input NAND off or on. Additional drive at
the logic HIGH level (max. $I_{OH} = -6.5$ mA instead of the usual -1
mA) allows the 134 to drive bus lines without external pull-up resistors.

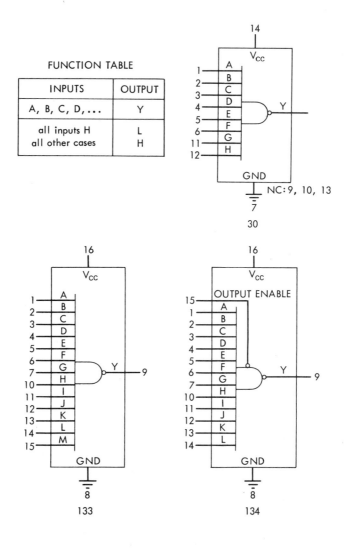

FUNCTION TABLE

INPUTS	OUTPUT
A, B, C, D, ...	Y
all inputs H	L
all other cases	H

NC: 9, 10, 13

30

133

134

30   8-input NAND gates

MANUFACTURERS	TTL	LS	HC	HCT	ALS/A	S	AS	F	AC	ACT
Fairchild	■		■			■				
Motorola		■	D							
National Semiconductor	■	■	■		■			■		
RCA			■							
SGS		■	▓							
Signetics	■	■				■		■		
Texas Instruments	■	■		■	■	■				
Toshiba			▓							

KEY PARAMETERS		TTL	LS	HC	HCT	ALS/A	S	AS	F	AC	ACT
I_{OH} (max.)	mA	-0.4^a	-0.4	-4	-4	-0.4	-1	-2	-1		
I_{OL} (max.)	mA	16	8	4	4	8	20	20	20		
I_{CCH} (max.)	mA	2	0.5			0.36	5	1.5	1.5		
I_{CCL} (max.)	mA	6	1.1			0.9	10	4.9	4		
I_{CC} (quiesc.)	mA			0.02^d	0.02						
C_{pd} (typ.)	pF/gate			25^e	26^g						
t_{PLH} (max.)	ns	22	15^b	33^f	35^h	10	6	5	5		
t_{PHL} (max.)	ns	15	20^c	33^f	35^h	12^i	7	4.5	5.5		

[a] Fairchild -0.8 mA

[b] Fairchild and National 12 ns

[c] Fairchild and National 15 ns

[d] SGS and Toshiba 0.01 mA

[e] RCA; TI 22 pF/gate, Signetics 23 pF/gate, Toshiba 30 pF/gate, National 34 pF/gate

[f] Toshiba 31 ns, National 42 ns

[g] RCA; Signetics 24 pF

[h] RCA; Signetics 39 ns

[i] National; TI has t_{PHL} (max.) = 20 ns

---

134   12-input NAND gates with 3-state outputs

MANUFACTURERS	TTL	LS	HC	HCT	ALS	S	AS	F	AC	ACT
National Semiconductor						■				
Signetics						■				
Texas Instruments						■				

KEY PARAMETERS		TTL	LS	HC	HCT	ALS	S	AS	F	AC	ACT
I_{OH} (max.)	mA						-6.5				
I_{OL} (max.)	mA						20				
I_{CCH} (max.)	mA						13				
I_{CCL} (max.)	mA						16				
I_{CCZ} (max.)	mA						25				
t_{PLH} (max.)	ns						6				
t_{PHL} (max.)	ns						7.5				
t_{PZH} (max.)	ns						19.5				
t_{PZL} (max.)	ns						21				
t_{PHZ} (max.)	ns						8.5				
t_{PLZ} (max.)	ns						14				

133   13-input NAND gates

MANUFACTURERS	TTL	LS	HC	HCT	ALS	S	AS	F	AC	ACT
Fairchild		■				■				
Motorola			D							
National Semiconductor			■		■					
SGS		■				■				
Signetics			■		■					
Texas Instruments			■		■					
Toshiba			■							

KEY PARAMETERS		TTL	LS	HC	HCT	ALS	S	AS	F	AC	ACT
I_{OH} (max.)	mA		-0.4	-4		-0.4	-1				
I_{OL} (max.)	mA		8	4		8	20				
I_{CCH} (max.)	mA		0.5			0.34	5				
I_{CCL} (max.)	mA		1.1			0.8	10				
I_{CC} (quiesc.)	mA			0.02^b							
C_{pd} (typ.)	pF			34^c							
t_{PLH} (max.)	ns		15	38^d		11	6				
t_{PHL} (max.)	ns		38^a	38^d		25	7				

[a] Motorola 59 ns   [c] TI 24 pF

[b] SGS and Toshiba 0.01 mA   [d] SGS 36 ns, National 42 ns

AND GATES

An AND gate is a device that normally has a LOW output except in the special case where all its inputs are HIGH, in which case its output is also HIGH. In other words, the AND gate signals the fact that all its inputs have just attained the HIGH state by a high-going transition at its output. A low-going transition at the output signals that one or more of its inputs has just broken this uniformity by going LOW.

The AND gate is logically simpler and more natural than the NAND, which can be thought of as an AND gate with an inverter at the output. Electrically, however, the NAND is usually the simpler circuit, with the AND derived from the NAND by adding an inverting stage.

The 2-input AND gates are listed here; AND gates with three or four inputs follow as a separate group.

2-Input AND Gates

08   *Quad 2-input AND gates.* Standard AND gates, available in most technologies. 14-pin DIP.

1008   *Quad 2-input AND buffers/drivers.* High-current ALS and AS versions of the 08.

808   *Hex 2-input AND buffers/drivers*. Basically the 1008 in a hex
           package with 20 pins. Max. I_{OH} increased to -15 mA in ALS
           version, and -1 option raises ALS max. I_{OL} to 48 mA. HC
           version available.

09   *Quad 2-input AND gates with open-collector or open-drain out-
           puts*. Standard OC/OD versions of the 08.

[7001]   *Quad 2-input Schmitt-trigger AND gates*. An HC circuit cur-
           rently in the design stage. (Texas Instruments)

FUNCTION TABLE

INPUTS		OUTPUT
A	B	Y
L	L	L
L	H	L
H	L	L
H	H	H

08, 1008

808

09

*Open-collector outputs

08   Quad 2-input AND gates

MANUFACTURERS	TTL	LS	HC	HCT	ALS	S	AS	F	AC	ACT
Fairchild	■		■			■		■		▒
Motorola		■	▒							
National Semiconductor	■			▒	■					
RCA			■						▒	▒
SGS		■	▒							
Signetics	■		■			■		■		
Texas Instruments	■		■		■	■		■		
Toshiba			■						D	

KEY PARAMETERS		TTL	LS	HC	HCT	ALS	S	AS	F	AC	ACT
I_{OH} (max.)	mA	-0.8	-0.4	-4	-4	-0.4	-1	-2	-1	-24	-24
I_{OL} (max.)	mA	16	8	4	4	8	20	20	20	24	24
I_{CCH} (max.)	mA	21	4.8			2.4	32	9.3	8.3		
I_{CCL} (max.)	mA	33	8.8			4	57	24	12.9		
I_{CC} (quiesc.)	mA			0.02[c]	0.02					0.04	0.04
C_{pd} (typ.)	pF/gate			38[d]	38[f]					20[h]	20[j]
t_{PLH} (max.)	ns	27	15[a]	23[e]	30[g]	14	7	5.5	6.6	8.5[i]	12.9[k]
t_{PHL} (max.)	ns	19	20[b]	23[e]	30[g]	10	7.5	5.5	6.3	7.5[i]	12.9[k]

[a] Fairchild and National 13 ns
[b] Fairchild and National 11 ns
[c] SGS and Toshiba 0.01 mA
[d] Motorola, National; Signetics 10 pF/gate, TI 20 pF/gate, Toshiba 21 pF/gate, RCA 37 pF/gate
[e] SGS 22 ns, TI 25 ns, Motorola 30 ns, National t_{PLH} (max.) = 30 ns and t_{PHL} (max.) = 23 ns
[f] National; Signetics 20 pF/gate, RCA 51 pF/gate
[g] RCA 31 ns
[h] RCA 100 pF/gate
[i] RCA 8.1 ns
[j] RCA 115 pF/gate
[k] RCA; comparable Fairchild data not available

1008   Quad 2-input AND buffers/drivers

MANUFACTURERS	TTL	LS	HC	HCT	ALS/A	S	AS	F	AC	ACT
National Semiconductor					■		■			
Texas Instruments					■		■			

KEY PARAMETERS		TTL	LS	HC	HCT	ALS/A	S	AS	F	AC	ACT
I_{OH} (max.)	mA					-2.6		-48			
I_{OL} (max.)	mA					24		48			
I_{CCH} (max.)	mA					3		9.5			
I_{CCL} (max.)	mA					9.3		22			
t_{PLH} (max.)	ns					9		5[a]			
t_{PHL} (max.)	ns					9		5[a]			

[a] National 6 ns

808 Hex 2-input AND buffers/drivers

MANUFACTURERS	TTL	LS	HC	HCT	ALS	S	AS/B	F	AC	ACT
National Semiconductor			■		■		■			
Texas Instruments			■		■		■			

Added change control symbol: TI 74ALS808A. ALS -1 version available from National

KEY PARAMETERS		TTL	LS	HC	HCT	ALS	S	AS/B	F	AC	ACT
I_{OH} (max.)	mA			-6		-15		-48			
I_{OL} (max.)	mA			6		24^a		48			
I_{CCH} (max.)	mA					7^b		13^d			
I_{CCL} (max.)	mA					16		33^e			
I_{CC} (quiesc.)	mA			0.08							
C_{pd} (typ.)	pF/gate			20							
t_{PLH} (max.)	ns			25		9^c		6^f			
t_{PHL} (max.)	ns			25		8		6^f			

^{a} -1 option has I_{OL} (max.) = 48 mA
^{b} National 6 mA
^{c} National 8 ns
^{d} National 11 mA
^{e} National 32 mA
^{f} National 5 ns

09 Quad 2-input AND gates with open-collector or open-drain outputs

MANUFACTURERS	TTL	LS	HC	HCT	ALS	S	AS	F	AC	ACT
Fairchild	■	■								
Motorola	■	■								
National Semiconductor	■	■			■	■				
SGS	■	■								
Signetics	■	■								
Texas Instruments	■	■	■		■	■				

KEY PARAMETERS		TTL	LS	HC	HCT	ALS	S	AS	F	AC	ACT
I_{OH} (max.)	mA	0.25^a	0.1^a	0.005		0.1	0.25				
I_{OL} (max.)	mA	16	8	4		8	20				
I_{CCH} (max.)	mA	21	4.8			2.4	32				
I_{CCL} (max.)	mA	33	8.8			4	57				
I_{CC} (quiesc.)	mA			0.02							
C_{pd} (typ.)	pF/gate			20							
t_{PLH} (max.)	ns	32^b	35^c	31		54	10				
t_{PHL} (max.)	ns	24	35^d	25		15	10				

^{a} Fairchild data not available
^{b} R_L = 400 Ωs for t_{PLH} as well as for t_{PHL}
^{c} Fairchild and National 20 ns
^{d} Fairchild and National 15 ns

3-Input AND Gates

11   *Triple 3-input AND gates.* Standard 3-input ANDs, available in most technologies. 14-pin DIP.

1011   *Triple 3-input AND buffers/drivers.* High-current ALS versions of the 11.

15   *Triple 3-input AND gates with open collector outputs.* OC versions of the 11.

4-Input AND Gates

21   *Dual 4-input AND gates.* 14-pin DIP with 2 pins inactive.

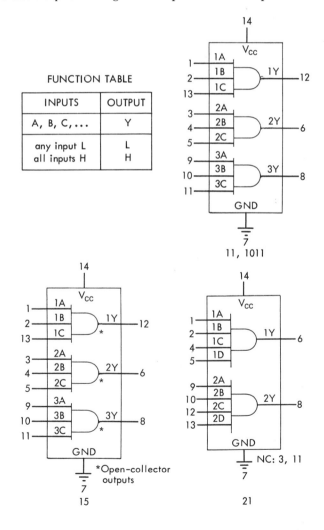

FUNCTION TABLE

INPUTS	OUTPUT
A, B, C, ...	Y
any input L	L
all inputs H	H

11   Triple 3-input AND gates

MANUFACTURERS	TTL	LS	HC	HCT	ALS	S	AS	F	AC	ACT
Fairchild		██				██			██	
Motorola		██	▒▒							
National Semiconductor					██					
RCA			██							
SGS		██	▒▒							
Signetics		██				██		██		
Texas Instruments		██		██	██					
Toshiba			██						D	

Added change control symbol: National 74ALS11A

KEY PARAMETERS		TTL	LS	HC	HCT	ALS	S	AS	F	AC	ACT
I_{OH} (max.)	mA	-0.8^a	-0.4^d	-4	-4	-0.4	-1^d	-2	-1	-24	
I_{OL} (max.)	mA	16^a	8^d	4	4	8	20^d	20	20	24	
I_{CCH} (max.)	mA	15^b	3.6			1.8	24	7	6.2		
I_{CCL} (max.)	mA	22^c	6.6			3	42	18	9.7		
I_{CC} (quiesc.)	mA			0.02^g	0.02					0.04	
C_{pd} (typ.)	pF/gate			28^h	28					20	
t_{PLH} (max.)	ns	27	15^e	28^i	35	20^j	7	6	6.6	8.5	
t_{PHL} (max.)	ns	19	20^f	28^i	35	10	7.5	5.5	6.5	7.5	

[a] National; Signetics data not available in this form

[b] Signetics 12 mA

[c] Signetics 20 mA

[d] Signetics data not available in this form

[e] Fairchild and National 13 ns

[f] Fairchild and National 11 ns

[g] SGS and Toshiba 0.01 mA

[h] SGS and Toshiba; TI 25 pF/gate, RCA and Signetics 26 pF/gate, Motorola and National 35 pF/gate

[i] RCA and Toshiba; Signetics and TI 25 ns, SGS 26 ns, Motorola and National 31 ns

[j] National 13 ns

1011   Triple 3-input AND buffers/drivers

MANUFACTURERS	TTL	LS	HC	HCT	ALS/A	S	AS	F	AC	ACT
National Semiconductor					██					
Texas Instruments					██					

KEY PARAMETERS		TTL	LS	HC	HCT	ALS/A	S	AS	F	AC	ACT
I_{OH} (max.)	mA					-2.6					
I_{OL} (max.)	mA					24					
I_{CCH} (max.)	mA					2.3					
I_{CCL} (max.)	mA					7					
t_{PLH} (max.)	ns					10					
t_{PHL} (max.)	ns					9					

15   Triple 3-input AND gates with open-collector outputs

MANUFACTURERS	TTL	LS	HC	HCT	ALS	S	AS	F	AC	ACT
Fairchild		■								
Motorola		■								
National Semiconductor		■			■					
SGS		■								
Texas Instruments		■			■					

Added change control symbol: TI 74ALS15A

KEY PARAMETERS		TTL	LS	HC	HCT	ALS	S	AS	F	AC	ACT
I_{OH} (max.)	mA		0.1[a]			0.1	0.25				
I_{OL} (max.)	mA		8			8	20				
I_{CCH} (max.)	mA		3.6			1.8	19.5				
I_{CCL} (max.)	mA		6.6			3	42				
t_{PLH} (max.)	ns		35[b]			45	8.5				
t_{PHL} (max.)	ns		35[c]			20	9				

[a] Fairchild data not available
[b] Fairchild and National 20 ns
[c] Fairchild and National 15 ns

21   Dual 4-input AND gates

MANUFACTURERS	TTL	LS	HC	HCT	ALS	S	AS	F	AC	ACT
Fairchild		■								
Motorola		■								
National Semiconductor		■			■		■			
RCA			■							
SGS			▨							
Signetics	■		D	D						
Texas Instruments	■				■		■			
Toshiba			■							

KEY PARAMETERS		TTL	LS	HC	HCT	ALS	S	AS	F	AC	ACT
I_{OH} (max.)	mA	-0.8	-0.4	-4	-4	-0.4		-2			
I_{OL} (max.)	mA	16	8	4	4	8		20			
I_{CCH} (max.)	mA	8	2.4			1.2		4.6			
I_{CCL} (max.)	mA	13	4.4			2		12			
I_{CC} (quiesc.)	mA			0.02[c]	0.02						
C_{pd} (typ.)	pF/gate			29[d]	42						
t_{PLH} (max.)	ns	27	15[a]	28[e]	34	26		6			
t_{PHL} (max.)	ns	19	20[b]	28[e]	34	10		6			

[a] National 13 ns
[b] National 11 ns, Fairchild 15 ns
[c] SGS and Toshiba 0.01 mA
[d] TI 25 pF/gate, RCA 36 pF/gate
[e] SGS 26 ns

NOR GATES

The logic of NOR is that the gate will put out a logic HIGH only if neither of its inputs is HIGH. In other words, a NOR gate is equivalent to an AND gate with an inverter at each input. The output of an AND gate is normally LOW but goes to logic HIGH in the special case where all its inputs are HIGH, whereas a NOR gate goes to HIGH when all its inputs are LOW.

NOR gates with two inputs are listed here, and 3-input, 4-input, and 5-input types follow separately.

2-Input NOR Gates

02   *Quad 2-input NOR gates.* Standard components available in nearly all technologies. 14-pin DIP.

36§   *Quad 2-input NOR gates.* Variant-pinout HC version of the 0.2. (Texas Instruments)

I_{OH} (max.)	−4 mA	t_{PHL} (max.)	25 ns
I_{OL} (max.)	4 mA	t_{PHL} (max.)	25 ns
I_{CC} (quiesc.)	0.02 mA		
C_{pd} (max.)	20 pF/gate		

28   *Quad 2-input NOR buffers/drivers.* High-current versions of the 02 with max. I_{OL} of 24-48 mA.

1002   *Quad 2-input NOR buffers/drivers.* ALS versions of the 28.

1036   *Quad 2-input NOR buffers/drivers.* AS versions of the 36 with I_{OH} = −48 mA and I_{OL} = mA.

805   *Hex 2-input NOR buffers/drivers.* Basically the 1002 or 1036 in a hex package with 20 pins. Max. I_{OH} increased to −15 mA in ALS version, and −1 option raises ALS max. I_{OL} to 48 mA. HC version available.

128   *Quad 2-input NOR buffers/drivers.* TTL versions of the 28 with max. I_{OH} increased from −2.4 mA to −42.4 mA.

33   *Quad 2-input NOR buffers/drivers with open-collector outputs.* OC versions of the 28.

[7002]   *Quad 2-input Schmitt-trigger NOR gates.* An HC circuit currently in the design stage. (Texas Instruments)

FUNCTION TABLE

INPUTS		OUTPUT
A	B	Y
L	L	H
L	H	L
H	L	L
H	H	L

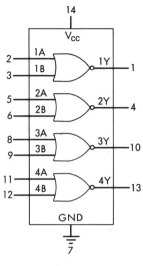

02, 28, 1002, 128

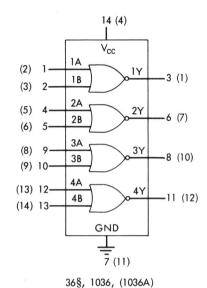

36§, 1036, (1036A)

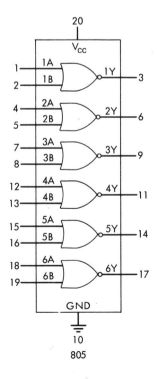

10

805

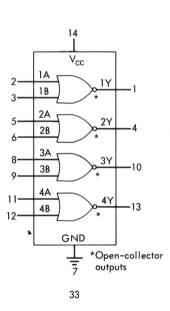

*Open-collector
outputs

33

Quad 2-input NOR gates

MANUFACTURERS	TTL	LS	HC	HCT	ALS	S	AS	F	AC	ACT
Fairchild	■	■				■		■		
Motorola		■	▒							
National Semiconductor		■	■		■					
RCA									▒	▒
SGS		■	▒							
Signetics	■	■				■		■		
Texas Instruments	■	■			■		■		■	
Toshiba		■							▒	

KEY PARAMETERS		TTL	LS	HC	HCT	ALS	S	AS	F	AC	ACT
I_{OH} (max.)	mA	-0.4[a]	-0.4	-4	-4	-0.4	-1	-2	-1	-24	-24
I_{OL} (max.)	mA	16	8	4	4	8	20	20	20	24	24
I_{CCH} (max.)	mA	16	3.2			2.2	29	5.9	5.6		
I_{CCL} (max.)	mA	27	5.4			4	45	20.1	13		
I_{CC} (quiesc.)	mA			0.02[e]	0.02					0.04	0.04
C_{pd} (typ.)	pF/gate			22[f]	26[h]					30[j]	n/a
t_{PLH} (max.)	ns	22[b]	15[c]	23[g]	28[i]	12	5.5	4.5	6.5	6.5[k]	10.9
t_{PHL} (max.)	ns	15	15[d]	23[g]	28[i]	10	5.5	4.5	5.3	7[k]	10.9

[a] Fairchild -0.8 mA
[b] Fairchild 15 ns
[c] Fairchild 10 ns, National 13 ns
[d] Fairchild and National 10 ns
[e] SGS and Toshiba 0.01 mA
[f] Signetics and TI; Motorola and National 20 pF/gate, RCA 26 pF/gate, SGS and Toshiba 27 pF/gate
[g] SGS 22 ns
[h] RCA; Signetics 24 pF/gate
[i] RCA; Signetics 24 ns
[j] Toshiba 24 pF/gate; RCA data not available
[k] RCA 10.3 ns; Toshiba data not available

Quad 2-input NOR buffers/drivers

MANUFACTURERS	TTL	LS	HC	HCT	ALS/A	S	AS	F	AC	ACT
Fairchild		■								
Motorola		■								
National Semiconductor					■					
SGS		■								
Signetics	■									
Texas Instruments	■				■					

KEY PARAMETERS		TTL	LS	HC	HCT	ALS/A	S	AS	F	AC	ACT
I_{OH} (max.)	mA	-2.4	-1.2			-2.6					
I_{OL} (max.)	mA	48	24			24					
I_{CCH} (max.)	mA	21	3.6			2.8					
I_{CCL} (max.)	mA	57	13.8			9					
t_{PLH} (max.)	ns	9	24[a]			8					
t_{PHL} (max.)	ns	12	24[a]			7					

[a] Fairchild 20 ns

1002   Quad 2-input NOR buffers/drivers

MANUFACTURERS	TTL	LS	HC	HCT	ALS/A	S	AS	F	AC	ACT
National Semiconductor					■					
Texas Instruments					■					

KEY PARAMETERS		TTL	LS	HC	HCT	ALS/A	S	AS	F	AC	ACT
I_{OH} (max.)	mA					-2.6					
I_{OL} (max.)	mA					24					
I_{CCH} (max.)	mA					2.8					
I_{CCL} (max.)	mA					9					
t_{PLH} (max.)	ns					8					
t_{PHL} (max.)	ns					7					

1036   Quad 2-input NOR buffers/drivers

MANUFACTURERS	TTL	LS	HC	HCT	ALS	S	AS/A	F	AC	ACT
National Semiconductor							■			
Texas Instruments							■			

KEY PARAMETERS		TTL	LS	HC	HCT	ALS	S	AS/A	F	AC	ACT
I_{OH} (max.)	mA							-48			
I_{OL} (max.)	mA							48			
I_{CCH} (max.)	mA							7			
I_{CCL} (max.)	mA							23			
t_{PLH} (max.)	ns							4.3			
t_{PHL} (max.)	ns							4.3			

805   Hex 2-input NOR buffers/drivers

MANUFACTURERS	TTL	LS	HC	HCT	ALS	S	AS/B	F	AC	ACT
National Semiconductor			■		■		■			
Texas Instruments			■		■		■			

Added change control symbol: TI 74ALS805A. ALS -1 version available from National

KEY PARAMETERS		TTL	LS	HC	HCT	ALS	S	AS/B	F	AC	ACT
I_{OH} (max.)	mA			-6		-15		-48			
I_{OL} (max.)	mA			6		24^a		48			
I_{CCH} (max.)	mA					4		10^d			
I_{CCL} (max.)	mA					14		32			
I_{CC} (quiesc.)	mA			0.08							
C_{pd} (typ.)	pF/gate			40							
t_{PLH} (max.)	ns			24		7^b		4.3^e			
t_{PHL} (max.)	ns			24		8^c		4.3^e			

^{a} -1 option has I_{OL} (max.) = 48 mA
^{b} National 6 ns
^{c} National 7 ns

^{d} National 9 mA
^{e} National 4 ns

128   Quad 2-input NOR buffers/drivers

MANUFACTURERS	TTL	LS	HC	HCT	ALS	S	AS	F	AC	ACT
Signetics	■									
Texas Instruments	■									

KEY PARAMETERS		TTL	LS	HC	HCT	ALS	S	AS	F	AC	ACT
I_{OH} (max.)	mA	-42.4									
I_{OL} (max.)	mA	48									
I_{CCH} (max.)	mA	21									
I_{CCL} (max.)	mA	57									
t_{PLH} (max.)	ns	9									
t_{PHL} (max.)	ns	12									

33	Quad 2-input NOR buffers/drivers with open-collector outputs

MANUFACTURERS	TTL	LS	HC	HCT	ALS/A	S	AS	F	AC	ACT
Fairchild		■								
Motorola		■								
National Semiconductor					■					
SGS	■	■								
Signetics	■									
Texas Instruments		■			■					

KEY PARAMETERS		TTL	LS	HC	HCT	ALS/A	S	AS	F	AC	ACT
I_{OH} (max.)	mA	0.25	0.25^a			0.1					
I_{OL} (max.)	mA	48	24			24					
I_{CCH} (max.)	mA	21	3.6			2.8					
I_{CCL} (max.)	mA	57	13.8			9					
t_{PLH} (max.)	ns	15	32^b			33					
t_{PHL} (max.)	ns	18	28^b			12					

[a] SGS 0.1 mA
[b] Fairchild 22 ns

3-Input, 4-Input, and 5-Input NOR Gates

The basic logic of the multiple-input NOR is that the output is HIGH if, and only if, all the inputs are LOW.

27   *Triple 3-input NOR gates.* Standard components available in most technologies. 14-pin DIP.

4002   *Dual 4-input NOR gates.* HC and HCT only. 14-pin DIP with two pins inactive.

25   *Dual 4-input NOR gates with independent strobe.* The two pin positions left unused on the 4002 are used for two strobe inputs in this TTL circuit. Each strobe input is ANDed internally with the four inputs to one of the NOR gates. Holding a strobe input LOW will disable that NOR gate's inputs and thus keep the output of the NOR gate HIGH regardless of its input conditions. A HIGH at the strobe input enables the gate inputs, and the 25 then becomes a simple 4-input NOR gate like the 4002.

260   *Dual 5-input NOR gates.* 5-input LS and S circuits in 14-pin DIPs.

FUNCTION TABLE

INPUTS	OUTPUT
A, B, C, ...	Y
all inputs L	H
any input H	L

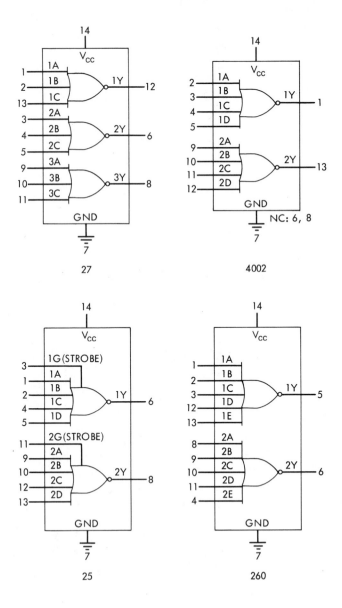

27

4002

25

260

27   Triple 3-input NOR gates

MANUFACTURERS	TTL	LS	HC	HCT	ALS	S	AS	F	AC	ACT
Fairchild	■	■	■							
Motorola		■	▨							
National Semiconductor			■		■		■			
RCA			■							
SGS		■	▨							
Signetics			■					■		
Texas Instruments	■	■	■	■						
Toshiba		■	■							

KEY PARAMETERS		TTL	LS	HC	HCT	ALS	S	AS	F	AC	ACT
I_{OH} (max.)	mA	-0.8	-0.4	-4	-4	-0.4		-2	-1		
I_{OL} (max.)	mA	16	8	4	4	8		20	20		
I_{CCH} (max.)	mA	16	4			1.8		6.4	5.5		
I_{CCL} (max.)	mA	26	6.8			4		17.1	12		
I_{CC} (quiesc.)	mA			0.02[e]	0.02						
C_{pd} (typ.)	pF/gate			27[f]	28[h]						
t_{PLH} (max.)	ns	15[a]	15[c]	23[g]	29[i]	15		5.5	5.5		
t_{PHL} (max.)	ns	11[b]	15[d]	23[g]	29[i]	9		4.5	5		

[a] National 11 ns
[b] National 15 ns
[c] Fairchild and National 13 ns
[d] National 10 ns, Fairchild 13 ns
[e] SGS and Toshiba 0.01 mA

[f] SGS and Toshiba; Signetics 24 pF/gate, TI 25 pF/gate, RCA 26 pF/gate, Motorola and National 36 pF/gate
[g] SGS 22 ns, RCA 24 ns
[h] RCA; Signetics 30 pF/gate
[i] RCA; Signetics 26 ns

4002   Dual 4-input NOR gates

MANUFACTURERS	TTL	LS	HC	HCT	ALS	S	AS	F	AC	ACT
Motorola			▨							
National Semiconductor			■							
RCA			■	■						
SGS			▨							
Signetics			■	■						
Texas Instruments			■							
Toshiba			■							

KEY PARAMETERS		TTL	LS	HC	HCT	ALS	S	AS	F	AC	ACT
I_{OH} (max.)	mA			-4	-4						
I_{OL} (max.)	mA			4	4						
I_{CC} (quiesc.)	mA			0.02[a]	0.02						
C_{pd} (typ.)	pF/gate			25[b]	22						
t_{PLH} (max.)	ns			26[c]	28						
t_{PHL} (max.)	ns			26[c]	28[d]						

[a] SGS and Toshiba 0.01 mA
[b] National and TI; Signetics 16 pF/gate, RCA 22 pF/gate, Motorola and SGS 26 pF/gate

[c] SGS and Toshiba; RCA and Signetics 25 ns, TI 28 ns, Motorola and National 30 ns
[d] RCA 36 ns

25   Dual 4-input NOR gates with independent strobes

MANUFACTURERS	TTL	LS	HC	HCT	ALS	S	AS	F	AC	ACT
Fairchild	■									
National Semiconductor	■									
Signetics	■									
Texas Instruments	■									

KEY PARAMETERS		TTL	LS	HC	HCT	ALS	S	AS	F	AC	ACT
I_{OH} (max.)	mA	-0.8									
I_{OL} (max.)	mA	16									
I_{CCH} (max.)	mA	16									
I_{CCL} (max.)	mA	19									
t_{PLH} (max.)	ns	22									
t_{PHL} (max.)	ns	15									

260   Dual 5-input NOR gates

MANUFACTURERS	TTL	LS	HC	HCT	ALS	S	AS	F	AC	ACT
Fairchild		■								
Motorola		■								
SGS		■								
Signetics		■				■		■		
Texas Instruments						■				

KEY PARAMETERS		TTL	LS	HC	HCT	ALS	S	AS	F	AC	ACT
I_{OH} (max.)	mA		-0.4				-1		-1		
I_{OL} (max.)	mA		8				20		20		
I_{CCH} (max.)	mA		4				29		6.5		
I_{CCL} (max.)	mA		5.5				45		9.5		
t_{PLH} (max.)	ns		15[a]				5.5		6.5		
t_{PHL} (max.)	ns		15[b]				6		4.5		

[a] Fairchild 10 ns, Signetics 12 ns

[b] Fairchild and Signetics 12 ns

OR GATES

The difference between OR gates and AND gates (and the difference between NOR gates and NAND gates) is the difference between "all" and "any." Whereas an AND gate puts out a logic HIGH when *all* of its inputs are HIGH, an OR gate puts out a logic HIGH when *any* of its inputs is HIGH, and thus puts out a logic LOW only when all of its inputs are LOW.

An OR gate behaves like a NOR gate with an inverter attached to the output, just as a NOR gate can be thought of as an OR followed by an inverter. Also, an OR gate is equivalent to a NAND gate with an inverter at each input.

32	*Quad 2-input OR gates.* Standard components, available in nearly all technologies. 14-pin DIP.
1032	*Quad 2-input OR buffers/drivers.* High-current ALS and AS versions of the 32.
832	*Hex 2-input OR buffers/drivers.* Basically the 1032 in a hex package with 20 pins. Max. I_{OH} increased to -15 mA in ALS version, and -1 option raises ALS max. I_{OL} to 48 mA. HC version available.
[7032]	*Quad 2-input Schmitt-trigger OR gates.* An HC circuit currently in the design stage. (Texas Instruments)
4075	*Triple 3-input OR gates.* HC and HCT circuits. 14-pin DIP.
4072	*Dual 4-input OR gates.* HC circuits. 14-pin DIP.

FUNCTION TABLE

INPUTS	OUTPUT
A, B, C, ...	Y
all inputs L	L
any input H	H

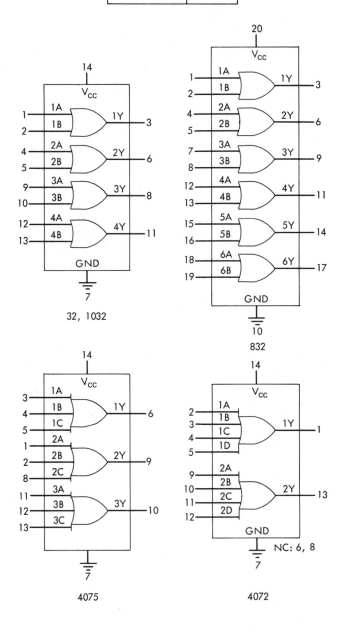

32   Quad 2-input OR gates

MANUFACTURERS	TTL	LS	HC	HCT	ALS	S	AS	F	AC	ACT
Fairchild	■	■	■		■		■			▨
Motorola			D							
National Semiconductor	■	■	■	□	■					
RCA	■	■							▨	▨
SGS		■	▨							
Signetics	■	■				■	■			
Texas Instruments	■	■	■		■	■	■			
Toshiba			■						D	

KEY PARAMETERS		TTL	LS	HC	HCT	ALS	S	AS	F	AC	ACT
I_{OH} (max.)	mA	-0.8	-0.4	-4	-4	-0.4	-1	-2	-1	-24	-24
I_{OL} (max.)	mA	16	8	4	4	8	20	20	20	24	24
I_{CCH} (max.)	mA	22	6.2			4	32	12	9.2		
I_{CCL} (max.)	mA	38	9.8			4.9	68	26.6	15.5		
I_{CC} (quiesc.)	mA			0.02^b	0.02					0.04	0.04
C_{pd} (typ.)	pF/gate			22^c	22^e					20^f	20^h
t_{PLH} (max.)	ns	15	22^a	23^d	30	14	7	5.8	6.6	8.5	10.9
t_{PHL} (max.)	ns	22	22^a	23^d	30	12	7	5.8	6.3	7.5^g	10.9

[a] National 11 ns, Fairchild 15 ns
[b] SGS and Toshiba 0.01 mA
[c] RCA; Signetics 16 pF/gate, TI 20 pF/gate, SGS and Toshiba 23 pF/gate, National 50 pF/gate
[d] SGS 22 ns, National and TI 25 ns

[e] RCA; Signetics 28 pF/gate
[f] RCA 47 pF/gate
[g] RCA 8.5 ns
[h] RCA 67 pF/gate

1032   Quad 2-input OR buffers/drivers

MANUFACTURERS	TTL	LS	HC	HCT	ALS/A	S	AS/A	F	AC	ACT
National Semiconductor					■		■			
Texas Instruments					■		■			

KEY PARAMETERS		TTL	LS	HC	HCT	ALS/A	S	AS/A	F	AC	ACT
I_{OH} (max.)	mA					-2.6		-48			
I_{OL} (max.)	mA					24		48			
I_{CCH} (max.)	mA					5		11.5			
I_{CCL} (max.)	mA					10.6		24			
t_{PLH} (max.)	ns					9		5.5^a			
t_{PHL} (max.)	ns					12		5.5^a			

[a] National 6.3 ns

832   Hex 2-input OR buffers/drivers

MANUFACTURERS	TTL	LS	HC	HCT	ALS	S	AS	F	AC	ACT
National Semiconductor					■		■			
Texas Instruments			■		■		■			

Added change control symbol: National 74AS832B. ALS -1 versions available from National and TI

KEY PARAMETERS		TTL	LS	HC	HCT	ALS	S	AS	F	AC	ACT
I_{OH} (max.)	mA			-6		-15		-48			
I_{OL} (max.)	mA			6		24^a		48			
I_{CCH} (max.)	mA					8		15^b			
I_{CCL} (max.)	mA					16		36			
I_{CC} (quiesc.)	mA			0.08							
C_{pd} (typ.)	pF/gate			20							
t_{PLH} (max.)	ns			25		8		5.5			
t_{PHL} (max.)	ns			25		8		5.5			

[a] -1 option has I_{OL} (max.) = 48 mA
[b] National 17 mA

4075   Triple 3-input OR gates

MANUFACTURERS	TTL	LS	HC	HCT	ALS	S	AS	F	AC	ACT
Motorola			▨							
National Semiconductor			■	■						
RCA			■							
SGS			▨							
Signetics			■	■						
Texas Instruments			■	■						
Toshiba			■							

KEY PARAMETERS		TTL	LS	HC	HCT	ALS	S	AS	F	AC	ACT
I_{OH} (max.)	mA			-4	-4						
I_{OL} (max.)	mA			4	4						
I_{CC} (quiesc.)	mA			0.02^a	0.02						
C_{pd} (typ.)	pF/gate			30^b	32^d						
t_{PLH} (max.)	ns			25^c	30						
t_{PHL} (max.)	ns			25^c	30						

[a] SGS and Toshiba 0.01 mA
[b] RCA and TI 26 pF/gate, Signetics 28 pF/gate; Motorola data not available
[c] SGS 24 ns, Motorola and National 29 ns
[d] Signetics; RCA 28 pF/gate

4072   Dual 4-input OR gates

MANUFACTURERS	TTL	LS	HC	HCT	ALS	S	AS	F	AC	ACT
SGS			▓							
Toshiba			■							

KEY PARAMETERS		TTL	LS	HC	HCT	ALS	S	AS	F	AC	ACT
I_{OH} (max.)	mA			-4							
I_{OL} (max.)	mA			4							
I_{CC} (quiesc.)	mA			0.01^a							
C_{pd} (typ.)	pF/gate			28							
t_{PLH} (max.)	ns			26^b							
t_{PHL} (max.)	ns			26^b							

[a] Toshiba 0.04 mA

[b] SGS; Toshiba 28 ns

OR/NOR GATES

The user who needs an occasional OR gate or NOR gate can implement nearly any form of OR/NOR logic by keeping on hand a single component, the 4078. This HC circuit is the equivalent of an 8-input OR gate with output Y or an 8-input NOR gate with output X.

FUNCTION TABLE

INPUTS	OUTPUTS	
A, B, C, ...	X	Y
all inputs L	H	L
any input H	L	H

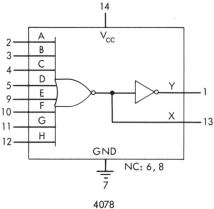

4078

4078   8-input OR/NOR gates

MANUFACTURERS	TTL	LS	HC	HCT	ALS	S	AS	F	AC	ACT
Motorola			D							
National Semiconductor			■							
SGS			▒							
Texas Instruments			■							
Toshiba			■							

Added change control symbol: Texas Instruments 74HC4078A

KEY PARAMETERS		TTL	LS	HC	HCT	ALS	S	AS	F	AC	ACT
I_{OH} (max.)	mA			-4							
I_{OL} (max.)	mA			4							
I_{CC} (quiesc.)	mA			0.02^a							
C_{pd} (typ.)	pF			73^b							
t_{PLH} output Y (max.)	ns			33^c							
t_{PHL} output Y (max.)	ns			33^c							
t_{PLH} output X (max.)	ns			35^d							
t_{PHL} output X (max.)	ns			35^d							

^{a} SGS and Toshiba 0.01 mA ^{c} SGS 36 ns

^{b} Toshiba; TI 25 pF, SGS 66 pF, National 100 pF ^{d} National; Toshiba and TI 33 ns, SGS 36 ns

EXCLUSIVE-NOR GATES

The ordinary NOR gate puts out a LOW logic level if either of its inputs is active or HIGH. "Either" includes the case where both of the inputs are HIGH. An exclusive-NOR gate acts like an ordinary NOR gate except that it excludes the case where both of the inputs are HIGH. That is, it puts out a HIGH level if both inputs are LOW and falls to a LOW level if either of the inputs goes HIGH, but comes back up to a HIGH level again if both of the inputs go HIGH. In general terms, the output of an exclusive-NOR is HIGH if all of its inputs have the same logic value and LOW if they don't.

One common application of the exclusive-NOR gate is as a switchable inverter/noninverter, with one input designated the control line and the other designated the data line. If the control line is held HIGH, the input to the data line is reproduced unchanged at the gate output. If the control line is held LOW, the gate acts as an inverter, producing the logical complement of the input data.

Other applications of the exclusive-NOR present themselves in the implementation of parity-checking schemes. Parity is a common way of building error detection into data transmission and storage procedures. A word is said to have even parity if its bits add up to an even number and odd parity if they don't. An added bit (the parity bit) attached to each word in the process of sending or storing it tells whether the word has even or odd parity, and this information can be used in receiving or retrieving the word to check the validity of the data. Each exclusive-NOR gate can function as a parity element, producing a logic HIGH if an even number of inputs (0 or 2) are active (HIGH) or a logic LOW if an odd number (one of the two) is active.

The quad package as a unit can also be used as a 4-bit digital comparator. If a 4-bit word is placed at the four A inputs, and another 4-bit word is placed at the four B inputs, the four Y outputs will all be HIGH if and only if the two words are identical. If the gates have open-collector outputs, they can be wired together to produce a single output bit that indicates equality or inequality of the two input words.

Exclusive-NOR logic has a kind of symmetry (see function table) that allows for an unusual flexibility in how it is represented. The following four logic symbols are equally valid ways of representing the action of this logic element. The designer should pick the form of representation most suitable to the given application.

As shown in Figure 1.1, an odd consequence of exclusive-NOR logic is that an exclusive-NOR gate can be made from an exclusive-OR gate and an inverter not only in the usual way, by adding the inverter to the gate output, but also by adding the inverter to either gate input. If inverters are added to both inputs, however, the gate's logic remains unchanged.

An unfortunate numbering conflict complicates a basically simple picture of the commercially available exclusive-NOR gates. The original exclusive-NOR gate was an open-collector LS circuit, the 266. As just noted, the open-collector outputs allowed the circuit to be used as a simple digital comparator.

When HC versions of the 266 with totem-pole (push-pull) outputs were developed, several manufacturers (RCA, Signetics, Texas Instruments, and Toshiba) wisely chose a different designation, 7266, to mark the important distinction between output types. But Motorola, National, and SGS chose instead to recycle the designation 266 for their totem-pole HC versions, creating some confusion between these standard-output components and the open-collector LS266. Worse yet, Texas Instruments quite properly used 266 to designate its open-drain HC version of the original open-collector circuit, so that HC 266 can at present refer to a component with either kind of output. In this book, the symbol "%" is used to mark aberrant part numbers in case of numbering conflicts, so the HC circuits with totem-pole outputs marketed as the 266 are here referred to as the 266%.

All currently available 7400-series exclusive-NOR gates are quad 2-input types supplied in 14-pin DIPs.

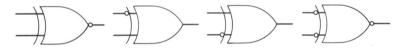

Figure 1.1

7266   *Quad 2-input exclusive-NOR gates.* HC circuits with ordinary to-tem-pole outputs.

266%   *Quad 2-input exclusive-NOR gates.* Same as 7266 with (depre-cated) variant type number.

810§   *Quad 2-input exclusive-NOR gates.* ALS and AS circuits. Pinout differs from that of the 7266 and 266. Data below is for ALS version; most data on AS version not available. (National Semi-conductor)

I_{OH} (max.)	−0.4 mA	t_{PLH} o.i. LO (max.)	20 ns
I_{OL} (max.)	8 mA	t_{PHL} o.i. LO (max.)	14 ns
I_{CC} (max.)	7.5 mA	t_{PLH} o.i. HI (max.)	18 ns
		t_{PHL} o.i. HI (max.)	14 ns

266   *Quad 2-input exclusive-NOR gates with open-collector or open-drain outputs.* Standard LS circuits with OC outputs plus an HC version with OD outputs.

811   *Quad 2-input exclusive-NOR gates with open collector outputs.* Open-collector versions of the 810§.

Note.   In some exclusive-NOR implementations, the propagation delay between a given input and the gate output can change markedly depending on whether the other input happens to be LOW or HIGH. In the following data sheets, the propagation delay times have been broken down into the two cases where the other input is LOW ("o.i. LO") or HIGH ("o.i. HI") when such information has been provided by the manufacturers.

FUNCTION TABLE

INPUTS		OUTPUT
A	B	Y
L	L	H
L	H	L
H	L	L
H	H	H

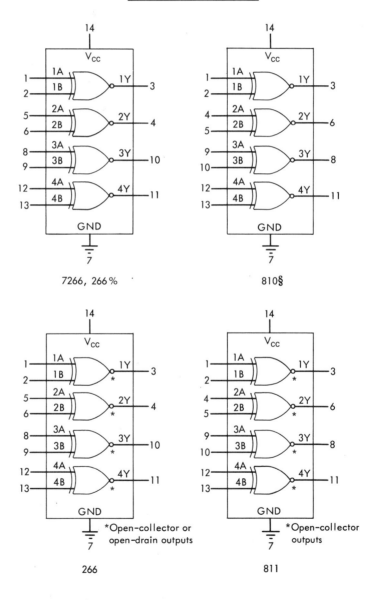

7266, 266%

810§

266

*Open-collector or
open–drain outputs

811

*Open-collector
outputs

7266   Quad 2-input exclusive-NOR gates

MANUFACTURERS	TTL	LS	HC	HCT	ALS	S	AS	F	AC	ACT
RCA			■							
Signetics			■							
Texas Instruments			■							
Toshiba			■							

KEY PARAMETERS		TTL	LS	HC	HCT	ALS	S	AS	F	AC	ACT
I_{OH} (max.)	mA			-4							
I_{OL} (max.)	mA			4							
I_{CC} (quiesc.)	mA			0.02^a							
C_{pd} (typ.)	pF/gate			33^b							
t_{PLH} (max.)	ns			29^c							
t_{PHL} (max.)	ns			29^c							

[a] Toshiba 0.01 mA
[b] RCA; Signetics 17 pF/gate, Toshiba 34 pF/gate, TI 35 pF/gate
[c] Signetics and TI 25 ns

266%   Quad 2-input exclusive-NOR gates

MANUFACTURERS	TTL	LS	HC	HCT	ALS	S	AS	F	AC	ACT
Motorola			▨							
National Semiconductor			■							
SGS			▨							

KEY PARAMETERS		TTL	LS	HC	HCT	ALS	S	AS	F	AC	ACT
I_{OH} (max.)	mA			-4							
I_{OL} (max.)	mA			4							
I_{CC} (quiesc.)	mA			0.02^a							
C_{pd} (typ.)	pF/gate			33^b							
t_{PLH} (max.)	ns			30^c							
t_{PHL} (max.)	ns			30^c							

[a] SGS 0.01 mA
[b] Motorola; National 25 pF/gate, SGS 34 pF/gate
[c] SGS 26 ns

266 Quad 2-input exclusive-NOR gates with OC or OD outputs

MANUFACTURERS	TTL	LS	HC	HCT	ALS	S	AS	F	AC	ACT
Fairchild		■								
Motorola		■								
National Semiconductor		■								
SGS		■								
Signetics		■								
Texas Instruments		■	■							
Toshiba			D							

KEY PARAMETERS		TTL	LS	HC	HCT	ALS	S	AS	F	AC	ACT
I_{OH} (max.)	mA		0.1[a]	0.005							
I_{OL} (max.)	mA		8	4							
I_{CC} (max.)	mA		13								
I_{CC} (quiesc.)	mA			0.02							
C_{pd} (typ.)	pF/gate			35							
t_{PLH} (max.)	ns		30[b]	31							
t_{PHL} (max.)	ns		30[b]	25							

[a] Fairchild data not available
[b] Fairchild 23 ns

811 Quad 2-input exclusive-NOR gates with open-collector outputs

MANUFACTURERS	TTL	LS	HC	HCT	ALS	S	AS	F	AC	ACT
National Semiconductor					■		▒			
Texas Instruments					■					

KEY PARAMETERS		TTL	LS	HC	HCT	ALS	S	AS	F	AC	ACT
I_{OH} (max.)	mA					0.1		0.1			
I_{OL} (max.)	mA					8		20			
I_{CC} (max.)	mA					7.5		n/a			
t_{PLH} o.i. LO (max.)	ns					55		n/a			
t_{PHL} o.i. LO (max.)	ns					28		n/a			
t_{PLH} o.i. HI (max.)	ns					50		n/a			
t_{PHL} o.i. HI (max.)	ns					23		n/a			

EXCLUSIVE-OR GATES

The exclusive-OR gate can best be understood as an inverted form of the exclusive-NOR: the gate's output is LOW if both inputs have the same logic value and HIGH if they don't. Applications are the same as for the exclusive-NOR, and there are likewise several equally valid ways of symbolizing this circuit, as shown in Figure 1.2.

An odd consequence of exclusive-OR logic is that an exclusive-OR gate can be made from an exclusive-NOR gate and an inverter not only in the usual way, by adding the inverter to the gate output, but also, as shown below, by adding the inverter to either gate *input*. However, the addition of inverters to *both* inputs leaves the gate's logic unchanged because the common inversion does not affect the equality or inequality of the inputs, which is the condition that determines the gate's output.

All currently available 7400-series exclusive-OR gates are quad 2-input types supplied in 14-pin DIPs.

86   *Quad 2-input exclusive-OR gates.* Standard components available in nearly all technologies.

386   *Quad 2-input exclusive-OR gates.* LS and HC versions of the 86 with a variant pinout.

136   *Quad 2-input exclusive-OR gates with open-collector outputs.* OC versions of the 86.

Note.   In some exclusive-OR implementations, the propagation delay between a given input and the gate output can change markedly depending on whether the other input happens to be LOW or HIGH. In the following data sheets, the propagation delay times have been broken down into the two cases where the other input is LOW ("o.i. LO") or HIGH ("o.i. HI") when such information has been provided by the manufacturers.

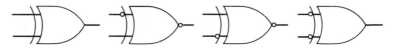

Figure 1.2

FUNCTION TABLE

INPUTS		OUTPUTS
A	B	Y
L	L	L
L	H	H
H	L	H
H	H	L

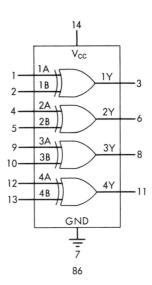

86

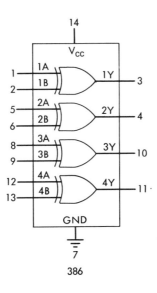

386

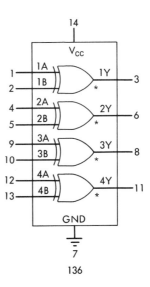

136

MANUFACTURERS	TTL	LS	HC	HCT	ALS	S	AS	F	AC	ACT
Fairchild										
Motorola										
National Semiconductor										
RCA										
SGS										
Signetics										
Texas Instruments										
Toshiba				D					D	

Added change control symbol: Texas Instruments 74LS86A

KEY PARAMETERS		TTL	LS	HC	HCT	ALS	S	AS	F	AC	ACT
I_{OH} (max.)	mA	-0.8	-0.4	-4	-4	-0.4	-1	-2	-1	-24	-24
I_{OL} (max.)	mA	16	8	4	4	8	20	20	20	24	24
I_{CCH} (max.)	mA	50	10			5.9	75[i]	n/a	23[j]		
I_{CCL} (max.)	mA	57[a]	15[a]			5.9	75[a]	n/a	28[k]		
I_{CC} (quiesc.)	mA			0.02[e]	0.02					0.04	0.04
C_{pd} (typ.)	pF/gate			30[f]	30[h]					57[l]	81
t_{PLH} o.i. LO (max.)	ns	23	23[b]	30[g]	40	17	10.5	n/a	6.5	9.7[l]	13
t_{PHL} o.i. LO (max.)	ns	17	17	30[g]	40	12	10	n/a	6.5	9.7[l]	13
t_{PLH} o.i. HI (max.)	ns	30	30[c]	30[g]	40	17	10.5	n/a	8.0	9.7[l]	13
t_{PHL} o.i. HI (max.)	ns	22	22[d]	30[g]	40	10	10	n/a	7.5	9.7[l]	13

[a] National (I_{CCL} for others not available)
[b] Fairchild 12 ns, National 18 ns
[c] National 10 ns, Fairchild 13 ns
[d] Fairchild and National 12 ns
[e] SGS and Toshiba 0.01 mA
[f] Signetics; RCA 22 pF/gate, National 25 pF/gate, Motorola 33 pF/gate, SGS and Toshiba 34 pF/gate, TI 35 pF/gate

[g] Signetics 18 ns, TI 25 ns, SGS 33 ns
[h] Signetics; RCA 27 pF/gate
[i] National 50 mA
[j] Fairchild 28 mA
[k] Fairchild 18 mA
[l] RCA; Fairchild data not available

386   Quad 2-input exclusive-OR gates

MANUFACTURERS	TTL	LS	HC	HCT	ALS	S	AS	F	AC	ACT
Motorola		■								
National Semiconductor		■								
SGS		■	▨							
Texas Instruments		■								
Toshiba		■								

Added change control symbol: Texas Instruments 74LS386A

KEY PARAMETERS		TTL	LS	HC	HCT	ALS	S	AS	F	AC	ACT
I_{OH} (max.)	mA		-0.4	-4							
I_{OL} (max.)	mA		8	4							
I_{CCH} (max.)	mA		10[a]								
I_{CC} (quiesc.)	mA			0.02[e]							
C_{pd} (typ.)	pF/gate			33[f]							
t_{PLH} o.i. LO (max.)	ns		23[b]	30[g]							
t_{PHL} o.i. LO (max.)	ns		17	30[g]							
t_{PLH} o.i. HI (max.)	ns		30[c]	30[g]							
t_{PHL} o.i. HI (max.)	ns		22[d]	30[g]							

[a] National adds I_{CCL} = 15 mA
[b] National 18 ns
[c] National 10 ns
[d] National 12 ns

[e] SGS and Toshiba 0.01 mA
[f] TI 35 pF/gate
[g] Toshiba; TI 25 ns, SGS 33 ns

136   Quad 2-input exclusive-OR gates with open-collector outputs

MANUFACTURERS	TTL	LS	HC	HCT	ALS	S	AS	F	AC	ACT
Fairchild		■								
Motorola		■								
National Semiconductor					■	■	▨			
SGS		■								
Signetics		■								
Texas Instruments	■	■			■					

KEY PARAMETERS		TTL	LS	HC	HCT	ALS	S	AS	F	AC	ACT
I_{OH} (max.)	mA	0.25	0.1[a]			0.1	0.25	0.25			
I_{OL} (max.)	mA	16	8			8	20	20			
I_{CC} (max.)	mA	50	10			5.9	75	n/a			
t_{PLH} o.i. LO (max.)	ns	18	30[b]			50	12.5	n/a			
t_{PHL} o.i. LO (max.)	ns	50	30[b]			15	12	n/a			
t_{PLH} o.i. HI (max.)	ns	22	30[b]			50	12.5	n/a			
t_{PHL} o.i. HI (max.)	ns	55	30[b]			12	12	n/a			

[a] Fairchild data not available in this form
[b] Fairchild 23 ns

EXCLUSIVE-OR/NOR GATES

135   *Quad exclusive-OR/NOR gates.* A pair of C inputs allow these S
        components to perform as either quad exclusive-OR gates or quad
        exclusive-NOR gates. If a C input is held LOW, the associated pair
        of gates function as exclusive-ORs; if the C input is held HIGH, the
        same two gates become exclusive-NORs.

FUNCTION TABLE

INPUTS			OUTPUT	
A	B	C	Y	
L	L	L	L	
L	H	L	H	C is LOW
H	L	L	H	(exclusive-OR mode)
H	H	L	L	
L	L	H	H	
L	H	H	L	C is HIGH
H	L	H	L	(exclusive-NOR mode)
H	H	H	H	

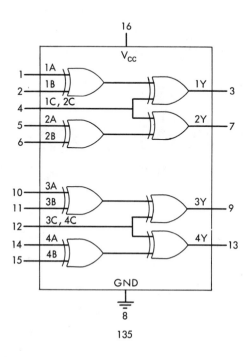

135   Quad exclusive-OR/NOR gates

MANUFACTURERS	TTL	LS	HC	HCT	ALS	S	AS	F	AC	ACT
National Semiconductor						■				
Signetics						■				
Texas Instruments						■				

KEY PARAMETERS		TTL	LS	HC	HCT	ALS	S	AS	F	AC	ACT
I_{OH} (max.)	mA						-1				
I_{OL} (max.)	mA						20				
I_{CC} (max.)	mA						99				
t_{PLH} (max.)	ns						15				
t_{PHL} (max.)	ns						15				

AND-OR-INVERT GATES

Logically, the AND-OR-invert (AOI) gate is the AND-OR gate (see below) with an inverted output. In practice, the AOI is much more common than the noninverted AND-OR gate.

The AOI output comes from a NOR gate with inputs provided by two or more AND gates. (Commercially available versions have either two AND gates or four AND gates in the input stage). A logic LOW at the AOI output signifies that one or more of the AND gates has inputs that are all HIGH. A logic HIGH at the AOI output signifies that none of the AND gates has attained this condition. The AOI gate is useful when several distinct sets of inputs must be monitored for the simultaneous occurrence of a given condition within any one of them.

All 7400-series AND-OR-invert gates are supplied in 14-pin DIPs.

51   *2-2-input and 3-3input AND-OR-invert gates.* An IC consisting of dual independent 2-wide AOI gates, available in two basic configurations. In the original TTL and S versions, both AOI gates receive input through a pair of 2-input ANDs, an arrangement that leaves two of the DIP's 14 pins unconnected. In the later LS and HC versions, these two pins (11 and 12) provide a third input to each of the two AND gates in one of the dual AOIs making up the package, so that one of the AOIs receives input through a pair of 3-input ANDs rather than 2-input ANDs. The 51 is the most widely supported of the AOI gates and is the only one available in HC versions.

55   *4-4-input AND-OR-invert gates.* Single AOI gates with two 4-input ANDs. LS only.

54   *2-2-2-2-input or 2-2-3-3 input AND-OR-invert gates.* Single AOI gates with four ANDs, available (like the 51) in two basic configura-

tions. In the original TTL version, each AND has just two inputs, and three pins are left unconnected. In the LS version, two of the AND gates again have two inputs, but the other two have three inputs, leaving just one of the 14 pins still unconnected.

64   *2-2-3-4-input AND-OR-invert gates.* These S and F versions of the 54 make one of the 3-input ANDs into a 4-input AND, thus using all 14 pins.

65   *2-2-3-4-input AND-OR-invert gates with open-collector outputs.* OC versions of the 64. S only.

FUNCTION TABLE
(AOI with 2-input ANDs)

INPUTS				OUTPUT
A	B	C	D	Y
H	H	.	.	L
.	.	H	H	L
all other cases				H

FUNCTION TABLE
(AOI with 3-input ANDs)

A	B	C	D	E	F	Y
INPUTS						OUTPUT
H	H	H	.	.	.	L
.	.	.	H	H	H	L
all other cases						H

· = Either LOW or HIGH logic level

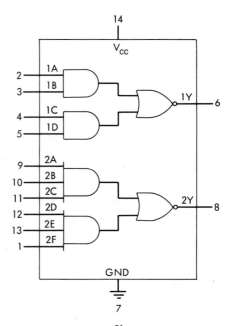

51
(Pins 11, 12 are NC in
the TTL and S versions)

FUNCTION TABLE

INPUTS								OUTPUT
A	B	C	D	E	F	G	H	Y
H	H	H	H	.	.	.	.	L
.	.	.	.	H	H	H	H	L
all other cases								H

· = Either-LOW or HIGH logic level

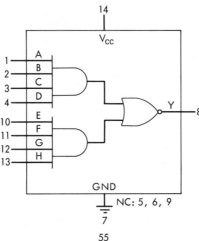

55

FUNCTION TABLE

INPUTS								OUTPUT
A	B	C	D	E	F	G	H	Y
H	H	.	.	.	.	.	.	L
.	.	H	H	.	.	.	.	L
.	.	.	.	H	H	.	.	L
.	.	.	.	.	.	H	H	L
all other cases								H

• = Either LOW or HIGH logic level

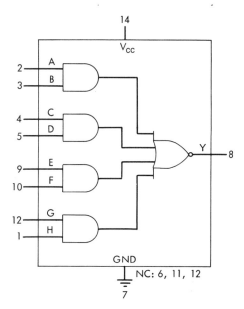

54 (TTL versions)

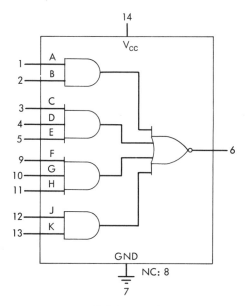

54 (LS versions)

FUNCTION TABLE

INPUTS										OUTPUT
A	B	C	D	E	F	G	H	J	K	Y
H	H	.	.	.	.	.	.	.	.	L
.	.	H	H	H	.	.	.	.	.	L
.	.	.	.	.	H	H	H	.	.	L
.	.	.	.	.	.	.	.	H	H	L
all other cases										H

• = Either LOW or HIGH logic level

FUNCTION TABLE

INPUTS											OUTPUT
A	B	C	D	E	F	G	H	J	K	L	Y
H	H	.	.	.	.	.	.	.	.	.	L
.	.	H	H	H	.	.	.	.	.	.	L
.	.	.	.	.	H	H	.	.	.	.	L
.	.	.	.	.	.	.	H	H	H	H	L
all other cases											H

• = Either LOW or HIGH logic level

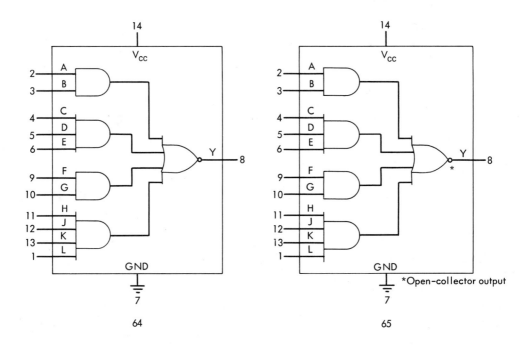

*Open-collector output

64   65

51   2-2-input and 3-3-input AND-OR-invert gates

MANUFACTURERS	TTL	LS	HC	HCT	ALS	S	AS	F	AC	ACT
Fairchild	■	■				■				
Motorola		■	▨			■				
National Semiconductor		■	▨			■				
SGS		■	▨							
Signetics	■					■		■		
Texas Instruments	■		■			■				
Toshiba			■							

KEY PARAMETERS		TTL	LS	HC	HCT	ALS	S	AS	F	AC	ACT
I_{OH} (max.)	mA	-0.4^a	-0.4	-4			-1		-1		
I_{OL} (max.)	mA	16	8	4			20		20		
I_{CCH} (max.)	mA	8	1.6				17.8		3		
I_{CCL} (max.)	mA	14	2.8				22		7.5		
I_{CC} (quiesc.)	mA			0.02^d							
C_{pd} (typ.)	pF/gate			25^e							
t_{PLH} (max.)	ns	22	20^b	32^f			5.5		6.5		
t_{PHL} (max.)	ns	15	20^c	32^f			5.5		4.5		

[a] Fairchild -0.8 mA
[b] National 13 ns
[c] National 12 ns
[d] SGS and Toshiba 0.01 mA

[e] TI; National 20 pF/gate, SGS and Toshiba 33 pF/gate. Motorola data not available
[f] Motorola and National; SGS and Toshiba 26 ns, TI 35 ns

55   4-4-input AND-OR-invert gates

MANUFACTURERS	TTL	LS	HC	HCT	ALS	S	AS	F	AC	ACT
Fairchild		■								
Motorola		■								
National Semiconductor		■								
SGS		■								
Texas Instruments		■								

KEY PARAMETERS		TTL	LS	HC	HCT	ALS	S	AS	F	AC	ACT
I_{OH} (max.)	mA		-0.4								
I_{OL} (max.)	mA		8								
I_{CCH} (max.)	mA		0.8								
I_{CCL} (max.)	mA		1.3								
t_{PLH} (max.)	ns		20^a								
t_{PHL} (max.)	ns		20^b								

[a] National 13 ns, Fairchild 15 ns
[b] National 12 ns, Fairchild 15 ns

54   2-2-2-2-input or 2-2-3-3-input AND-OR-invert gates

MANUFACTURERS	TTL	LS	HC	HCT	ALS	S	AS	F	AC	ACT
Fairchild		▓								
Motorola		▓								
National Semiconductor		▓								
SGS	▓	▓								
Signetics	▓	▓								
Texas Instruments	▓	▓								

KEY PARAMETERS		TTL	LS	HC	HCT	ALS	S	AS	F	AC	ACT
I_{OH} (max.)	mA	-0.4	-0.4								
I_{OL} (max.)	mA	16	8								
I_{CCH} (max.)	mA	8	1.6								
I_{CCL} (max.)	mA	9.5	2								
t_{PLH} (max.)	ns	22	20[a]								
t_{PHL} (max.)	ns	15	20[b]								

[a] Fairchild 15 ns
[b] National 13 ns, Fairchild 15 ns

64   2-2-3-4-input AND-OR-invert gates

MANUFACTURERS	TTL	LS	HC	HCT	ALS	S	AS	F	AC	ACT
Fairchild						▓		▓		
Motorola								▓		
National Semiconductor						▓				
Signetics						▓		▓		
Texas Instruments						▓				

KEY PARAMETERS		TTL	LS	HC	HCT	ALS	S	AS	F	AC	ACT
I_{OH} (max.)	mA						-1		-1		
I_{OL} (max.)	mA						20		20		
I_{CCH} (max.)	mA						12.5		2.8		
I_{CCL} (max.)	mA						16		4.7		
t_{PLH} (max.)	ns						5.5		7[a]		
t_{PHL} (max.)	ns						5.5		5.5		

[a] Fairchild 7.5 ns

65	**2-2-3-4-input AND-OR-invert gates with open-collector outputs**

MANUFACTURERS	TTL	LS	HC	HCT	ALS	S	AS	F	AC	ACT
National Semiconductor						■				
Texas Instruments						■				

KEY PARAMETERS		TTL	LS	HC	HCT	ALS	S	AS	F	AC	ACT
I_{OH} (max.)	mA						0.25				
I_{OL} (max.)	mA						20				
I_{CCH} (max.)	mA						11				
I_{CCL} (max.)	mA						16				
t_{PLH} (max.)	ns						7.5				
t_{PHL} (max.)	ns						8.5				

AND-OR GATES

The AND-OR gate is a noninverted version of the more commonly found AND-OR-invert gate. In an AND-OR gate, a number of AND gates supply input to a single OR gate that provides the overall AND-OR output. This output is HIGH if all the inputs to any one (or more) of the AND gates are HIGH. The output is LOW otherwise, that is, in the case where none of the AND gates has uniformly HIGH inputs. Only one kind of 7400-series AND-OR gate is commercially available.

58   *2-2-input and 3-3-input AND-OR gates.* A 14-pin DIP houses two independent 2-wide AND-OR gates, one with a pair of 2-input ANDs and the other with a pair of 3-input ANDs. These HC circuits are identical to the HC51 (see above) except that the outputs are not inverted.

FUNCTION TABLE
(2-input ANDs)

INPUTS	OUTPUT
A B C D	Y
H H . .	H
. . H H	H
all other cases	L

FUNCTION TABLE
(3-input ANDs)

INPUTS	OUTPUT
A B C D E F	Y
H H H . . .	H
. . H H H	H
all other cases	L

• = Either LOW or HIGH logic level

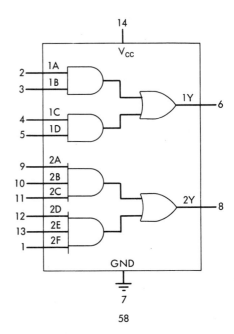

58

58   2-2-input and 3-3-input AND-OR gates

MANUFACTURERS	TTL	LS	HC	HCT	ALS	S	AS	F	AC	ACT
Motorola			▓							
National Semiconductor			▓							
Signetics			D							

KEY PARAMETERS		TTL	LS	HC	HCT	ALS	S	AS	F	AC	ACT
I_{OH} (max.)	mA			-4							
I_{OL} (max.)	mA			4							
I_{CC} (quiesc.)	mA			0.02							
C_{pd} (typ.)	pF/gate			20[a]							
t_{PLH} (max.)	ns			32							
t_{PHL} (max.)	ns			32							

[a] National; Motorola data not available

COMPLEMENTARY OUTPUT GATES

These single-sourced components (two of them still in the planning stage) are specifically designed for use as true/complement drivers. See Texas Instruments data for further information and application notes.

265§   *Quad complementary-output elements.* These TTL circuits solve the problem of how to provide both inverted and noninverted versions of a signal, such as the output from a clock generator, without introducing the time lag or "skew" between the two forms that would result if an extra inverter were inserted into one of the lines. They can also be used as differential line drivers, as components of complementary input sections for decoders and code converters, and in switch debouncing. The quad package (14-pin DIP) provides two 1-input inverter/noninverter gates and two 2-input AND/NAND gates.

I_{OH} (max.)	−0.8 mA	t_{PLH} (max.)	18 ns
I_{OL} (max.)	16 mA	t_{PHL} (max.)	18 ns
I_{CC} (max.)	34 mA	Skew (max.)	3 ns

[800]   *Triple 4-input AND/NAND buffers/drivers.* AS components presently in the design stage. Will have high-current (−48/48 mA) complementary outputs with typical skew less than 0.5 ns, suitable for complementary output applications such as clock generators, decoder inputs, and switch debouncing.

[802]   *Triple 4-input OR/NOR buffers/drivers.* AS components presently in the design stage. Will have high-current (−48/48 mA) complementary outputs with typical skew less than 0.5 ns, suitable for complementary output applications such as clock generators, decoder inputs, and switch debouncing.

CURRENT-SENSING INTERFACE GATES

63§   *Hex current-sensing interface gates.* LS gates that translate logic based on the difference between a high current level and a low current level into common logic based on the difference between a high voltage level and a low voltage level. Specifically, low-level input current (0.05 mA or less) is translated into a logic LOW, and high-level input current (0.2 mA or more) is translated into a logic HIGH. These devices are designed to interface common logic circuits to PLAs or other logic elements that source current but do not sink current. Supplied in a 14-pin DIP that operates from the standard 5 V power supply. See manufacturer's data for further information. (Texas Instruments)

I_{OH} (max.)	-0.4 mA	t_{PLH} (max.)	45 ns ($C_L = 15$ pF)
I_{OL} (max.)	8 mA	t_{PHL} (max.)	25 ns ($C_L = 15$ pF)
I_{CC} (max.)	16 mA		

2

Latches and Flip-Flops

Unlike simple gates, which are combinational devices, latches and flip-flops implement *sequential* logic, that is, logic that comprehends not just the present combination of inputs but also their past history. Simply put, latches and flip-flops can "remember" data after their inputs are removed.

Each latch and flip-flop in this section can retain just one bit of data, a logic 1 or a logic 0 (represented by a HIGH or LOW voltage level). Sets of four to eight latches or flip-flops with common clocks can retain four to eight bits of data at once; these essential interface components, classified as parallel latches and registers, are treated in Part II, Parallel Circuits. Larger arrays of flip-flops are classified as memories; computer memories, which are beyond the scope of this book, are made up of thousands or millions of flip-flops, each retaining just one bit of data.

The individually accessible latches and flip-flops listed in this section are grouped as follows:

$\overline{S}$-$\overline{R}$ (Set-Reset) latches
Transparent latches
D flip-flops
J-K flip-flops
J-$\overline{K}$ flip-flops

AND-gated J-K and J-$\overline{\text{K}}$ flip-flops

Multifunction circuits with flip-flops

Latches and flip-flops are ordinarily activated by incoming pulses representing clock signals or data. Timing requirements associated with input pulses introduce three new key parameters:

t_W (min.)	The (maximum) minimum input pulse width, that is, the worst-case figure for the minimum width (in nanoseconds) that an input pulse must have to guarantee proper operation.
t_{SU} (min.)	The (maximum) minimum setup time, that is, the worst-case figure for the minimum amount of time that valid data must be present at an input before the clock or enable transition that causes the device to read or hold data.
t_H (min.)	The (maximum) minimum hold time, that is, the worst-case figure for the minimum amount of time that valid data must be held at an input after the clock or enable transition in order to be properly read by the device.

Another new key parameter is f_{MAX}, the (minimum) maximum clock frequency at which the device is guaranteed to operate properly. The minimum f_{MAX} of flip-flops and other sequential devices is widely used as a figure of merit for comparing relative speeds of operation.

All 7400-series latches and flip-flops have standard totem-pole outputs with standard output-current ratings for the given technology. Concerning inputs, however, it should be noted that in TTL, LS, ALS, S, AS, and F technologies the data, clock, and other inputs may differ from one another in their maximum input current ratings I_H and I_L. See manufacturer's data sheets for the particular component involved.

$\overline{S}$-$\overline{R}$ (SET-RESET) LATCHES

$\overline{S}$-$\overline{R}$ latches are the simplest of all electronic sequential logic devices. An $\overline{S}$-$\overline{R}$ latch is made from two NAND gates by cross-coupling them as shown in Figure 2.1.

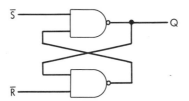

Figure 2.1

In this circuit, the $\overline{S}$ and $\overline{R}$ inputs are normally HIGH. If a low-going pulse arrives at the $\overline{S}$ or "set" input, the output at Q is forced HIGH (because any LOW input to a NAND will cause the output to be. HIGH) and then stays HIGH regardless of any further changes at the $\overline{S}$ input. That is, once the latch is set (Q is HIGH), it stays set until specifically reset by a low-going $\overline{R}$ input. This is because the HIGH level at Q is also applied to the lower NAND gate and combines with the normal HIGH level at $\overline{R}$ to produce a LOW output at the lower NAND. This LOW output is in turn fed back up to the top NAND, maintaining the HIGH output regardless of the input at $\overline{S}$, which is why further changes at the $\overline{S}$ input have no effect.

When the $\overline{R}$ (reset) input goes LOW at some later time, the dual HIGH condition at the lower NAND inputs is broken and the output of the lower NAND goes HIGH. This output, fed back to the upper NAND, combines with the normal HIGH level at $\overline{S}$ to force the latch output at Q back to LOW. The cross-coupled feedback now keeps the Q output LOW regardless of further changes at $\overline{R}$ until such time as a low-going signal at $\overline{S}$ sets the latch again. In summary, a low-going pulse at $\overline{S}$ sets the latch, which then provides a HIGH level at the Q output (presumably signifying that Q = 1) until a low-going pulse at $\overline{R}$ resets the latch back to Q = 0. In effect, the latch holds or "remembers" a LOW input until it is specifically instructed to change by a LOW signal at the other input.

There is only one form of commercially available common-logic $\overline{S}$-$\overline{R}$ latch, the 279. This provides four independent $\overline{S}$-$\overline{R}$ latches in a single 16-pin DIP. Two of the four latches in each package implement the circuit shown above; the other two have dual $\overline{S}$ inputs that are ANDed together to form the internal $\overline{S}$ input.

Simple $\overline{S}$-$\overline{R}$ latches cannot be synchronized with a system clock without a significant amount of additional circuitry, and consequently they have become obsolete for most applications. One of the very few exceptions is switch debouncing. An ordinary contact switch will commonly experience jittery contact or "bounce" in the brief instant that the contacts are coming together or pulling apart. The resulting momentary voltage fluctuations can seriously affect the performance of devices reading the switch, which might, for example, mistake a series of erratic voltage changes for a train of pulses.

A bounceless switch can be made by tying both the $\overline{S}$ and $\overline{R}$ inputs of an $\overline{S}$-$\overline{R}$ latch to V_{CC} through appropriate resistors and then connecting a single-pole, double-throw switch so that in one switch position the $\overline{S}$ input is connected to LOW (i.e., grounded) and in the other position the $\overline{R}$ input is connected to LOW instead. When $\overline{S}$ is selected, Q will be HIGH, and when $\overline{R}$ is selected, Q will be LOW. The ability of the latch to reject changes in input after the first voltage transition means that momentary fluctuations in voltage caused by the switch opening or closing will have no effect. Also, the latched switch is faster than the purely mechanical version.

FUNCTION TABLE
(Latches 1 and 3)

INPUTS			OUTPUT
$\overline{S}_1$	$\overline{S}_2$	$\overline{R}$	Q
H	H	H	Q_0
H	H	L	L
L	•	H	H
•	L	H	H
L	•	L	H*
•	L	L	H*
L	L	L	H*

No change
Reset latch
Set latch

(Transient)

• = Either LOW or HIGH logic level
Q_0 = The level at Q before establishment
　　of the present input conditions
* = Nonstable state (may not persist
　　when $\overline{S}$ and $\overline{R}$ return to their normal
　　HIGH level)

FUNCTION TABLE
(Latches 2 and 4)

INPUTS		OUTPUT
$\overline{S}$	$\overline{R}$	Q
H	H	Q_0
H	L	L
L	H	H
L	L	H*

No change
Reset latch
Set latch

(Transient)

Q_0 = The level at Q before establishment
　　of the present input conditions
* = Nonstable state (may not persist
　　when $\overline{S}$ and $\overline{R}$ return to their normal
　　HIGH level)

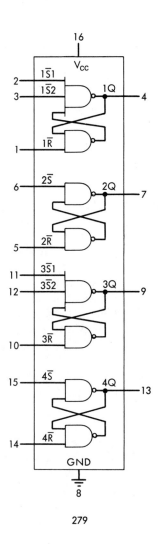

279

279 Quad $\overline{S}$-$\overline{R}$ latches

MANUFACTURERS	TTL	LS	HC	HCT	ALS	S	AS	F	AC	ACT
Fairchild	■									
Motorola		■								
National Semiconductor		■								
SGS		■	▨							
Signetics	■									
Texas Instruments	■	■								
Toshiba			■							

Added change control symbol: Texas Instruments 74LS279A

KEY PARAMETERS		TTL	LS	HC	HCT	ALS	S	AS	F	AC	ACT
I_{OH} (max.)	mA	-0.8	-0.4	-4							
I_{OL} (max.)	mA	16	8	4							
I_{CC} (max.)	mA	30	7								
I_{CC} (quiesc.)	mA			0.02							
C_{pd} (typ.)	pF/latch			26							
t_{pd} $\overline{S}$ to Q (max.)	ns	22	22	24^a							
t_{pd} $\overline{R}$ to Q (max.)	ns	27	27	29^b							

[a] SGS; Toshiba 25 ns. In these HC versions, the two latches with dual gated $\overline{S}$ inputs have t_{pd} (max.) = 32 ns (SGS) or 33 ns (Toshiba)

[b] SGS; Toshiba 30 ns

TRANSPARENT LATCHES

A transparent latch can be thought of as a low-current buffer with data input D and data output Q. The Q output follows the D input as in an ordinary noninverting buffer, with just the propagation delay t_{pd} separating changes in D from changes in Q.

In addition, this circuit has the ability to "latch" or lock in its current D state at a given moment under control of an additional enable input, E. When E is HIGH, the latch is enabled to act like a buffer: it is "transparent," simply reproducing its current input. But when E goes LOW, the bit then present at D is stored at the Q output, and thereafter the latch ignores its input and maintains its Q output unchanged for as long as E is held LOW. When E goes HIGH again, data then at D will appear at Q after the normal propagation delay, and the latch will be back in its transparent mode.

For proper latching operation the D input must have valid data for at least one setup time t_{SU} before the HIGH-to-LOW transition at E that actuates the latching operation, and it must stay at the same level for at least one hold time t_H after this transition. Since the propagation delay t_{pd} between D and Q in the transparent mode is greater than the setup time t_{SU}, it is possible in marginal cases for the level at output Q during the HIGH-to-LOW transition of E to be momentarily different from the level that will, after a short delay, appear at Q and end up stored in the latch. In normal cases, however, the latching operation can be considered to have zero propagation delay, because under ordinary conditions the information at D has been valid for some time, and the enable transition simply locks out further changes that would disturb an already existing condition at Q.

Transparent latches are commonly used to provide temporary storage of data between processing units and exterior devices. They are particularly useful in asynchronous data transfers, where the output of one unit must be held steady until another unit is ready to read the data.

There are three commercially available 7400-series transparent latches that allow direct control at the bit level (or, to be more accurate, control over pairs of bits): the 75, 375, and 77. Each package contains two independent devices, and each device consists of a pair of one-bit transparent latches controlled by a common enable (E) input. Thus, each IC can be considered to provide a single 4-bit transparent latch, two independent 2-bit transparent latches, or four 1-bit latches that must be enabled in pairs. Transparent latches that handle groups of four or more bits as a single unit will be found in Chapter 5, Parallel Latches.

75   *Quad transparent latches (enabled in pairs).* The basic transparent latch, available in TTL, LS, HC, and HCT. Each of the four latches provides both a noninverted Q output and, for convenience, an inverted $\overline{Q}$ output. In the TTL and LS versions the propagation delay between the D input and the inverted output $\overline{Q}$ is slightly less than the delay between D and the noninverted output Q shown in the table of key parameters. Supplied in 16-pin DIP.

375   *Quad transparent latches (enabled in pairs).* Same as the 75 but with supply voltage and ground assigned to the corner pins to simplify PC board layout. Currently available only in LS and HC versions.

77   *Quad transparent latches (enabled in pairs).* In these LS and HC components the four $\overline{Q}$ outputs provided by the 75 and 375 are omitted, allowing the same functions to be supplied in a 14-pin package (two pins, 7 and 10, are unused).

FUNCTION TABLE

INPUTS		OUTPUTS	
D	E	Q	$\overline{Q}$
L	H	L	H
H	H	H	L
•	L	Q_0	$\overline{Q}_0$

• = Either LOW or HIGH
    logic level
Q_0= Level at D (and normally
    at Q) before the HIGH – to
    LOW transition of E

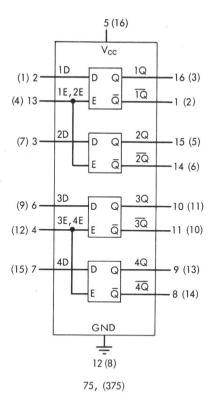

75, (375)

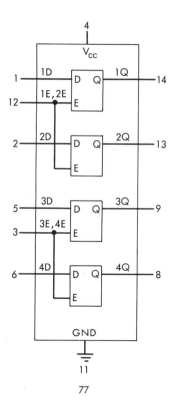

77

149

MANUFACTURERS	TTL	LS	HC	HCT	ALS	S	AS	F	AC	ACT
Fairchild	■									
Motorola		■	▒							
National Semiconductor										
RCA			■	■						
SGS		▒	▒							
Signetics	■	■	■	■	■					
Texas Instruments	■	■								
Toshiba			■							

KEY PARAMETERS		TTL	LS	HC	HCT	ALS	S	AS	F	AC	ACT
I_{OH} (max.)	mA	-0.4	-0.4	-4							
I_{OL} (max.)	mA	16	8	4							
I_{CC} (max.)	mA	53[a]	12								
I_{CC} (quiesc.)	mA			0.04[e]	0.04						
C_{pd} (typ.)	pF/latch			46[f]	46[l]						
t_W (min.)	ns	20	20[c]	20[g]	20						
t_{SU} (min.)	ns	20	20[c]	15[h]	15						
t_H (min.)	ns	5[b]	0[c]	3[i]	3						
t_{pd} D to Q (max.)	ns	30	27[d]	28[j]	38[m]						
t_{pd} E to Q (max.)	ns	30	27[d]	33[k]	38[m]						

[a] National 50 mA
[b] Fairchild 0 ns
[c] Motorola data not available
[d] t_{pd} E to $\overline{Q}$ (max.) = 30 ns
[e] SGS and Toshiba 0.02 mA
[f] RCA, TI; Motorola 40 pF/pair, Signetics 42 pF/latch, SGS and Toshiba 48 pF/latch
[g] SGS 18 ns, Toshiba 19 ns, National 24 ns

[h] SGS 12 ns, Toshiba 13 ns, National and TI 25 ns
[i] National 0 ns, TI 5 ns, Toshiba and SGS 6 ns
[j] RCA, Signetics; TI and Toshiba 30 ns, Motorola and National 32 ns, SGS 35 ns
[k] RCA, TI; Signetics 30 ns, Motorola, National, Toshiba 36 ns, SGS 39 ns
[l] RCA; Signetics 42 pF/latch
[m] RCA; Signetics 35 ns

Quad transparent latches (enabled in pairs)

MANUFACTURERS	TTL	LS	HC	HCT	ALS	S	AS	F	AC	ACT
Fairchild		■								
Motorola		■								
SGS			▒							
Signetics		■								
Texas Instruments			▒							
Toshiba			■							

KEY PARAMETERS		TTL	LS	HC	HCT	ALS	S	AS	F	AC	ACT
I_{OH} (max.)	mA		-0.4	-4							
I_{OL} (max.)	mA		8	4							
I_{CC} (max.)	mA		12								
I_{CC} (quiesc.)	mA			0.04[b]							
C_{pd} (typ.)	pF/latch			48[c]							
t_W (min.)	ns		20	19[d]							
t_{SU} (min.)	ns		20	13[e]							
t_H (min.)	ns		0	5[f]							
t_{pd} D to Q (max.)	ns		27	30[g]							
t_{pd} E to Q (max.)	ns		27[a]	39[h]							

[a] t_{pd} E to $\overline{Q}$ (max.) = 30 ns
[b] SGS 0.01 mA
[c] Toshiba; SGS 49 pF/latch. TI data not available
[d] Toshiba; SGS 18 ns, TI 20 ns

[e] Toshiba; SGS 12 ns, TI 25 ns
[f] TI, Toshiba; SGS 6 ns
[g] TI, Toshiba; SGS 29 ns
[h] SGS; TI 33 ns, Toshiba 40 ns

77   Quad transparent latches (enabled in pairs)

MANUFACTURERS	TTL	LS	HC	HCT	ALS	S	AS	F	AC	ACT
Motorola		■								
National Semiconductor		■								
SGS			▨							
Texas Instruments			■							
Toshiba			■							

KEY PARAMETERS		TTL	LS	HC	HCT	ALS	S	AS	F	AC	ACT
I_{OH} (max.)	mA		-0.4	-4							
I_{OL} (max.)	mA		8	4							
I_{CC} (max.)	mA		13								
I_{CC} (quiesc.)	mA			0.02^c							
C_{pd} (typ.)	pF/latch			27^d							
t_W (min.)	ns		20	19^e							
t_{SU} (min.)	ns		20	13^f							
t_H (min.)	ns		0	6^g							
t_{pd} D to Q (max.)	ns		27^a	30^h							
t_{pd} E to Q (max.)	ns		27^b	36^i							

[a] National; Motorola 19 ns
[b] National; Motorola 18 ns
[c] SGS, Toshiba; TI 0.04 mA
[d] SGS and Toshiba; TI data not available
[e] Toshiba; SGS 18 ns, TI 20 ns

[f] Toshiba; SGS 12 ns, TI 25 ns
[g] SGS and Toshiba; TI 5 ns
[h] Toshiba, TI; SGS 35 ns
[i] Toshiba; TI 33 ns, SGS 39 ns

D FLIP-FLOPS

The D (Data) flip-flop somewhat resembles a transparent latch but is triggered differently. Like the latch, the D flip-flop has a single D input plus a triggering control input, a Q output, and sometimes a $\overline{Q}$ output. The control input, however, is not level-triggering like the enable or E input of the latch, but is rather an edge-triggering signal usually provided by a clock pulse (CP). As a result, the D flip-flop has no transparent mode, that is, no extended period during which the output at Q follows or tracks the input at D. Instead, the flip-flop is sensitive to its data input only at the instant that its clock input is going from LOW to HIGH. At the moment that the voltage level at CP passes a certain threshold, the level at D is read by the flip-flop and transferred to the output at Q. This level is then held at Q until the next high-going clock transition.

In operation, therefore, the D flip-flop acts something like a camera with a fast shutter, taking a snapshot of its input when triggered by the positive edge of the clock pulse. At all other times during the clock cycle (including the transition from HIGH back to LOW), the flip-flop is blind to its D input and maintains the level stored at Q at the last high-going clock transition. For proper functioning, the level to be stored in the flip-flop must be present at D for at least the setup time t_{SU} before the positive edge of the clock pulse arrives and must be held at D for at least the hold time t_H after the transition. Outside of this brief span of time, changes at the D input have no effect.

In 7400-series logic there is just one standard D flip-flop that allows control over individual bits of data, the 74. This very basic and useful IC, available in nearly all technologies, provides two independent D flip-flops in one 14-pin DIP.

Each of the two flip-flops in the 74 has the usual D and clock pulse (CP) inputs and the standard Q outputs, plus, for maximum versatility, an inverted $\overline{Q}$ output and two additional active-LOW control inputs, set or $\overline{S}$ (also known as preset) and reset or $\overline{R}$ (also known as clear). The $\overline{S}$ input, when taken LOW, will set the Q output to HIGH; the $\overline{R}$ input, when taken LOW, will reset the Q output to LOW. The $\overline{S}$ and $\overline{R}$ inputs are, in effect, the inputs of an $\overline{S}$-$\overline{R}$ latch that forms the output stage of the D flip-flop. By providing direct access to this internal latch, the $\overline{S}$ and $\overline{R}$ inputs bypass the circuitry that handles the D and clock inputs, overriding these signals. From a logical standpoint (assuming positive logic), the set (preset) operation directly loads the flip-flop with a logic 1, and the reset (clear) operation puts it back at logic 0.

In the 74, triggering occurs when the clock input reaches a certain voltage level and is not directly related to the rise time (transition time) of the clock pulse. Even so, certain restrictions on the transition time must be observed. As Signetics notes for their TTL, LS, and S versions of the 74, "Although the clock input is level-sensitive, the positive transition of the clock pulse between the 0.8 V and 2.0 V levels should be equal to or less than the clock-to-output delay time for reliable operation."* While HC and HCT versions provide some Schmitt-trigger hysteresis at the clock input for improved tolerance to slow rise and fall times, the maximum recommended input rise or fall time (specified at 4.5 V for HC or 5 V for HCT) is put at 500 ns by all manufacturers.

The 74 is the only D flip-flop included in this section because it is the only one that allows clock control over individual bits of data. A number of other important edge-triggered D flip-flops are provided in groups of four, six, or eight flip-flops triggered by a common clock input. These common multiple packages, categorized as 4-bit, 6-bit, and 8-bit *registers*, are described in Chapter 6.

* *Signetics TTL Data Manual* (1984), p. 4-99.

FUNCTION TABLE

INPUTS				OUTPUTS		
$\overline{S}$	$\overline{R}$	CP	D	Q	$\overline{Q}$	
L	H	•	•	H	L	Direct set or reset
H	L	•	•	L	H	
L	L	•	•	H*	H*	
H	H	↑	H	H	L	Read and store D
H	H	↑	L	L	H	
H	H	H	•	NO CHANGE		Do nothing
H	H	↓	•	NO CHANGE		
H	H	L	•	NO CHANGE		

• = Either LOW or HIGH logic level
* = Nonstable state (may not persist
   when $\overline{S}$ and $\overline{R}$ return to their
   normal HIGH level)

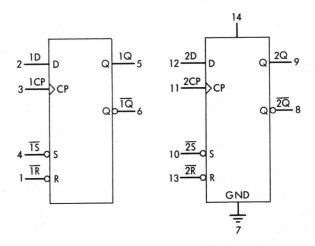

154

MANUFACTURERS	TTL	LS/A	HC	HCT	ALS	S	AS	F	AC	ACT
Fairchild		■						■		
Motorola	■		▨				■			
National Semiconductor		■		▨				■		
RCA			■	■					▨	▨
SGS			▨	■						
Signetics	■	■		■		■	■			
Texas Instruments	■	■		■	■	■	■	■		
Toshiba									D	

Added change control symbol: National 74ALS74A

KEY PARAMETERS		TTL	LS/A	HC	HCT	ALS	S	AS	F	AC	ACT
I_{OH} (max.)	mA	-0.4^a	-0.4	-4	-4	-0.4	-1	-2	-1	-24	-24
I_{OL} (max.)	mA	16	8	4	4	8	20	20	20	24	24
I_{CC} (max.)	mA	30^b	8			4	50^w	16	16		
I_{CC} (quiesc.)	mA			0.04^i	0.04					0.04	0.04
C_{pd} (typ.)	pF/FF			35^j	30^q					35^{aa}	35^{aa}
t_W (min.)	ns	37	25^d	20^k	25^r	15	7.3^x	5.5	5	5^{bb}	6^{ff}
t_{SU} (min.)	ns	20	20^e	25^l	25^s	15	3	4.5	3	3^{cc}	3.5^{gg}
t_H (min.)	ns	5	5^f	3^m	3^t	0	2^y	0	1	0	1^{hh}
f_{MAX} (min.)	MHz	15^c	25^g	21^n	20^u	34	75	105	100	125^{dd}	125^{ii}
t_{pd} CP (max.)	ns	40	40^h	44^o	44	18	9^z	9	9.2	10.5^{ee}	13^{ee}
t_{pd} $\overline{S}$ or $\overline{R}$ (max.)	ns	40	40^h	50^p	50^v	15	13.5	10.5	10.5	10.5^{ee}	11.5^{ee}

[a] Fairchild -0.8 mA

[b] Texas Instruments 15 mA

[c] National 20 MHz

[d] Fairchild and National 18 ns

[e] SGS 25 ns

[f] National 0 ns

[g] Fairchild 30 MHz

[h] National 30 ns, Fairchild 35 ns

[i] SGS and Toshiba 0.02 mA

[j] TI; Signetics 24 pF/FF, RCA 25 pF/FF, TI 35 pF/FF, SGS and Toshiba 53 pF/FF, Motorola 60 pF/FF, National 80 pF/FF

[k] SGS, TI, Toshiba 25 ns for preset and clear inputs

[l] Signetics 15 ns

[m] RCA and Signetics; Motorola, National, SGS, TI, Toshiba 0 ns

[n] Motorola and National; SGS 20 MHz, Toshiba 22 MHz, Signetics 24 MHz, RCA and TI 25 MHz

[o] SGS 39 ns, Toshiba 40 ns

[p] Toshiba 51 ns, Motorola, National, TI 58 ns

[q] Signetics 29 pF/FF

[r] National and Signetics 20 ns

[s] Signetics 15 ns

[t] National 0 ns

[u] National 21 MHz, Signetics 22 MHz

[v] National and Signetics 44 ns

[w] Texas Instruments 25 mA

[x] National 9 ns

[y] Fairchild 0 ns

[z] Fairchild 11 ns

[aa] RCA 86 pF/FF

[bb] RCA t_W (max.) from $\overline{R}$ or $\overline{S}$ is 5.1 ns

[cc] RCA 5.5 ns

[dd] RCA 100 MHz

[ee] RCA 9.4 ns

[ff] RCA t_W (max.) at CP is 5.6 ns, at $\overline{R}$ or $\overline{S}$ is 5.1 ns

[gg] RCA 6.5 ns

[hh] RCA 0 ns

[ii] RCA 90 MHz

J-K FLIP-FLOPS

The versatile J-K flip-flop can be thought of as a more highly evolved form of the set-reset latch that adds edge-triggering from a clock pulse and the ability to emulate other kinds of flip-flops.

Like the $\overline{S}$-$\overline{R}$ latch, the basic idea of the J-K flip-flop is that an appropriate signal at one data input (here an active-HIGH input J) will set the flip-flop, producing a HIGH level at the Q output, while the same signal applied instead to the other input (K) will reset or clear the flip-flop, storing a LOW level at Q. But unlike the $\overline{S}$-$\overline{R}$ latch, which is activated simply by the appearance of the triggering level at $\overline{S}$ or $\overline{R}$, the J-K will "see" its inputs at J and K only at the moment that a triggering clock transition occurs at a separate clock input. In short, the J-K is edge-triggered by a clock (like the D flip-flop) rather than level-triggered by the data itself.

All the J-K flip-flops listed here trigger on the *negative* edge of the clock pulse, that is, on the HIGH-to-LOW transition of the clock. In operation, a HIGH level placed at the J or K input at least one setup time prior to the HIGH-to-LOW transition at $\overline{CP}$ will set or reset the output at Q to store a 1 or 0 in the flip-flop for the remainder of the clock cycle.

In the clocked set-reset mode just described, where J and K ordinarily have the opposite level from one another, the action of the flip-flop can be pictured in a rather different way that is more useful in understanding certain applications. Since all the commercially available J-K flip-flops listed here have both a Q and an inverted $\overline{Q}$ output, in the normal mode they can be thought of simply as memory devices that respond to the negative clock transition by reading their inputs at J and K and storing these inputs at Q and $\overline{Q}$, respectively. In this regard the J-K behaves much like the D flip-flop. In fact, any J-K flip-flop can be converted into a D flip-flop by connecting the K input to the J input through an inverter, guaranteeing that the J and K inputs will have opposite levels. Thus modified, the J input becomes the D input of a negative-edge-triggered by flip-flop (see Fig. 2.2).

In addition to the ordinary clocked set-reset function, the J-K has a further ability that substantially expands its range of application. In a set-reset latch, activating both inputs at once produces a transient HIGH level at the Q output, and the eventual state of the latch if both inputs are removed simultaneously is indeterminate. Thus, the input condition where both inputs are active has no practical use. In the J-K flip-flop, this condition is used to implement a new

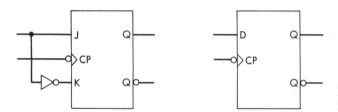

Figure 2.2 Converting a J-K flip-flop into a D flip-flop.

function. If HIGH levels are placed at both J and K, the subsequent arrival of a low-going clock transition will cause the Q output to "toggle" or switch to the state opposite the one it currently has.

The toggle mode has important applications in the construction of counters and dividers. A J-K flip-flop can be permanently held in this mode by tying both inputs, J and K, to V_{CC} (logic HIGH). Thus configured, the device is often called a T (Toggle) flip-flop, represented by the simplified diagram below. The J and K inputs are deleted, and the clock input is relabeled with a T. Such a device will simply switch states every time the T input goes from HIGH to LOW (see Figure 2.3).

The J-K flip-flops listed in this section provide both Q and $\overline{Q}$ outputs for the convenience of the designer. Also provided is an active-LOW set (preset) input, $\overline{S}$, or an active-LOW reset (clear) input, $\overline{R}$, or both. As with the D flip-flop, the $\overline{S}$ and $\overline{R}$ inputs bypass the flip-flop's ordinary input circuitry and thus override the other inputs to directly load the flip-flop with a logic 1 or 0.

It should be noted that while most J-K flip-flops are true edge-triggered devices, as explained in the description of the D flip-flop, some of the TTL versions use the older master-slave configuration, forming what is technically a positive pulse-triggered device. A master-slave flip-flop consists internally of two simpler level-triggered flip-flops, a "master" and a "slave" wired in series and connected to a common clock input but enabled by opposite clock levels. When the clock goes HIGH, data at J and K are loaded into the master flip-flop and internally provided as input to the slave, but are inhibited by the HIGH clock level from affecting the state of the slave, which continues to put out previously stored data. When the clock goes LOW again, further changes of data to the master are locked out, the slave is enabled to read its input from the master, and the new data are provided to external devices at the slave outputs Q and $\overline{Q}$.

This sequence of operations is set in motion by a *positive* clock pulse, but devices reading the master-slave flip-flop see its outputs change on the *negative* clock edge that triggers the slave. The reason that the newer edge-triggered J-K flip-flops all trigger on the negative edge of the clock is to more closely emulate the behavior of the original master-slave versions.

Although their logical functions are similar, the master-slave differs fundamentally from the edge-triggered flip-flop in that the J and K inputs must be valid for the entire time the clock is HIGH, not only at the HIGH-to-LOW transition.

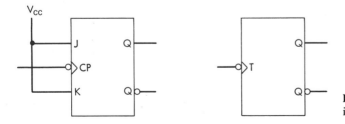

Figure 2.3 Converting a J-K flip-flop into a T flip-flop.

In other words, the master-slave's set-up time is the entire width of the positive clock pulse. While a newer edge-triggered component can usually replace an older master-slave version, the reverse is almost never advisable. New designs should use the edge-triggered versions.

Triggering of the J-K flip-flop generally occurs at a certain voltage level of the clock input and is not directly related to the transition time of the clock pulse. Nevertheless, certain restrictions on the transition time must be observed. See the note on this in the description of the D flip-flop above.

73   *Dual J-K flip-flops with individual clocks and clears.* Two independent J-K flip-flops, each with its own $\overline{CP}$ and $\overline{R}$ inputs. LS, HC, and HCT versions are edge-triggered, TTL versions are master-slave. Supplied in 14-pin DIP.

107   *Dual J-K flip-flops with individual clocks and clears.* Virtually identical to the corresponding versions of the 73, but with supply voltage and ground assigned to the corner pins to simplify PC board layout.

113   *Dual J-K flip-flops with individual clocks and presets.* Two $\overline{S}$ inputs, one for each flip-flop, replace the two $\overline{R}$ inputs provided by the 73 and 107 in the same 14-pin DIP. No TTL versions, but available in most of the other technologies. All versions are edge-triggered. LS versions with the "A" suffix (74LS113A) draw slightly less current and have slightly shorter maximum propagation delays than those without the "A" suffix.

76   *Dual J-K flip-flops with individual clocks, presets, and clears.* A 16-pin DIP allows each flip-flop to have both an $\overline{S}$ and an $\overline{R}$ input for maximum versatility. Available in TTL, LS, HC, and HCT versions; TTL versions are master-slave, others are edge-triggered.

112   *Dual J-K flip-flops with individual clocks, presets, and clears.* Virtually identical to the corresponding versions of the 76, but with supply voltage and ground assigned to the corner pins to simplify PC board layout. No TTL versions, but available in most of the other technologies. All versions are edge-triggered. LS versions with the "A" suffix (74LS112A) draw slightly less current and have slightly shorter maximum propagation delays than those without the "A" suffix.

111§   *Dual J-K variable-skew flip-flops with individual clocks, presets, and clears.* A 16-pin master-slave TTL design from Texas Instruments that loads data into the master on the rising edge of the clock, but then locks out changes at the J and K inputs during the remainder of the HIGH level clock interval. According to the manufacturer, these flip-flops are called variable-skew because "the system designer can set the maximum allowable clock skew needed by varying the clock pulse width. Thus system design is made easier and the requirements for sophisticated clock distribution systems are minimized or, in some cases, entirely eliminated."*

I_{OH} (max.)	−0.8 mA	t_H (min.)	30 ns
I_{OL} (max.)	16 mA	f_{MAX} (min.)	20 MHz
I_{CC} (max.)	20.5 mA	t_{pd} $\overline{CP}$ (max.)	30 ns
t_W (min.)	25 ns	t_{pd} $\overline{S}$ or $\overline{R}$ (max.)	30 ns
t_{SU} (min.)	0 ns		

78   *Dual J-K flip-flops with individual presets and common clock and clear.* Edge-triggered LS and HC components that squeeze dual J-K flip-flops with both presets and clears into a 14-pin DIP by using common $\overline{CP}$ and $\overline{R}$ inputs that trigger or clear both flip-flops at once.

114   *Dual J-K flip-flops with individual presets and common clock and clear.* Virtually identical to the corresponding versions of the 78 but with supply voltage and ground assigned to the corner pins to simplify PC board layouts. No TTL versions, but available in most of the other technologies. All versions are edge-triggered. LS versions with the "A" suffix (74LS114A) draw slightly less current and have slightly shorter maximum propagation delays than those without the "A" suffix.

* *Texas Instruments TTL Data Book, Vol. 2* (1985), p. 3-441.

FUNCTION TABLE

	$\overline{S}$	$\overline{R}$	$\overline{CP}$	J	K	Q	$\overline{Q}$	
$\overline{S}$ or $\overline{R}$ active	L	H	.	.	.	H	L	Direct set or reset
	H	L	.	.	.	L	H	
	L	L	.	.	.	H*	H*	
Negative clock edge arrives	H	H	↓	L	L	NO CHANGE		Maintain state
	H	H	↓	H	L	H	L	Read and store
	H	H	↓	L	H	L	H	J and K inputs
	H	H	↓	H	H	TOGGLE		T (Toggle) mode
All other clock levels	H	H	H	.	.	NO CHANGE		Do nothing
	H	H	↑	.	.	NO CHANGE		
	H	H	L	.	.	NO CHANGE		

. = Either LOW or HIGH logic level
* = Nonstable state (may not persist when $\overline{S}$ and $\overline{R}$ return to their normal HIGH level)

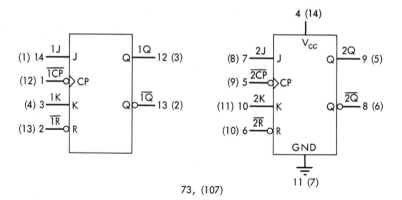

73, (107)

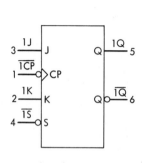

113

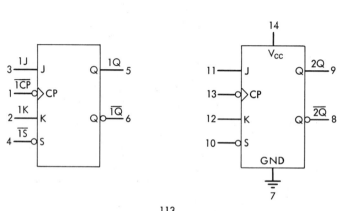

76, (112)

160

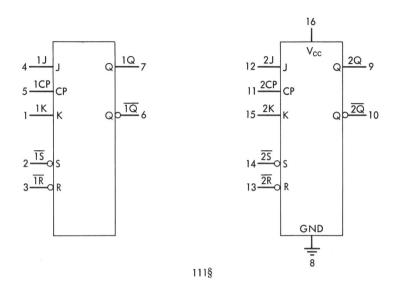

111§

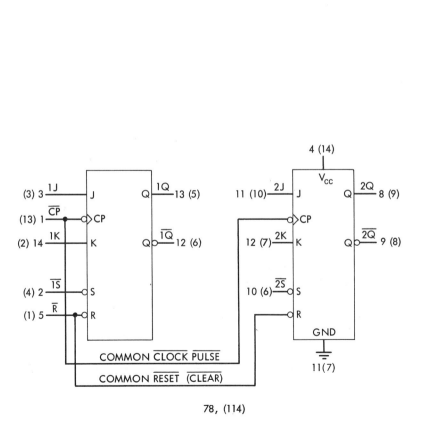

COMMON CLOCK PULSE

COMMON RESET (CLEAR)

78, (114)

MANUFACTURERS	TTL	LS/A	HC	HCT	ALS	S	AS	F	AC	ACT
Fairchild	■									
Motorola		■	▒							
National Semiconductor	■	■		■						
RCA			■	■						
SGS		▒	▒	■						
Signetics	■	■	■	■						
Texas Instruments	■	■	▒							
Toshiba	■		■							

Missing change control symbol "A": Signetics 74LS73

KEY PARAMETERS		TTL	LS/A	HC	HCT	ALS	S	AS	F	AC	ACT
I_{OH} (max.)	mA	-0.4	-0.4	-4	-4						
I_{OL} (max.)	mA	16	8	4	4						
I_{CC} (max.)	mA	40^a	6^b								
I_{CC} (quiesc.)	mA			0.04^e	0.04						
C_{pd} (typ.)	pF/FF			42^f	30^m						
t_W (min.)	ns	47	25^c	20^g	29^n						
t_{SU} (min.)	ns	0	20	24^h	20^o						
t_H (min.)	ns	0	0	0^i	5^p						
f_{MAX} (min.)	MHz	15	30	24^j	22^q						
t_{pd} $\overline{CP}$ (max.)	ns	40	20^d	32^k	48						
t_{pd} $\overline{R}$ (max.)	ns	40	20^d	39^l	43						

[a] TI 20 mA, National 34 mA

[b] Signetics 8 mA

[c] TI 20 ns

[d] Signetics 30 ns. Motorola data for t_{pd} from $\overline{R}$ not available

[e] Toshiba 0.01 mA, SGS 0.02 mA

[f] SGS and Toshiba; RCA 28 pF/FF, Signetics and TI 30 pF/FF, Motorola and National 80 pF/FF

[g] SGS 18 ns, Toshiba 19 ns

[h] SGS; RCA, Signetics, and TI 20 ns, Motorola, National, and Toshiba 25 ns

[i] RCA 3 ns, Signetics 4 ns

[j] Signetics and Toshiba; Motorola and National 21 MHz, RCA, SGS, and TI 25 MHz

[k] Toshiba 38 ns, RCA 40 ns, Signetics 43 ns. TI data not available

[l] RCA and Signetics 36 ns, Toshiba 46 ns. TI data not available

[m] Signetics; RCA 28 pF/FF

[n] Signetics; RCA $\overline{CP}$ 20 ns, $\overline{R}$ 23 ns

[o] RCA; Signetics 15 ns

[p] Signetics; RCA 3 ns

[q] Signetics; RCA 25 MHz

107   Dual J-K flip-flops with individual clocks and clears

MANUFACTURERS	TTL	LS/A	HC	HCT	ALS	S	AS	F	AC	ACT
Motorola		■	▨							
National Semiconductor			■	■						
RCA			■	■						
SGS		▨	▨							
Signetics	■	■	■	■						
Texas Instruments	■	■	■	■						
Toshiba	■	■	■	■						

Missing change control symbol "A": Signetics 74LS107

KEY PARAMETERS		TTL	LS/A	HC	HCT	ALS	S	AS	F	AC	ACT
I_{OH} (max.)	mA	-0.4	-0.4	-4	-4						
I_{OL} (max.)	mA	16	8	4	4						
I_{CC} (max.)	mA	40[a]	6[b]								
I_{CC} (quiesc.)	mA			0.04[f]	0.04						
C_{pd} (typ.)	pF/FF			46[g]	30						
t_W (min.)	ns	47	25[c]	20[h]	30[n]						
t_{SU} (min.)	ns	0	20	25[i]	25						
t_H (min.)	ns	0	0[d]	0[j]	5						
f_{MAX} (min.)	MHz	15	30	24[k]	22[o]						
t_{pd} $\overline{CP}$ (max.)	ns	40	20[e]	40[l]	54[p]						
t_{pd} $\overline{R}$ (max.)	ns	40	20[e]	39[m]	48						

[a] TI 20 mA, National 34 mA

[b] Signetics 8 mA

[c] TI 20 ns

[d] TI 20 ns

[e] Signetics 30 ns. Motorola and SGS data for t_{pd} from $\overline{R}$ not available

[f] SGS and Toshiba 0.02 mA

[g] SGS; Signetics 30 pF/FF, RCA 31 pF/FF, TI 35 pF/FF, Toshiba 42 pF/FF, Motorola and National 80 pF/FF

[h] SGS 18 ns, Toshiba and TI 25 ns

[i] Toshiba 19 ns, Signetics 20 ns, SGS 24 ns

[j] RCA 3 ns, Signetics 5 ns

[k] Signetics and Toshiba; Motorola and National 21 MHz, RCA, SGS, and TI 25 MHz

[l] SGS and Signetics; Motorola, National, and TI 32 ns, Toshiba 41 ns, RCA 43 ns

[m] SGS 47 ns, Toshiba 55 ns

[n] Signetics; RCA 23 ns

[o] RCA; Signetics 24 MHz

[p] RCA; Signetics 45 ns

113 Dual J-K flip-flops with individual clocks and presets

MANUFACTURERS	TTL	LS/A	HC	HCT	ALS/A	S	AS	F	AC	ACT
Fairchild		■						■		
Motorola			▨					▨		
National Semiconductor		■			■		▨			
SGS		■	▨							
Signetics		■					■			
Texas Instruments		■					D			
Toshiba			■							

Missing change control symbol "A": Fairchild, Signetics, and SGS 74LS113. SGS also makes a 74LS113A

KEY PARAMETERS		TTL	LS/A	HC	HCT	ALS/A	S	AS	F	AC	ACT
I_{OH} (max.)	mA		-0.4	-4		-0.4	-1	-2	-1		
I_{OL} (max.)	mA		8	4		8	20	20	20		
I_{CC} (max.)	mA		6[a]			4.5	50[l]	n/a	19		
I_{CC} (quiesc.)	mA			0.04[e]							
C_{pd} (typ.)	pF/FF			38[f]							
t_W (min.)	ns		25[b]	20[g]		16.5	8	n/a	5[m]		
t_{SU} (min.)	ns		20[c]	25[h]		22	3	n/a	5[n]		
t_H (min.)	ns		0[c]	0		0	0	n/a	0		
f_{MAX} (min.)	MHz		30	25[i]		30	80	n/a	100[o]		
t_{pd} $\overline{CP}$ (max.)	ns		20[d]	33[j]		19	7	n/a	7		
t_{pd} $\overline{S}$ (max.)	ns		20[d]	41[k]		16	7	n/a	7.5		

[a] 74LS113 (Fairchild, SGS, Signetics) 8 mA
[b] Fairchild 74LS113 20 ns, SGS 74LS113 18 ns
[c] TI 20 ns
[d] Fairchild 74LS113 and SGS 74LS113 24 ns, Signetics 74LS113 30 ns
[e] SGS and Toshiba 0.02 mA
[f] SGS and Toshiba; TI 35 pF/FF, Motorola and National 80 pF/FF
[g] SGS 18 ns, Toshiba 19 ns, TI 25 ns
[h] SGS 12 ns, Toshiba 19 ns

[i] TI; Motorola and National 21 MHz, Toshiba 27 MHz, SGS 29 MHz
[j] SGS; Motorola and National 32 ns, Toshiba 34 ns, TI 35 ns
[k] SGS 39 ns, Toshiba 40 ns
[l] TI 25 mA
[m] Motorola 4.5 ns
[n] Motorola 4 ns
[o] Motorola data over temperature range not available

Dual J-K flip-flops with individual clocks, presets, and clears

MANUFACTURERS	TTL	LS/A	HC	HCT	ALS	S	AS	F	AC	ACT
Fairchild	███									
Motorola		███	░░░							
National Semiconductor			███	░░░						
SGS		░░░	░░░							
Signetics	███		███							
Texas Instruments	███	███								
Toshiba			███	███						

Missing change control symbol "A": Signetics 74LS76

KEY PARAMETERS		TTL	LS/A	HC	HCT	ALS	S	AS	F	AC	ACT
I_{OH} (max.)	mA	-0.4^a	-0.4	-4	-4						
I_{OL} (max.)	mA	16	8	4	4						
I_{CC} (max.)	mA	20^b	6^c								
I_{CC} (quiesc.)	mA			0.04^f	0.04						
C_{pd} (typ.)	pF			47^g	n/a						
t_W (min.)	ns	47	25	20^h	20^i						
t_{SU} (min.)	ns	0	20^d	25^i	25						
t_H (min.)	ns	0	0	0	0						
f_{MAX} (min.)	MHz	15	30	24^j	22						
t_{pd} $\overline{CP}$ (max.)	ns	40	20^e	36^k	52						
t_{pd} $\overline{S}$ or $\overline{R}$ (max.)	ns	40	20^e	41^l	52						

[a] Fairchild -0.8 mA

[b] National 34 mA, Signetics 40 mA

[c] Signetics 8 mA

[d] TI 25 ns

[e] Signetics 30 ns

[f] SGS and Toshiba 0.02 mA

[g] SGS and Toshiba; TI 36 pF/FF, Motorola and National 80 pF/FF

[h] SGS 18 ns, Toshiba 19 ns, TI 25 ns

[i] SGS 18 ns, Toshiba 19 ns, TI 38 ns

[j] Toshiba; Motorola and National 21 MHz, SGS and TI 25 MHz

[k] National 31 ns, Motorola 32 ns

[l] TI 39 ns, Toshiba 45 ns, SGS 46 ns

MANUFACTURERS	TTL	LS/A	HC	HCT	ALS/A	S	AS	F	AC	ACT
Fairchild		■	■			■		■		
Motorola			▨	▨				▨		
National Semiconductor				▨	▨		▨			
RCA		□							D	D
SGS		■	▨							
Signetics		■					■		D	
Texas Instruments		■		■			D			
Toshiba		■		■					D	

Missing change control symbol "A": Fairchild and Signetics 74LS112

KEY PARAMETERS		TTL	LS/A	HC	HCT	ALS/A	S	AS	F	AC	ACT
I_{OH} (max.)	mA		-0.4	-4	-4	-0.4	-1	-2	-1	[-24]	[-24]
I_{OL} (max.)	mA		8	4	4	8	20	20	20	[24]	[24]
I_{CC} (max.)	mA		6^a			4.5	50^r	n/a	19		
I_{CC} (quiesc.)	mA			0.04^e	0.04					[0.08]	[0.08]
C_{pd} (typ.)	pF/FF			54^f	30^l					n/a	n/a
t_W $\overline{CP}$ (min.)	ns		25^b	20^g	20	16.5	8	n/a	5^t	n/a	n/a
t_{SU} (min.)	ns		20^c	25^h	20^m	22	3^s	n/a	5^u	n/a	n/a
t_H (min.)	ns		0	0	0^n	0	0	n/a	0	n/a	n/a
f_{MAX} (min.)	MHz		30	24^i	22^o	30	80	n/a	100^v	n/a	n/a
t_{pd} $\overline{CP}$ (max.)	ns		20^d	38^j	44^p	19	7	n/a	7.5	[9.2]	[9.2]
t_{pd} $\overline{S}$ or $\overline{R}$ (max.)	ns		20^d	41^k	46^q	18	7	n/a	7.5	[10.9]	[12.1]

[a] 74LS112 (Fairchild and Signetics) 8 mA

[b] Fairchild 74LS112 20 ns

[c] TI 25 ns

[d] Fairchild 74LS112 24 ns, Signetics 74LS112 30 ns

[e] SGS and Toshiba 0.02 mA

[f] SGS and Toshiba; RCA 12 pF/FF, Signetics 27 pF/FF, TI 35 pF/FF, Motorola and National 80 pF/FF

[g] SGS 18 ns, Toshiba 19 ns, TI 25 ns

[h] SGS 18 ns, Toshiba 19 ns, RCA and Signetics 20 ns

[i] Signetics and Toshiba; TI 20 MHz, Motorola and National 21 MHz, RCA 25 MHz, SGS 29 MHz

[j] Toshiba; TI 31 ns, Motorola and National 32 ns, SGS 33 ns, RCA and Signetics 44 ns

[k] SGS 40 ns, RCA, Signetics, and Toshiba 45 ns

[l] Signetics; RCA 20 pF/FF. National data not available

[m] National 25 ns

[n] RCA 3 ns

[o] RCA 25 MHz

[p] Signetics 50 ns

[q] National 44 ns

[r] TI 25 mA

[s] Fairchild 7 ns

[t] Motorola 4.5 ns

[u] Motorola 4 ns

[v] Motorola data over temperature range not available

MANUFACTURERS	TTL	LS/A	HC	HCT	ALS	S	AS	F	AC	ACT
Motorola		■								
National Semiconductor		■								
SGS		▒								
Texas Instruments		■	▒							

KEY PARAMETERS		TTL	LS/A	HC	HCT	ALS	S	AS	F	AC	ACT
I_{OH} (max.)	mA		-0.4	-4							
I_{OL} (max.)	mA		8	4							
I_{CC} (max.)	mA		6								
I_{CC} (quiesc.)	mA			0.04							
C_{pd} (typ.)	pF/FF			30							
t_W (min.)	ns		25	a							
t_{SU} (min.)	ns		20	a							
t_H (min.)	ns		0	a							
f_{MAX} (min.)	MHz		30	a							
t_{pd} $\overline{CP}$ (max.)	ns		20	a							
t_{pd} $\overline{S}$ or $\overline{R}$ (max.)	ns		20	a							

[a] Data over temperature range not available

MANUFACTURERS	TTL	LS/A	HC	HCT	ALS/A	S	AS	F	AC	ACT
Fairchild		■						▒		
Motorola		■						▒		
National Semiconductor		■			■		■	▒		
SGS		■								
Signetics								■		
Texas Instruments					■		D			

Missing change control symbol "A": Fairchild 74LS114. SGS makes both a 74LS114 and a 74LS114A

KEY PARAMETERS		TTL	LS/A	HC	HCT	ALS/A	S	AS	F	AC	ACT
I_{OH} (max.)	mA		-0.4	-4		-0.4	-1	-2	-1		
I_{OL} (max.)	mA		8	4		8	20	20	20		
I_{CC} (max.)	mA		6[a]			4.5	50[e]	n/a	19		
I_{CC} (quiesc.)	mA			0.04							
C_{pd} (typ.)	pF/FF			50							
t_W (min.)	ns		25[b]	25		16.5	8	n/a	5[f]		
t_{SU} (min.)	ns		20[c]	25		22	3	n/a	5[g]		
t_H (min.)	ns		0	0		0	0	n/a	0		
f_{MAX} (min.)	MHz		30	20		30	80	n/a	90[h]		
t_{pd} $\overline{CP}$ (max.)	ns		20[d]	44		19	7	n/a	8.5		
t_{pd} $\overline{S}$ or $\overline{R}$ (max.)	ns		20[d]	44		18	7[j]	n/a	7.5		

[a] 74LS114 (Fairchild and SGS) 8 mA
[b] SGS 74LS114 18 ns, Fairchild 74LS114 20 ns
[c] TI 25 ns
[d] 74LS114 (Fairchild and SGS) 24 ns
[e] TI 25 mA
[f] Motorola 4.5 ns
[g] Motorola 4 ns
[h] Motorola data over temperature range not available

J-K̄ FLIP-FLOPS

A J-K̄ flip-flop is simply a J-K flip-flop with an inverted K input. This makes it especially easy to convert to a D flip-flop when necessary by connecting the J input directly to the K̄ input without needing to add a separate inverter.

Since the J-K̄ flip-flop can still be converted into a T flip-flop without further circuitry (by grounding K̄ and tying J HIGH), it is the most versatile of all flip-flops and the one to have on hand if only one kind of flip-flop can be carried in stock.

The standard J-K̄ flip-flop is the 109, a dual package that is very widely supported and available in almost all technologies. Two other J-K̄ types, the 276§ and 376§, are single-sourced quad packages available only in TTL; they are described briefly below. Refer to Texas Instruments data for further details.

All of the J-K̄ flip-flops listed below, including TTL versions, are edge-triggered, not master-slave. Like other edge-triggered designs, they are activated when the clock voltage reaches a certain level, and triggering is not directly related to the transition time of the clock pulse. As with other edge-triggered circuits, however, this is true only within certain limits. For TTL, LS, and F versions of the 109, Signetics cautions that "the positive transition of the clock pulse between the 0.8 V and 2.0 V levels should be equal to or less than the clock to output delay time for reliable operation."* Maximum input rise or fall time for HC and HCT versions is generally 500 ns at 4.5 V (HC) or 5 V (HCT).

109 *Dual J-K̄ flip-flops with individual clocks, presets, and clears.* The standard J-K̄. Two independent flip-flops, each with its own J, K̄, CP, S̄, and R̄ inputs and complementary Q and Q̄ outputs, in a 16-pin DIP. All versions are positive-edge-triggered.

276§ *Quad J-K̄ flip-flops with individual clocks and common preset and clear.* Eliminating the Q̄ outputs of the 109 and sharing single preset (S̄) and clear (R̄) pins allow four TTL J-K̄ flip-flops to be housed in a single 20-pin DIP. External control over each flip-flop is maintained through individual *negative*-edge-triggered clock inputs. A moderate amount of hysteresis (typ. 0.2 V) is provided at clock and preset (S̄) inputs. Key parameters are listed below. (Texas Instruments)

376§ *Quad J-K̄ flip-flops with common clock and clear.* Sharing a single clock input and omitting the preset capability of the 276 reduces the size of this TTL quad package to a 16-pin DIP. The clock input is positive-edge-triggered like the clock input of the 109 and provides a moderate amount (typ. 0.2 V) of input hysteresis. (Texas Instruments)

* *Signetics TTL Data Manual* (1984), p. 4-161; *Signetics FAST Data Manual* (1986), p. 6-66.

	276	376
I_{OH} (max.)	−0.8 mA	−0.8 mA
I_{OL} (max.)	16 mA	16 mA
I_{CC} (max.)	81 mA	74 mA
t_W (min.)	15 ns	22 ns
t_{SU} (min.)	10 ns	10 ns
t_H (min.)	10 ns	20 ns
f_{MAX} (min.)	35 MHz	30 MHz
t_{pd} Clock (max.)	30 ns	35 ns
t_{pd} Clear (max.)	30 ns	30 ns

FUNCTION TABLE (109)

	INPUTS					OUTPUTS		
	$\bar{S}$	$\bar{R}$	CP	J	$\bar{K}$	Q	$\bar{Q}$	
$\bar{S}$ or $\bar{R}$ active	L	H	.	.	.	H	L	Direct set or reset
	H	L	.	.	.	L	H	
	L	L	.	.	.	H*	H*	
Positive clock edge arrives	H	H	↟	L	H	NO CHANGE		Maintain state
	H	H	↟	H	H	H	L	Clock set
	H	H	↟	L	L	L	H	or reset
	H	H	↟	H	L	TOGGLE		T (toggle) mode
All other clock levels	H	H	H	.	.	NO CHANGE		Do nothing
	H	H	↡	.	.	NO CHANGE		
	H	H	L	.	.	NO CHANGE		

• = Either LOW or HIGH logic level
* = Nonstable state (may not persist when
$\bar{S}$ and $\bar{R}$ return to their normal HIGH level)

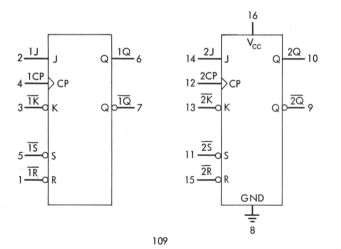

109   Dual J-$\overline{\text{K}}$ flip-flops with individual clocks, presets, and clears

MANUFACTURERS	TTL	LS/A	HC	HCT	ALS/A	S	AS	F	AC	ACT
Fairchild		■				■			■	
Motorola		▨					■		■	
National Semiconductor	■			▨				■		
RCA			■						D	D
SGS		■	▨							
Signetics		■						■		
Texas Instruments	■		■	■						
Toshiba		■							D	

Concurrent change control symbols: SGS makes both a 74LS109 and a 74LS109A

KEY PARAMETERS		TTL	LS/A	HC	HCT	ALS/A	S	AS	F	AC	ACT
I_{OH} (max.)	mA	-0.8[a]	-0.4	-4	-4	-0.4	-1	-2	-1	-24	-24
I_{OL} (max.)	mA	16	8	4	4	8	20	20	20	24	24
I_{CC} (max.)	mA	30[b]	8			4	52	17	17		
I_{CC} (quiesc.)	mA			0.04[l]	0.04					0.04	0.04
C_{pd} (typ.)	pF/FF			47[m]	33[t]					35	35
t_W (min.)	ns	20	25[g]	20[n]	23[u]	15	7	5.5	5	3.5	6
t_{SU} (min.)	ns	10[c]	35[h]	25[o]	23[v]	15	6	5.5	3	5	2.5
t_H (min.)	ns	6[d]	5[i]	0[p]	3[w]	0	0	0	1	0.5	2
f_{MAX} (min.)	MHz	25[e]	25[j]	25[q]	22	34	75	105	90	125	125
t_{pd} CP (max.)	ns	28	40[k]	44[r]	44[x]	18	11	9	9.2	10.5	13
t_{pd} $\overline{\text{S}}$ or $\overline{\text{R}}$ (max.)	ns	35[f]	40[k]	48[s]	44[y]	15	12	10.5	10.5	10.5	11.5

[a] National -1.2 mA
[b] Texas Instruments 15 mA
[c] National 15 ns
[d] National 10 ns
[e] National 30 MHz
[f] National 29 ns
[g] National 74LS109A and SGS 74LS109 18 ns, Fairchild 20 ns
[h] Fairchild 18 ns, Signetics 74LS109A and SGS 74LS109 20 ns, National 30 ns
[i] Fairchild and National 74LS109A and SGS 74LS109 have t_H (min.) = 0 ns
[j] Fairchild 74LS109A and SGS 74LS109 30 MHz
[k] National 30 ns, Fairchild 35 ns
[l] SGS and Toshiba 0.02 mA
[m] SGS; Signetics 20 pF/FF, RCA 30 pF/FF, TI 35 pF/FF,

National 80 pF/FF. Motorola data not available
[n] SGS 18 ns, Toshiba 19 ns
[o] SGS and Signetics 18 ns, Toshiba 19 ns, RCA 23 ns
[p] Motorola, RCA, Signetics 5 ns
[q] Motorola and National 21 MHz, Signetics and Toshiba 24 MHz
[r] SGS 40 ns, Toshiba 41 ns
[s] Toshiba; RCA, SGS, and Signetics 46 ns, Motorola, National, and TI 58 ns
[t] RCA; Signetics 22 pF/FF. National data not available
[u] National 20 ns
[v] National 25 ns
[w] National 0 ns
[x] RCA 50 ns
[y] RCA 56 ns

AND-GATED J-K AND J-$\overline{\text{K}}$ FLIP-FLOPS

In this last category of J-K flip-flops are five ICs that provide various AND-gated combinations of multiple J, K, $\overline{\text{J}}$, and/or $\overline{\text{K}}$ inputs. The basic idea in each case is to provide multiple gated inputs in a 14-pin DIP without requiring the user to add external gating. These specialized components are available only in TTL versions, and three of the five are available only from Texas Instruments. The other two, the 72 and the 70, are second-sourced by National Semiconductor. Because of their relative obscurity, these circuits are described only briefly, and functional diagrams are provided only for the second-sourced 72 and 70. Refer to manufacturer's data for further information.

72   *AND-gated J-K flip-flops with preset and clear.* Master-slave TTL J-K flip-flops with three AND-gated J inputs and three AND-gated K inputs. Positive-pulse-triggered clock (output data appears on negative edge of clock pulse). Active-LOW direct preset and clear inputs, complementary Q and $\overline{\text{Q}}$ outputs.

110§   *AND-gated variable-skew J-K flip-flops with preset and clear.* TTL master-slave devices, functionally the same as the 72 but with "data lockout" to prevent changes in J or K during the HIGH level of the clock pulse from affecting the operation of the flip-flop. See note on "variable-skew" feature in the description of the 111§ under "J-K flip-flops." (Texas Instruments)

70   *AND-gated J-$\overline{\text{J}}$-K-$\overline{\text{K}}$ flip-flops with preset and clear.* Positive-edge-triggered TTL flip-flops that combine two J inputs and an inverted $\overline{\text{J}}$ input at a three-input AND gate and two K inputs and an inverted $\overline{\text{K}}$ input at a second three-input AND gate. Also has active-LOW direct preset and clear inputs, complementary Q and $\overline{\text{Q}}$ outputs.

104§   *AND-gated J-K flip-flops with preset, clear, and common JK inputs.* Positive-pulse-triggered TTL master-slave flip-flops similar to the 72, but with an additional JK input that provides a common signal to both the J gate and the K gate. (Texas Instruments)

105§   *AND-gated J-$\overline{\text{J}}$-K-$\overline{\text{K}}$ flip-flops with preset, clear, and common JK input.* Positive-pulse-triggered TTL master-slave flip-flop similar to the 104, but with one of the three J inputs replaced by an inverted $\overline{\text{J}}$ and one of the three K inputs replaced by an inverted $\overline{\text{K}}$. (Texas Instruments)

	72	110§	70	104§	105§
I_{OH} (max.)	−0.4 mA	−0.8 mA	−0.4 mA	−1 mA	−1 mA
I_{OL} (max.)	16 mA	16 mA	16 mA	16 mA	16 mA
I_{CC} (max.)	20* mA	34 mA	26 mA	24 mA	28 mA
t_W (min.)	47 mA	25 ns	30 ns	20 ns	20 ns
t_{SU} (min.)	0 ns	20 ns	20 ns	35 ns	10 ns
t_H (min.)	0 ns	5 ns	5 ns	0 ns	0 ns
f_{MAX} (min.)	15 MHz	20 MHz	20 MHz	n/a	n/a
t_{pd} Clock (max.)	40 ns	30 ns	50 ns	25 ns	25 ns
t_{pd} Clear (max.)	40 ns	25 ns	50 ns	n/a	n/a

* National 17 mA

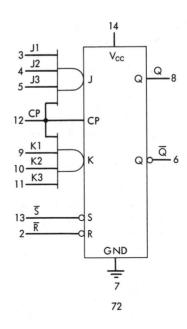

72

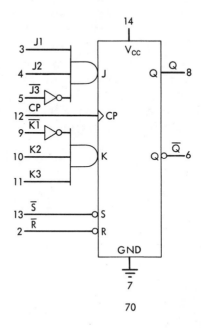

70

172

72   AND-gated J-K flip-flops with preset and clear

MANUFACTURERS	TTL	LS	HC	HCT	ALS	S	AS	F	AC	ACT
National Semiconductor	■									
Texas Instruments	■									

KEY PARAMETERS		TTL	LS	HC	HCT	ALS	S	AS	F	AC	ACT
I_{OH} (max.)	mA	-0.4									
I_{OL} (max.)	mA	16									
I_{CC} (max.)	mA	20[a]									
t_W (min.)	ns	47									
t_{SU} (min.)	ns	0									
t_H (min.)	ns	0									
f_{MAX} (min.)	MHz	15									
t_{pd} $\overline{CP}$ (max.)	ns	40									
t_{pd} $\overline{R}$ (max.)	ns	40									

[a] National 17 mA

70   AND-gated J-$\overline{J}$-K-$\overline{K}$ flip-flops with preset and clear

MANUFACTURERS	TTL	LS	HC	HCT	ALS	S	AS	F	AC	ACT
National Semiconductor	■									
Texas Instruments	■									

KEY PARAMETERS		TTL	LS	HC	HCT	ALS	S	AS	F	AC	ACT
I_{OH} (max.)	mA	-0.4									
I_{OL} (max.)	mA	16									
I_{CC} (max.)	mA	26									
t_W (min.)	ns	30									
t_{SU} (min.)	ns	20									
t_H (min.)	ns	5									
f_{MAX} (min.)	MHz	20									
t_{pd} CP (max.)	ns	50									
t_{pd} $\overline{R}$ (max.)	ns	50									

MULTIFUNCTION CIRCUITS WITH FLIP-FLOPS

The single-sourced item below is a "convenience package" that offers several common logic devices in a single IC. Such components can provide considerable flexibility in meeting the requirements of various simple logic applications while keeping a very small physical inventory of parts on hand. See manufacturer's data for logic diagram and function tables.

7074§   *Six-section multifunction circuits with flip-flops.* These HC circuits are provided in 24-pin DIPs that each contain one NAND gate, one NOR gate, two inverters, and two positive-edge-triggered D flip-flops with asynchronous active-LOW preset and clear. The flip-flops have f_{MAX} (min.) of 20 MHz at 4.5 V. (Texas Instruments)

3

One-Shots

A one-shot can be thought of as a special kind of flip-flop. An ordinary flip-flop has two stable states, a "set" state in which the Q output is HIGH and a "reset" state in which Q is LOW. The flip-flop will maintain either of these states indefinitely until an appropriate input signal makes it change to the other stable state.

A one-shot, on the other hand, has only one stable state—the reset state in which Q is LOW—and one quasi-stable state in which Q is HIGH, but only for a short period of time. When a one-shot is in the stable state, Q provides a LOW output indefinitely, just like a flip-flop in the reset state. When an appropriate triggering signal arrives, the one-shot changes to the set state (Q goes HIGH) just as a flip-flop would. But unlike a flip-flop, which would continue to hold this state until a subsequent reset signal arrived, the one-shot maintains the set condition for just a brief period of time and then automatically drops back to the reset state on its own, returning Q to LOW. In effect, the circuit responds to a triggering signal by emitting a single positive pulse, hence the name "one-shot" to describe this device.*

* Historically, one-shots and flip-flops developed from a class of relaxation oscillators called *multivibrators*. Because it has just one stable state, the one-shot is sometimes known as a *monostable multivibrator*, and the flip-flop, using the same terminology, is sometimes known as a *bistable multivibrator*. The word "multivibrator" was originally suggested by the large number of harmonic frequencies produced by simple free-running vacuum-tube oscillators (astable multivibrators). Though the old name still lingers in more formal contexts, it does not accurately describe the behavior of complex, normally non-oscillating digital devices, and for ordinary purposes the more descriptive terms "one-shot" and "flip-flop" should be used.

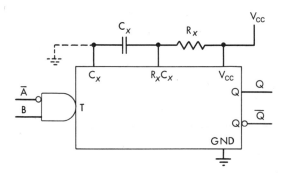

Figure 3.1

The most important variable parameter associated with a one-shot is the width of its output pulse at Q, t_{WQ}. The duration of this output pulse has nothing to do with the width of the input pulse; rather, it is a function of a resistance-capacitance time constant, normally chosen by connecting an appropriate external resistor and capacitor to the one-shot. Programming the timing constant is discussed in detail further on.

The features common to all 7400-series one-shots are summed up in the generalized diagram shown in Figure 3.1. All of them have at least one active-LOW input, usually designated $\overline{A}$, and at least one active-HIGH input, usually designated B. The B input generally has a Schmitt trigger and, in HC versions, the $\overline{A}$ input generally does, too. All commercially available one-shots have complementary Q and $\overline{Q}$ outputs. The external resistor R_X and external capacitor C_X are normally connected as shown; the optional external ground at C_X is discussed below.

TRIGGER INPUTS

The one-shot's trigger inputs are variously described as "level-triggered" or "edge-triggered." They are level-triggered (more properly, level-enabled) because each input serves as a complementary level-sensitive enable/inhibit control for the other. If, for example, the B input is held at its inactive (LOW) level, the $\overline{A}$ input is disabled and cannot trigger the one-shot. Conversely, if the $\overline{A}$ input is held HIGH, the B input is disabled.

These same inputs can also be described as edge-triggered, because it is a voltage *transition* at the triggering input that actually causes the one-shot to fire off a pulse; a stable combination of inputs can never trigger the device. The one-shot will produce an output pulse if either input undergoes an appropriate logic transition while the other input is enabled (B held HIGH or $\overline{A}$ held LOW). In many versions, simultaneous transitions at both trigger inputs may also fire the one-shot, but in published literature manufacturers often leave such cases of simultaneous triggering undefined, and in practice they are probably best avoided.

In all one-shots except the 4538, the "appropriate transition" that will trigger the one-shot is the transition *toward* the level that enables the one-shot: a low-going transition for the $\overline{\text{A}}$ input and a high-going transition for the B input. In the 4538 the rule is just the opposite: the trigger transition is the one *away* from the level that enables the device, so that a LOW-to-HIGH transition triggers A (note the omission of the overline to indicate positive-edge triggering) and a HIGH-to-LOW transition triggers $\overline{\text{B}}$ (which is correspondingly overlined to show the negative-edge triggering). Another way of stating this distinction is that, for most one-shots, a stable active level at one input enables the other, while in the 4538 the active level at one trigger input inhibits the other.

For the convenience of the designer, several types of one-shot offer multiple $\overline{\text{A}}$ and/or multiple B inputs that are gated internally to provide the $\overline{\text{A}}$ and B levels. The general rule for one-shots with gated inputs is that the device is active only when *all* of its B inputs are active (these inputs are internally AND-gated) and at least *one* of its $\overline{\text{A}}$ inputs is active (these inputs are internally OR-gated). As long as this multiple condition remains unsatisfied, the one-shot does nothing. When a particular set of input conditions finally satisfies the requirements, the one-shot fires, and the triggering transition is the last input change that finally makes this multiple condition true.

If the trigger condition remains true past the end of the output pulse, however, this does not cause the one-shot to fire again. The general rule for all one-shots is that the device is activated by its trigger conditions *becoming* satisfied, not by their *being* satisfied. It will trigger again only after one of its inputs is allowed to fall back to its inactive level and then pulsed active, whereupon it will fire on the edge of this pulse as before.

BASIC TYPES

One-shots can be classified according to physical organization or according to logical function. Physically, 7400-series one-shots come in two basic configurations:

1. A single device housed in a 14-pin DIP. The single packaging allows pins for dual OR-gated $\overline{\text{A}}$ and (in most cases) dual AND-gated B inputs. It also allows the inclusion of an internal timing resistor that in noncritical applications can take the place of the usual external resistor.
2. Dual one-shots housed in a 16-pin DIP. The dual configuration is achieved by limiting each device to a single $\overline{\text{A}}$ and B input and omitting the internal timing resistor.

Functionally, one-shots can be classified in three major categories, depending on how much control they provide over the length of the output pulse.

In the basic one-shot, an active-LOW or active-HIGH trigger causes the device to generate a single pulse the length of which (at a given supply voltage and temperature) is determined solely by the timing resistor and timing capacitor. Once triggered, changes at the one-shot's inputs during the timing cycle have no effect on the length of the output pulse. This category includes just the 121, a single TTL one-shot with a 2 kΩ internal timing resistor.

A *resettable* one-shot adds an active-LOW reset (clear) input, $\overline{R}$, to the basic set of external controls. If the $\overline{R}$ input, normally HIGH, is taken LOW during the timing cycle, it will cause premature termination of the output pulse. Thus, the resettable one-shot allows the output pulse to be shortened at will under the control of external input. In its pure form this second category includes only one type, the 221, a dual package without internal timing resistors.

In the third stage of evolution, the *retriggerable* one-shot adds the ability to extend the basic output pulse beyond the time determined by the R-C timing constant. Unlike the basic one-shot, which simply ignores any repeated appearances of the triggering pulse during the output pulse timing cycle, the retriggerable one-shot responds to the arrival of a new triggering transition during the timing cycle by "retriggering," that is, by beginning a new timing cycle without ever allowing the Q output to fall back to LOW. If triggering transitions keep arriving before the one-shot times out, the Q output can be kept at HIGH indefinitely.

Thus, the retriggerable one-shot provides nearly unlimited flexibility in shortening or lengthening the output pulse under external control. (In theory, a retriggerable one-shot need not have an $\overline{R}$ input, but in fact all 7400-series retriggerables are also resettable.) This category of retriggerable, resettable one-shots includes the 122 and 422§ (single, with internal 10 kΩ timing resistor) and the 123, 423, and 4538 (dual, without internal timing resistor).

TRIGGER FROM RESET (CLEAR)

In addition to the $\overline{A}$ and B inputs, many one-shots that have the direct reset (clear) capability provide a third avenue by which the device can be triggered—through the reset or $\overline{R}$ input itself. If the $\overline{R}$ input is held LOW while the $\overline{A}$ and B inputs are put at their active levels ($\overline{A}$ at LOW, B at HIGH), the direct reset will, in effect, inhibit the trigger, holding the one-shot's Q output LOW. When $\overline{R}$ is then brought HIGH, the enabling levels at the trigger inputs suddenly become operative and, in those one-shots that offer this capability, the high-going transition at $\overline{R}$ is applied through an internal connection to the internal trigger gate, producing an output pulse.

Note that at least one of the trigger inputs must be taken inactive ($\overline{A}$ to HIGH or B to LOW) and then returned to its active or enabling state sometime during the time of inactive (HIGH) level at $\overline{R}$ to set up an initial trigger from reset. If the input conditions at $\overline{A}$ and B then remain stable (both at their enabling levels), a one-shot that provides trigger from reset will continue to fire every time

TABLE 3.1 SUMMARY OF SALIENT FEATURES OF 7400-SERIES ONE-SHOTS

Type	Pkg. Count	Pin Count	Internal Resistor	Trigger From Reset	Active LOW Inputs	Active HIGH Inputs	Typical Pulse Width Min.	Typical Pulse Width Max.	Resistor (in kΩ) Min.	Resistor (in kΩ) Max.	Capacitor (in μF) Min.	Capacitor (in μF) Max.
Basic One-Shots												
121	Single	14	2 kΩ	—	TTL: 2	1 ⊓	30[a] ns	28 s	1.4	40	0	1000
Resettable One-Shots												
221	Dual	16	None	Yes	TTL: 1	1 ⊓	30[a] ns	28 s	1.4	40	0	1000
					LS: 1	1 ⊓	47[a] ns	70 s	1.4	100	0	1000
					HC, HCT: 1 ⊓[b]	1 ⊓	140[c] ns	inf.	500	inf.	0	inf.
Retriggerable, Resettable One-Shots												
122	Single	14	10 kΩ	Yes	TTL: 2	2 ⊓	45[d] ns	cont.	5	50	0	inf.
					LS: 2	2 ⊓[e]	116[d] ns	cont.	5	260	0	inf.
422§	Single	14	10 kΩ	No	LS: 2	2 ⊓[e]	116[d] ns	cont.	5	260	0	inf.
					TTL: 1	1	45[d] ns	cont.	5	50	0	inf.
123	Dual	16	None	Yes[f]	LS: 1	1 ⊓[e]	116[d] ns	cont.	5	260	0	inf.
					HC, HCT: 1 ⊓	1 ⊓	75[g] ns	cont.	2[h]	1000	0	inf.
423	Dual	16	None	No	LS: 1	1 ⊓[e]	116[d] ns	cont.	5	260	0	inf.
					HC, HCT: 1 ⊓	1 ⊓	75[g] ns	cont.	2[h]	1000	0	inf.
4538	Dual	16	None	No	HC, HCT: 1 ⊓	1 ⊓	147[i] ns	cont.	1	1000	0	inf.

[a] $C_X = 0$, $R_X = 2$ kΩ (the 1.4 kΩ minimum for R_X applies only if a certain amount of pulse-width jitter is allowed)

[b] As presently specified, the Signetics 74HC221 will have Schmitt-trigger circuitry only at its active-HIGH (B) input

[c] $C_X = 28$ pF, $R_X = 2$ kΩ (RCA); National 450 ns (typ.) under the same conditions. For its 74HC221 Toshiba specifies a typical minimum of 190 ns at $C_X = 0$ and $R_X = 1$ kΩ

[d] $C_X = 0$, $R_X = 5$ kΩ

[e] The active-HIGH (B) inputs of the 74LS122, 74LS123, 74LS422, and 74LS423 have only a moderate amount of Schmitt-trigger hysteresis (enough to ensure jitter-free triggering from transitions as slow as 0.1 V/μs)

[f] Except National 74123

[g] $C_X = 0$, $R_X = 5$ kΩ (Signetics); National has 450 ns (typ.) for $C_X = 28$ pF and $R_X = 2$ kΩ, Toshiba 118 ns (typ.) for $C_X = 0$ and $R_X = 1$ kΩ

[h] Signetics at $V_{CC} = 5$ V; RCA has minimum 5 kΩ (V_{CC} not specified), Toshiba states 1 kΩ for V_{CC} of 3 V or more

[i] $C_X = 12$ pF, $R_X = 1$ kΩ (Motorola, National); Toshiba 140 ns (typ.) under the same conditions

a pulse (negative or positive) is applied at the $\overline{R}$ input, the trigger itself always occurring on the high-going edge of the pulse. One-shots that can be triggered from $\overline{R}$ include the 221, 122, and 123 (except for the National 74123). One-shots that will not trigger from $\overline{R}$ include the 422, the 423, the 4538, National's TTL version of the 123, and, of course, the 121, which has no $\overline{R}$ input.

APPLICATIONS

Standing alone, one-shots are chiefly used to generate pulses of a specific duration, typically ranging from about 30 nanoseconds to a few microseconds. If the external timing capacitor is omitted and only the internal resistor or just a low-value external resistor is used, most one-shots will produce an output pulse in the neighborhood of 30 ns to 50 ns (a bit longer for HC and HCT versions) that can be used as a general-purpose asynchronous signal in applications that do not have stringent pulse width accuracy requirements. For example, a minimal one-shot can generate a dc-triggered reset pulse for initializing a collection of flip-flops or other devices. At the other end of the capability range, single retriggerable one-shots with external timing components can perform as frequency discriminators and envelope detectors, useful in cleaning up noisy waveforms and generating uniform pulses from irregular input signals.

Combined with other logic circuits, one-shots can form the basis for a wide variety of simple but extremely useful devices. For example, pairs of one-shots can be connected to form a delayed pulse generator, that is, a circuit that can introduce a timed delay into a particular signal. Further, pairs of retriggerable one-shots can be connected to form circuits that generate pulses at variable frequencies or detect missing pulses in an output train. Conjoined with a D flip-flop, the one-shot can function as a noise discriminator or pulse width detector.

For a readily available short discussion of these and other common uses of the versatile one-shot, see National Semiconductor Application Note 372, "Designer's Encyclopedia of Bipolar One-Shots," by Kern Wong (National *Logic Databook*, Vol. II).

OUTPUT PULSE TIMING

In 7400-series one-shots, the basic output pulse width is generally determined by the time it takes an external timing capacitor C_X, normally charged to V_{CC}, to suddenly discharge through an internal switching transistor to a certain internal reference voltage and then more slowly charge back up to a higher reference level through the timing resistor R_X. There is always a brief propagation delay, t_{pd}, between the arrival of the triggering edge and the beginning of the output pulse.

The width of the output pulse is calculated from a timing formula. In most cases, the time is given by the simple equation

$$T_W = K \, R_X \, C_X$$

The only exceptions are the TTL versions of the 122 and 123, which follow the equation

$$T_W = K \, R_X \, C_X \, (1 + 0.7/R_X)$$

In both equations, T_W is in seconds, R_X is in ohms, and C_X is in farads; or, more conveniently, T_W is expressed in nanoseconds, R_X in kilohms, and C_X in picofarads. K is a dimensionless constant, usually somewhere in the range 0.3 to 0.7; its value is different for different kinds of one-shots.

Specific values for K are given in the data sheets. In general, however, it should be noted that the given value is nominal and represents K exactly only at a specified C_X, usually 1000 pF for TTL and LS one-shots or 10,000 pF for HC and HCT one-shots. If the timing capacitance falls much below this threshold, K begins to rise from its nominal specification, generally reaching a value in the neighborhood of 1.0 to 1.2 by the time C_X has fallen to 10 pF.

K for any particular low value of C_X is found by referring to graphs published by the individual manufacturers. The pulse width when C_X is below the specified value can also be found directly from a more complex timing equation with several coefficients. For example, the pulse width of the Motorola and SGS 74LS122 and 74LS123 for C_X below 1000 pF is approximated by the expression $6 + 0.05 \, C_X + 0.45 \, R_X C_X + 11.6 \, R_X$ —but such equations are rarely published.

For C_X above 1000 (or 10,000) pF, K is still not necessarily constant. In some versions its curve with relation to C_X is indeed virtually flat, but in other versions K continues to slowly decrease with increasing C_X. K also varies with changes in supply voltage, as much as 2 percent above or below its nominal value over the TTL and LS V_{CC} range (4.5 V to 5.5 V) and considerably more than this over the usable HC range of 2 V to 6 V. Furthermore, while compensation circuitry in later designs has largely eliminated changes in the value of K due to differences in ambient temperature, in older TTL designs K can vary by as much as 5 percent above or below its nominal value over the rated temperature range.

The K coefficient is one of the few parameters in which meaningful differences can be found between components of the same kind produced by different manufacturers. In particular, the relation between K and V_{CC} can be radically different for differently sourced versions of exactly the same component, as can be seen by comparing different manufacturers' published graphs of K versus V_{CC} for the 74HC4538. Care should be taken to consult the graphs published by the individual manufacturer of the IC in question when calculating pulse widths at extremes of the capacitance, voltage, or temperature ranges. Changes in R_X, on the other hand, have no significant effect on K as long as R_X remains within the limits recommended for general operation.

PULSE WIDTH JITTER

Variation in width from pulse to pulse generated by the same one-shot is called *jitter*. In applications with stringent pulse width accuracy requirements, jitter can be held to a minimum by observing certain design practices.

First, jitter in any particular one-shot is independent of C_X but directly related to R_X, which should be as large as possible within the limits set for proper operation. While reasonably jitter-free triggering can be accomplished using the internal resistor in some one-shots, a larger external timing resistor should be used for maximum pulse width stability.

Another factor affecting pulse-width jitter is the transition time of the trigger pulse. As with flip-flops, one-shot triggering occurs at a voltage level and is not directly related to the rise or fall time of the input pulse. Even so, at ordinary logic inputs transition rates should be faster than 1 V/μs to minimize pulse width jitter. Signals with slower transition rates should be confined to Schmitt-trigger inputs (see table of features), which can generally provide jitter-free triggering with input transition rates as slow as 1 V/s in TTL and even slower in HC and HCT versions.

Though jitter in the output from a particular one-shot is independent of C_X, keeping C_X at 1000 pF or higher can help minimize variation in output pulse widths from different one-shots of the same kind. The range of jitter-free output pulse widths is extended if V_{CC} is kept at 5 V and the ambient temperature is maintained at 25°C.

DUTY CYCLE

In addition to the timing factors discussed above, the nonretriggerable devices (versions of the 121 and 221) require some consideration of duty cycle as well. This parameter, usually expressed as a percentage, gives the proportion of each complete operating cycle that is occupied by the output pulse, that is, the proportion of each cycle that the Q output spends at the HIGH level. Duty cycles above 50 percent will tend to reduce the output pulse width compared to the width predicted by the one-shot's timing formula. If the duty cycle varies between low and high values the output pulse width will also vary (jitter).

Since jitter is a function of R_X, the range of jitter-free duty cycles can be extended by using a relatively high value for the external resistor. If R_X is near the recommended maximum for a given component type, jitter will not be appreciable until the duty cycle approaches 90 percent. Even higher duty cycles can be achieved if a certain amount of pulse-width jitter is allowed. However, to take the final step it is necessary to use a retriggerable one-shot, which can achieve a 100 percent duty cycle (that is, its Q output will remain at the HIGH level indefinitely) if it continuously receives an input pulse train whose cycle time is shorter than the period of the output cycle.

RETRIGGER TIMING

When a trigger signal is repeated during the timing cycle, the retriggerable one-shot responds by quickly discharging the external timing capacitor back to the lower voltage reference level and beginning again its timed rise to the upper voltage reference level. Barring further retriggering, the new output pulse time, beginning with the retrigger transition, will include, first, the relatively constant propagation delay time t_{pd} between the input transition and the output transition (the output transition itself is invisible this time because Q is already HIGH, but the delay is still there), and second, the usually much larger timed pulse width determined by R_X and C_X. For one-shots that have the normal timing equation, therefore, the retriggerable pulse width is

$$t = t_W + t_{PLH} = K\ R_X\ C_X + t_{PLH}$$

where R_X and C_X are chosen by the designer and K and t_{PLH} (in the slightly more conservative form of t_{pd}) are key parameters given in the data pages.

DISCHARGE TIMING

The timing cycle, whether triggered or retriggered, always begins with a sudden discharge of the capacitor down to the lower reference voltage. The discharge time of the capacitor is very brief but not instantaneous, and during this time even the retriggerable one-shot will not respond to a repeated trigger signal.

Methods for calculating the discharge time vary with the technology. In LS one-shots the width of this "blind spot" is a function of C_X; the discharge time in nanoseconds is guaranteed to be less than $0.22\ C_X$ (pF) and is typically $0.05\ C_X$ (pF). In HC and HCT versions, t_{rr} is usually said to be a function of V_{CC} and C_X; for example, the discharge time of the National 74HC123A is approximated by the equation

$$t_{rr} = 20 + 187/(V_{CC} - 0.7) + (565 + 0.256\ V_{CC}\ C_X)/(V_{CC} - 0.7)^2 \text{ (typ.)}$$

and that of the Motorola 74HC4538 by the equation

$$t_{rr} = 72 + V_{CC}\ C_X/30.5 \text{ (typ.)}$$

where t_{rr} is in ns, V_{CC} is in V, and C_X is in pF. In some cases, however, the timing resistance R_X is also identified as a factor, as in the equation given for the Signetics 74HC123:

$$t_{rr} = 35 + 0.11\ C_X + 0.04\ R_X\ C_X \text{ (typ.)}$$

where $C_X > 10$ nF and $V_{CC} = 5$ V.

For some models, the manufacturer provides graphs or tables instead of formulas. For example, Toshiba specifies for the 74HC123 and 74HC423 a typical minimum t_{rr} of 74 ns if $C_X = 100$ pF or 1.1 μs if $C_X = 0.01\ \mu$F, assuming that $R_X =$

1 kΩ and V_{CC} = 4.5 V. In a lamentably high proportion of cases, however, such information is simply not available.

EXTERNAL TIMING COMPONENTS

For precise timing, use resistors and capacitors with a good temperature coefficient. Also, note that the one-shot's timing equation and other published data may fail to accurately predict its output pulse width if capacitor leakage exceeds 100 nA or stray capacitance to ground exceeds 50 pF.

Capacitors made of mica, glass, polystyrene, polypropylene, or polycarbonate are appropriate for small time constants; when larger values are called for, use solid tantalum or special aluminum capacitors. See note below on the use of electrolytic capacitors with certain TTL one-shots.

While the timing capacitance, C_X, can have almost any practical value, care should be taken to keep the timing resistance, R_X, within the limits specified in the table of features above. Failure to do so may result in malfunction or complete inoperation of the one-shot.

EXTERNAL CONNECTIONS

To physically program a one-shot's basic pulse width, the external timing capacitor (if any) should be connected between the pins labeled C_X and $R_X C_X$ in the corresponding functional diagram. If a polarized timing capacitor is used, the positive side of the capacitor should be connected to the $R_X C_X$ pin in the case of all one-shots except the 121 and the TTL and LS versions of the 221, where the positive side of the capacitor should be connected to C_X.

The external timing resistor (if any) should be connected between V_{CC} and the $R_X C_X$ pin. To use the optional internal resistor of the 121, 122, or 422, the pin labeled R_{int} in the appropriate diagram should be connected directly to V_{CC}. If an external resistor is used with these one-shots to improve pulse-width accuracy and repeatability, it should be connected between V_{CC} and $R_X C_X$ with the R_{int} pin left open-circuited.

The basic 121 can be configured to provide a brief, general-purpose reset pulse by simply connecting R_{int} to V_{CC} and leaving the C_X and $R_X C_X$ pins open. The 221 can be similarly configured by leaving C_X open and connecting a resistor between V_{CC} and $R_X C_X$.

TRIMMING RESISTORS

To trim the output pulse width, a variable resistor can be connected in series between the external timing resistor and V_{CC}, or an adjustable resistor can be chosen for R_X itself. In one-shots with internal resistors, the adjustable resistor is connected between V_{CC} and the R_{int} pin if the internal resistor is to be included in the timing circuit.

SWITCHING DIODES

When an electrolytic capacitor is used for C_X, the TTL versions of the various one-shots often require the addition of a switching diode to prevent high inverse leakage current. Such protection is indicated if the reverse voltage rating of the electrolytic (normally specified at 5 percent of the forward voltage rating) is less than 1 V. The diode, which can be any silicon switching diode such as the 1N916 or 1N3064, is connected to the $R_X C_X$ pin as shown in Figure 3.2.

When the switching diode is used, external resistance R_X should not exceed 60 percent of the maximum recommended external resistance specified in the table of features given previously. The timing coefficient, K, is lowered about 12 percent by the addition of the diode.

The switching diode is not needed, and should not be used, with LS, HC, and HCT implementations, and its use is discouraged in applications where the one-shot is retriggered.

CLAMPING DIODES

The timing capacitor C_X is generally kept fully charged in the quiescent state between output pulses. When the system's supply voltage is turned off, C_X discharges through the one-shot's $R_X C_X$ pin. If C_X is relatively large and the turn-off time of the power supply is relatively quick, the $R_X C_X$ pin may experience a momentary discharge current large enough to damage HC and HCT circuits.

The internal protection diodes of HC and HCT one-shots can safely handle a maximum current of about 30 mA, which limits the shut-off time of the power supply to

$$t = V_{CC} C_X/30$$

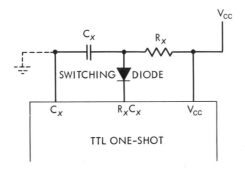

Figure 3.2

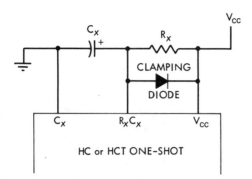

Figure 3.3

where C_X in μF and V_{CC} in volts give maximum shut-off time t in milliseconds. Thus, for example, if C_X is 15 μF and V_{CC} is 5 V then the maximum shut-off time is $5 \times 15/30 = 2.5$ ms. This is the formula suggested by Motorola and National. Toshiba puts the maximum input current at 20 mA and gives the formula as

$$t = (V_{CC} - 0.7)\, C_X/20$$

where t is the shortest time, in milliseconds, that the supply voltage can be allowed to fall to a level of 0.4 V_{CC}.

In practice, most power supplies are heavily filtered and do not shut off so quickly that "latch-up" becomes a problem. Under circumstances in which the shut-off time exceeds the limits given above, however, HC and HCT one-shots will need a clamping diode connected from the $R_X C_X$ pin to V_{CC} (in parallel with the external timing resistor) to prevent damage to the circuit (see Figure 3.3).

RECOMMENDED PRACTICES

For proper operation, the timing capacitor, timing resistor, and trimming resistor (if any) must be connected as closely as possible to the device pins to minimize noise pickup, stray capacitance, and stray inductance, especially inductance between the $R_X C_X$ junction and the one-shot's $R_X C_X$ pin. Such factors can interfere with the complete discharge of C_X in each cycle and therefore effect pulse width accuracy. Impedance added by lead lengths greater than 3 cm can cause a shift in the apparent capacitor voltage sensed by the one-shot and result in substantial differences between actual and calculated output pulse widths.

Power and ground wiring should conform to good high frequency standards so that switching transients in power and ground leads do not allow interaction between one-shots that would cause false triggering. In particular, a 0.001 μF to 0.1 μF bypass capacitor (disk ceramic or monolithic) should be connected, by the shortest possible path, between ground and the V_{CC} pin on each device. In cases of severe supply-line noise, decoupling should be provided by a local power supply voltage regulator.

The one-shot's trigger inputs should never be left uncommitted or "floating." All unused inputs should be hard-wired to their enabling levels, that is, all unused $\overline{A}$ (or A) inputs should be tied to ground, and all unused B (or $\overline{B}$) and $\overline{R}$ inputs should be tied to V_{CC}.

The C_X pin of most one-shots provides the external capacitor with an internal connection to ground. For improved noise immunity, however, most manufacturers recommend that the C_X pin(s) of LS and HC versions be externally wired to ground as well. The major exception to this is the LS221, in which the C_X pins are not pathways to ground and which will become inoperative if the C_X pins are externally grounded.

Basic one-shots

121 *Single one-shots with internal timing resistor.* Basic TTL one-shot with internal 2 kΩ timing resistor, two OR-gated active-LOW $\overline{A}$ inputs without Schmitt triggers, and one active-HIGH B input with Schmitt trigger that can provide jitter-free triggering from input transitions as slow as 1 V/s. The B input can conveniently function as an ENABLE/$\overline{INHIBIT}$ control for the two $\overline{A}$ inputs. There is no reset or clear input. Supplied in 14-pin DIP.

Resettable one-shots

221 *Dual resettable one-shots.* Two independent one-shots with additional separate direct reset (clear) inputs, $\overline{R}$, are fitted into a single 16-pin DIP by omitting the internal timing resistor and limiting each one-shot's trigger inputs to a single $\overline{A}$ and B. All versions have Schmitt-trigger circuitry at the B input; most of the HC and HCT versions (but none of the TTL or LS versions) also have Schmitt-trigger action at the $\overline{A}$ inputs. Each of the $\overline{A}$ and B inputs of a particular one-shot can function as an enable/inhibit control for the other, and the $\overline{R}$ input can also function as a trigger. A LOW level at the $\overline{R}$ input will terminate the output pulse prematurely, an ability that can be used to generate pulses significantly shorter than the minimum pulse width specification. The HC221 will reset on power-up. (See "Note on Interchangeability" below.)

Retriggerable one-shots

In addition to the retrigger capability, all of the following have an active-LOW reset input that can shorten the output pulse and disable the trigger inputs. In some types (noted below) $\overline{R}$ can also be used to trigger the device.

122 *Retriggerable, resettable one-shots with internal resistor.* Singly packaged TTL and LS one-shots with 10 kΩ internal timing resistor that can be used in place of the external resistor in applications

without stringent pulse width accuracy requirements. The one-shot provides two OR-gated active-LOW $\overline{A}$ inputs and two AND-gated active-HIGH B inputs, the latter doubling as convenient ENABLE/$\overline{\text{INHIBIT}}$ controls. In LS versions the B inputs provide moderate Schmitt-trigger action that will ensure jitter-free triggering from transition rates as slow as 0.1 V/μs (0.1 mV/ns). Can be triggered from $\overline{R}$.

422§   *Retriggerable, resettable one-shots with internal resistor.* Identical to 74LS122 except device will not trigger from the $\overline{R}$ input. Available only in an LS version from Texas Instruments.

123   *Dual retriggerable, resettable one-shots.* Retriggerable versions of the 221 (see "Note on Interchangeability" below). Two independent retriggerable and resettable one-shots with performance characteristics virtually identical to corresponding versions of the 122 are fitted into a single 16-pin DIP by omitting the internal timing resistor and limiting each one-shot's trigger inputs to a single active-LOW $\overline{A}$ input and a single active-HIGH B input. LS versions provide moderate Schmitt-trigger action (transition times to 0.1 V/μs) at the B inputs, while HC and HCT versions provide full Schmitt triggering at both $\overline{A}$ and B inputs to accommodate input transition rates down to 1 V/s. For each one-shot, each of the $\overline{A}$ and B inputs can conveniently function as an enable/inhibit control for the other input, and the $\overline{R}$ input can also function as a trigger in all versions except the National 74123.

423   *Dual retriggerable, resettable one-shots.* LS, HC, and HCT one-shots identical to corresponding versions of 123 except that the device will not trigger from the $\overline{R}$ input. The National 74123 (which will not trigger from $\overline{R}$) is really a misnamed TTL version of the 423.

4538   *Dual precision one-shots (retriggerable, resettable).* HC and HCT one-shots that provide maximum pulse-width accuracy. Inputs to the 4538 trigger on opposite edges of the input pulse from the corresponding inputs of the other one-shots, that is, A is positive-edge-triggered and $\overline{B}$ is negative-edge-triggered, although the enabling levels at these inputs are the same as for other types. The difference is caused by OR gating at the trigger input of the 4538 in place of the usual AND gating (see functional diagrams). As a result, a stable active level at either input will produce a steady HIGH from the OR gate that will prevent triggering transitions at the other input from being sensed by the one-shot. If the 4538 is to operate as a nonretriggerable device, Q should be connected to $\overline{B}$ if A is used for rising-edge triggering, or Q should be connected to A if $\overline{B}$ is used for falling-edge triggering. On power-up the device is reset.

Note on Interchangeability. Pin assignments on the nonretriggerable 221 are identical to those of the retriggerable 123 and 423. In fact, if retriggering is not needed, the 221 can be directly substituted for the 123, and can often be substituted for the 423 (if the ability of the 221 to trigger from reset causes no problems). In either case, however, the following details will have to be attended to:

(a) The values of R_X or C_X or both will probably have to be changed.

(b) If a TTL or LS version of the 221 is substituted for the corresponding version of the 123 or 423, the polarity of the timing capacitors must be reversed. The 74221 and 74LS221 require the positive leads of the capacitors, rather than the negative leads, to be connected to their corresponding C_X pins.

(c) In the retriggerable 123 and 423 the internal ground connection from the C_X pin is often augmented by an external ground connection to help improve noise immunity. In the LS221, however, C_X is not internally grounded, and if an LS221 is inserted where external ground connections have been added at the C_X pins, the device will not function.

Note on Function Tables. Published function tables for one-shots usually exclude certain atypical combinations of inputs, such as those in which more than one of the trigger inputs are experiencing a logic transition at exactly the same time. In effect, the manufacturers leave the circuit's response to such combinations undefined. All of the tables below assume that any output pulse caused by some previous set of conditions has already timed out, and therefore the device begins in the quiescent state.

FUNCTION TABLE
121

INPUTS			OUTPUTS	
$\overline{A1}$	$\overline{A2}$	B	Q	$\overline{Q}$
H	H	x	L	H
x	x	L	L	H
s	s	s	L	H
H	↓	H	⎍	⎍
↓	H	H	⎍	⎍
↓	↓	H	⎍	⎍
L	H	↑	⎍	⎍
H	L	↑	⎍	⎍
L	L	↑	⎍	⎍

HIGH at $\overline{A1}$ and $\overline{A2}$ inhibits trigger
LOW at B inhibits trigger
Steady state does not trigger

Appropriate transition
   triggers output pulse

x = Any voltage level or transition
s = Either stable level (HIGH or LOW)

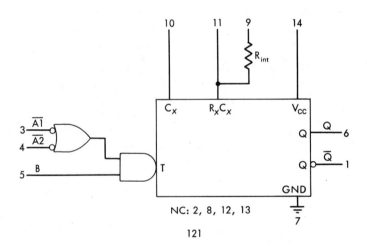

NC: 2, 8, 12, 13

121

FUNCTION TABLE
221, 123, 423

INPUTS			OUTPUTS	
$\overline{A}$	B	$\overline{R}$	Q	$\overline{Q}$
H	x	x	L	H
x	L	x	L	H
x	x	L	L	H
s	s	s	L	H
↓	H	H	⊓	⊔
L	↑	H	⊓	⊔
L	H	↑*	⊓	⊔

HIGH at A inhibits trigger
LOW at B inhibits trigger
LOW at $\overline{R}$ inhibits trigger
Steady state does not trigger

Appropriate transition
triggers output pulse

x = Any voltage level or transition
s = Either stable level (HIGH or LOW)
 * This line does not apply to the 423,
 which will not trigger from $\overline{R}$.

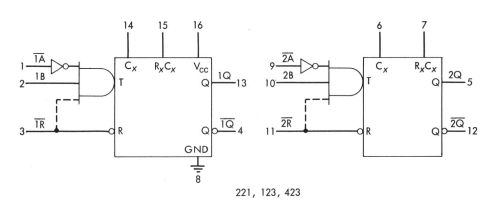

221, 123, 423

FUNCTION TABLE
122, 422

INPUTS					OUTPUTS		
$\overline{A1}$	$\overline{A2}$	B1	B2	$\overline{R}$	Q	$\overline{Q}$	
H	H	x	x	x	L	H	HIGH at $\overline{A1}$ and $\overline{A2}$ inhibits trigger
x	x	L	x	x	L	H	LOW at B1 or B2 inhibits trigger
x	x	x	L	x	L	H	
x	x	x	x	L	L	H	LOW at $\overline{R}$ inhibits trigger
s	s	s	s	s	L	H	Steady state does not trigger
↓	H	H	H	H	⊓	⊔	
H	↓	H	H	H	⊓	⊔	Triggers from $\overline{A}$ inputs
↓	↓	H	H	H	⊓	⊔	
L	s	↑	H	H	⊓	⊔	
s	L	↑	H	H	⊓	⊔	Triggers from B inputs
L	s	H	↑	H	⊓	⊔	
s	L	H	↑	H	⊓	⊔	
L	s	H	H	↑	⊓	⊔	Triggers from $\overline{R}$ input (122 only)
s	L	H	H	↑	⊓	⊔	

x = Any voltage level or transition
s = Either stable level (HIGH or LOW)

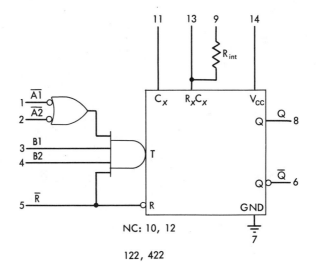

NC: 10, 12

122, 422

FUNCTION TABLE
4538

INPUTS			OUTPUTS	
A	$\overline{B}$	$\overline{R}$	Q	$\overline{Q}$
H	x	x	L	H
x	L	x	L	H
x	x	L	L	H
s	s	s	L	H
⬆	H	H	⊓ ⎍	
L	⬇	H	⊓ ⎍	

HIGH at A inhibits trigger
LOW at $\overline{B}$ inhibits trigger
LOW at $\overline{R}$ inhibits trigger
Steady state does not trigger

Appropriate transition triggers output pulse

x = Any voltage level or transition
s = Either stable level (HIGH or LOW)

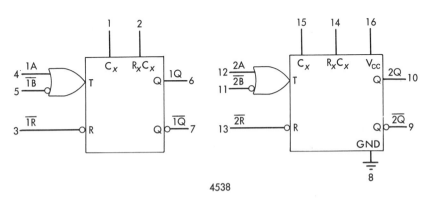

4538

121   One-shots with internal timing resistor

MANUFACTURERS	TTL	LS	HC	HCT	ALS	S	AS	F	AC	ACT
Fairchild	■									
National Semiconductor	■									
Signetics	■									
Texas Instruments	■									

KEY PARAMETERS		TTL	LS	HC	HCT	ALS	S	AS	F	AC	ACT
I_{OH} (max.)	mA	-0.4									
I_{OL} (max.)	mA	16									
I_{CC} (max.)	mA	40									
I_{CC} (quiesc. max.)	mA	25									
Timing constant, K		0.69									
t_W (min.)	ns	50									
$t_{WQ(min)}$ (max.)	ns	50[a]									
t_{WQ} (typ.)	ns	700[b]									
$t_{pd} \overline{A}$ (max.)	ns	80									
$t_{pd} B$ (max.)	ns	65									

[a] $C_X = 0$, R_{int} to V_{CC}. If $C_X = 80$ pF then $t_{WQ(min)}$ (typ.) [b] $C_X = 100$ pF, $R_X = 10$ kΩ
    = 110 ns

221   Dual resettable one-shots

MANUFACTURERS	TTL	LS	HC	HCT	ALS	S	AS	F	AC	ACT
Motorola		■	D							
National Semiconductor		■	■							
RCA			▨	▨						
SGS		▨	D							
Signetics	■		D	D						
Texas Instruments	■	■	D							
Toshiba			■							

Added change control symbol: National 74HC221A

KEY PARAMETERS		TTL	LS	HC	HCT	ALS	S	AS	F	AC	ACT
I_{OH} (max.)	mA	-0.8	-0.4	-4	-4						
I_{OL} (max.)	mA	16	8	4	4						
I_{CC} (max.)	mA	80	27	5.2^c	n/a						
I_{CC} (quiesc. max.)	mA	50	11	0.08^d	0.08						
C_{pd} (typ.)	pF/OS			166^e	166						
Timing constant, K		0.69	0.69	0.69^f	0.69						
t_W (min.)	ns	50	40	25^g	31						
$t_{WQ(min)}$ (max.)	ns	50^a	70^a	365^h	1						
t_{WQ} (typ.)	μs	0.7^b	0.74^b	7^i	7^b						
t_{pd} A (max.)	ns	80	80	60^j	68						
t_{pd} B (max.)	ns	65	65	60^j	68						
t_{pd} R̄ (max.)	ns	40	65	53^k	60						

^{a} C_X = 0, R_X = 2 kΩ

^{b} C_X = 100 pF, R_X = 10 kΩ

^{c} I.e., 2.6 mA per one-shot (National 74HC221A) at V_{CC} = 6 V

^{d} Toshiba 0.04 mA

^{e} RCA; Toshiba 109 pF/OS. National data not available

^{f} Toshiba 0.70, National 74HC221A 1.0

^{g} RCA; Toshiba 19 ns, National 74HC221A 37 ns

^{h} Toshiba at C_X = 0, R_X = 5 kΩ.; max. min. for others not available. At C_X = 28 pF and R_X = 2 kΩ a typical pulse width is 140 ns (RCA) or 450 ns (National 74HC221A)

^{i} C_X = 100 pF, R_X = 10 kΩ; Toshiba specifies 0.79 ms

^{j} RCA; National 74HC221A 51 ns, Toshiba 70 ns

^{k} National 74HC221A 42 ns

^{l} Max. min. not available. At C_X = 28 pF and R_X = 2 kΩ a typical pulse width is 140 ns (RCA)

MANUFACTURERS	TTL	LS	HC	HCT	ALS	S	AS	F	AC	ACT
Fairchild	■									
Motorola		■								
National Semiconductor										
SGS		▓								
Texas Instruments	■	■								

KEY PARAMETERS		TTL	LS	HC	HCT	ALS	S	AS	F	AC	ACT
I_{OH} (max.)	mA	-0.8	-0.4								
I_{OL} (max.)	mA	16	8								
I_{CC} (max.)	mA	36^a	11								
Timing constant, K		0.32^b	0.37^e								
t_W (min.)	ns	40	40								
$t_{WQ(min)}$ (max.)	ns	65^c	200^c								
t_{WQ} (typ.)	μs	3.42^d	4.5^d								
t_{pd} $\overline{A}$ (max.)	ns	40	45								
t_{pd} B (max.)	ns	36	56								
t_{pd} $\overline{R}$ (max.)	ns	40	45								

[a] Fairchild 28 mA

[b] Timing equation is $t_W = R_X C_X (1 + 0.7/R_X)$. If switching diode is used, K is 0.28 (TI)

[c] $C_X = 0$, $R_X = 5\ k\Omega$

[d] $C_X = 1000\ pF$, $R_X = 10\ k\Omega$

[e] National; TI 0.33, Motorola and SGS 0.45

MANUFACTURERS	TTL	LS	HC	HCT	ALS	S	AS	F	AC	ACT
Fairchild	■									
Motorola		■	D							
National Semiconductor			■							
RCA			▨	▨						
SGS		▨	D							
Signetics	■		■							
Texas Instruments	■	■	D							
Toshiba			■							

Added change control symbol: National 74HC123A

KEY PARAMETERS		TTL	LS	HC	HCT	ALS	S	AS	F	AC	ACT
I_{OH} (max.)	mA	-0.8	-0.4	-4	-4						
I_{OL} (max.)	mA	16	8	4	4						
I_{CC} (max.)	mA	66	20	0.8^f	n/a						
I_{CC} (quiesc. max.)	mA			0.08^g	0.08						
C_{pd} (typ.)	pF/OS			113^h	n/a						
Timing constant, K		0.28^a	0.37^d	0.45^i	0.45^o						
t_W (min.)	ns	40	40	25^j	25						
$t_{WQ(min)}$ (max.)	ns	65^b	200^b	k	p						
t_{WQ} (typ.)	μs	3.03^c	4.5^e	450^l	450^q						
t_{pd} $\overline{A}$ (max.)	ns	40	45	64^m	64^r						
t_{pd} B (max.)	ns	36	56	64^m	64^r						
t_{pd} $\overline{R}$ (max.)	ns	40	45	54^n	58^s						

a Fairchild, Signetics, TI; National 0.34. Timing equation is $t_W = K R_X C_X (1 + 0.7/R_X)$. If switching diode is used, K is 0.25 (Signetics, TI)

b $C_X = 0$, $R_X = 5$ kΩ

c $C_X = 1000$ pF, $R_X = 10$ kΩ. National has 3.42 μs under the same conditions. Typical values for Fairchild and Signetics not available

d National; TI 0.33, Motorola and SGS 0.45

e $C_X = 1000$ pF, $R_X = 10$ kΩ

f Toshiba (0.4 mA per one-shot); National 2.6 mA (1.3 mA per one-shot). Maximum active supply current for RCA and Signetics not available

g Toshiba 0.04 mA

h Toshiba; National, RCA, and Signetics data not available

i At $V_{CC} = 5$ V (RCA, Signetics); Toshiba 0.46 at 4.5 V. Signetics also gives K = 0.48 at $V_{CC} = 2$ V. National 74HC123A has K = 1.0

j National 37 ns

k Maximum value not available. At $C_X = 0$ and $R_X = 5$ kΩ

Signetics has output pulse width of 75 ns (typ.); at $C_X = 0$ and $R_X = 1$ kΩ Toshiba has 118 ns (typ.); at $C_X = 28$ pF and $R_X = 2$ kΩ National 74HC123A has 450 ns (typ.)

l $C_X = 100$ nF, $R_X = 10$ kΩ (Signetics). At $C_X = 10$ nF and $R_X = 10$ kΩ RCA has 38.7 μs (min.) and 51.3 μs (max.); at $C_X = 100$ pF and $R_X = 10$ kΩ Toshiba has 1.0 μs. Typical value for National not available

m Signetics; Toshiba and National 60 ns, RCA 80 ns. The delay when $\overline{R}$ is used as a trigger is the same as $\overline{A}$ or B except for Toshiba (66 ns instead of 60 ns)

n RCA and Signetics; National 42 ns, Toshiba 49 ns

o At $V_{CC} = 5$ V (RCA, Signetics)

p Maximum value not available. At $C_X = 0$ and $R_X = 5$ kΩ Signetics has typical output pulse width of 75 ns

q $C_X = 100$ nF, $R_X = 10$ kΩ (Signetics). At $C_X = 10$ nF and $R_X = 10$ kΩ RCA has 45 μs (typ.)

r Signetics; RCA 85 ns

s Signetics; RCA 60 ns

423   Dual retriggerable, resettable one-shots (no trigger from clear)

MANUFACTURERS	TTL	LS	HC	HCT	ALS	S	AS	F	AC	ACT
Motorola			D							
National Semiconductor			■	■						
RCA			▩	▩						
SGS			D							
Signetics			■							
Texas Instruments		■	D							
Toshiba			■							

Added change control symbol: National 74HC423A

KEY PARAMETERS		TTL	LS	HC	HCT	ALS	S	AS	F	AC	ACT
I_{OH} (max.)	mA		-0.4	-4	-4						
I_{OL} (max.)	mA		8	4	4						
I_{CC} (max.)	mA		20	0.8^c	n/a						
I_{CC} (quiesc. max.)	mA			0.08^d	0.08						
C_{pd} (typ.)	pF/OS			113^e	n/a						
Timing constant, K			0.33	0.45^f	0.45^l						
t_W (min.)	ns		40	25^g	25						
$t_{WQ(min)}$ (max.)	ns		200^a	h	m						
t_{WQ} (typ.)	μs		4.5^b	450^i	450^n						
t_{pd} $\overline{A}$ (max.)	ns		45	64^j	64^o						
t_{pd} B (max.)	ns		56	64^j	64^o						
t_{pd} $\overline{R}$ (max.)	ns		45	54^k	58^p						

^{a} C_X = 0, R_X = 5 kΩ

^{b} C_X = 1000 pF, R_X = 10 kΩ

^{c} Toshiba (0.4 mA per one-shot); National 2.6 mA (1.3 mA per one-shot). Maximum active supply current for RCA and Signetics not available

^{d} Toshiba 0.04 mA

^{e} Toshiba; National and Signetics data not available

^{f} At V_{CC} = 5 V (RCA, Signetics); Toshiba 0.46 at 4.5 V. Signetics also gives K = 0.48 at V_{CC} = 2 V. National 74HC123A has K = 1.0

^{g} RCA and Signetics; Toshiba 19 ns, National 37 ns

^{h} Maximum value not available. At C_X = 0 and R_X = 5 kΩ Signetics has output pulse width of 75 ns (typ.); at C_X = 0 and R_X = 1 kΩ Toshiba has 118 ns (typ.); at C_X = 28 pF

and R_X = 2 kΩ National 74HC423A has 450 ns (typ.)

^{i} C_X = 100 nF, R_X = 10 kΩ (Signetics). At C_X = 10 nF and R_X = 10 kΩ RCA has 38.7 μs (min.) and 51.3 μs (max.); at C_X = 100 pF and R_X = 10 kΩ Toshiba has 1.0 μs. Typical value for National not available

^{j} Signetics; Toshiba and National 60 ns, RCA 80 ns

^{k} RCA and Signetics; National 42 ns, Toshiba 49 ns

^{l} At V_{CC} = 5 V (RCA, Signetics)

^{m} Maximum value not available. At C_X = 0 and R_X = 5 kΩ Signetics has typical output pulse width of 75 ns

^{n} C_X = 100 nF, R_X = 10 kΩ (Signetics). At C_X = 10 nF and R_X = 10 kΩ RCA has 45 μs (typ.)

^{o} Signetics; RCA 85 ns

^{p} Signetics; RCA 60 ns

4538 Dual precision one-shots (retriggerable, resettable)

MANUFACTURERS	TTL	LS	HC	HCT	ALS	S	AS	F	AC	ACT
Motorola			▓							
National Semiconductor			■							
RCA			■	■						
SGS			D							
Signetics			D	D						
Toshiba			■							

KEY PARAMETERS		TTL	LS	HC	HCT	ALS	S	AS	F	AC	ACT
I_{OH} (max.)	mA			-4	-4						
I_{OL} (max.)	mA			4	4						
I_{CC} (max.)	mA			0.22^a	n/a						
I_{CC} (quiesc. max.)	mA			0.08^b	0.08						
C_{pd} (typ.)	pF/OS			150^c	134						
Timing constant, K				0.75^d	0.7^j						
t_W (min.)	ns			20^e	20						
$t_{WQ(min)}$ (max.)	ns			185^f	n/a						
t_{WQ} (typ.)	μs			10^g	k						
t_{pd} A (max.)	ns			63^h	69						
t_{pd} B (max.)	ns			63^h	69						
t_{pd} $\overline{R}$ (max.)	ns			63^i	63						

^{a} V_{CC} = 6 V (Motorola); Toshiba 0.8 mA at same voltage

^{b} RCA, Signetics; Toshiba 0.04 mA, Motorola 0.22 mA, National 0.25 mA

^{c} Motorola, National; Toshiba 90 pF/OS, RCA 136 pF/OS

^{d} National at V_{CC} = 4.5 V; Motorola 0.73 at 4.5 V if t_W > 100 μs; Toshiba 0.80 at V_{CC} > 4 V if C_X = 0.01 μF, 0.72 if C_X > 0.1 μF; RCA 0.7, V_{CC} not given

^{e} Toshiba 19 ns

^{f} C_X = 12 pF, R_X = 1 kΩ, V_{CC} = 5 V (Motorola, National). Toshiba has typical (not max.) minimum of 140 ns under the same conditions

^{g} C_X = 1000 pF, R_X = 10 kΩ, V_{CC} = 5 V (Motorola, National, Toshiba). At C_X = 0.1 μF, R_X = 10 Ω, RCA gives 0.602 ms for min. over temperature range, 0.798 ms for max. over temperature range

^{h} Motorola and National 69 ns

^{i} Toshiba 49 ns, Motorola and National 69 ns

^{j} RCA, V_{CC} not given

^{k} At C_X = 0.1 μF, R_X = 10 Ω, RCA gives 0.602 ms for min. over temperature range, 0.798 ms for max. over temperature range

PART II
Parallel Circuits

Part I of this book was devoted to those common digital logic circuits that are packaged in ways suited to the processing of individual bits of data, that is, data transmitted on single lines. Such one-line signal processing is particularly associated with the simple hard-wired logic or "hardware programming" found in high-speed control circuits. From an analytical standpoint, the chief characteristic of these "basic circuits" is that each device (gate, flip-flop, or one-shot) serving a particular line is individually controlled by its own enable or clock input.

In many other applications, however, the data being transmitted or processed are not individual control signals, but data "words" consisting of many bits transmitted in parallel along groups of parallel lines (buses). The most common example of the bus-oriented system is the digital computer. To meet the needs of bus-oriented applications efficiently there is a wide variety of devices for handling groups of bits in parallel, ranging in complexity from parallel line drivers all the way to microprocessors themselves.

The most elementary of the parallel devices are gathered in this section under the name *parallel circuits*. These are simply arrays of the basic circuits covered in Part I, supplied in a single package and activated by a common clock or enable line. Thus, arrays of buffers or line drivers sharing a common enable input constitute what in this book are called *parallel buffers/drivers*; arrays of transparent latches sharing a common enable input make up *parallel latches*; arrays of flip-flops sharing a common clock input form *registers*; and specially configured bidirectional arrays of bus drivers sharing common control inputs (often combined with internal latches or registers) are called *bus transceivers* or *bus interfaces*.

Part II is organized around these basic categories of parallel circuits as follows:

Chapter
    4 Parallel Buffers/Drivers
    5 Parallel Latches
    6 Registers
    7 Bus Registers
    8 Bus Transceivers

The inputs of some LS, ALS, S, AS, and F parallel devices are provided with PNP or NPN transistors that decrease input current requirements and thus reduce DC loading and increase fan-in. For ease of comparison, maximum I_{IH} and I_{IL} ratings for all non-CMOS parallel circuits are given directly in the data charts.

Since most parallel circuits are designed to drive buses or other relatively high-capacitance loads, their performance is usually specified under somewhat heavier test loads than the performance of the single-line components of Part I. Unless otherwise noted, the switching characteristics given under "key parameters" in Part II are valid under the following conditions:

TTL and LS circuits:

    C_L = 45 or 50 pF (except t_{PZH} and t_{PZL}, for which C_L = 5 pF)

    R_L = (TTL) 400 Ω
            (LS) 667 Ω

    T_A = 25°C

S buffers/drivers and bus transceivers:

    C_L = 50 pF (except t_{PZH} and t_{PZL}, for which C_L = 5 pF)

    R_L = 90 Ω

    T_A = 25°C

S latches and registers:

    C_L = 15 pF (except t_{PZH} and t_{PZL}, for which C_L = 5 pF)

    R_L = 280 Ω

    T_A = 25°C

ALS, AS, and F circuits:

   C_L = 50 pF

   R_L = 500 Ω (open-collector transceivers 680 Ω)

   T_A = 0 to 70°C

HC, HCT, AC, and ACT circuits:

   C_L = 50 pF

   R_L = (HC, HCT) 1 kΩ
           (AC, ACT) 500 Ω

   T_A = −40 to 85°C

It will be seen from the above that while there are still specification differences between the various families, these differences are for the most part significantly less than those found among the simpler devices of Part I. While it remains true that the older families (TTL, LS, and S) are specified somewhat less conservatively than the newer ones, the parameters given for various implementations of the same type can for most purposes be compared directly in assessing relative performance.

4

Parallel Buffers/Drivers

This chapter deals with inverting and noninverting buffer/driver circuits intended to be used in parallel to drive bus lines and other parallel arrays of devices, such as memory address registers.

The great majority of the parallel buffers/drivers have 3-state outputs, which can be controlled through separate output enable lines to either transmit input signals like the ordinary buffers/drivers of Chapter 1 or to present a high-impedance "Z" condition at their outputs, effectively disconnecting the drivers from the bus or other devices. The few parallel buffers/drivers that do not have 3-state outputs instead feature high-current open-collector outputs specifically designed for buffer/driver applications. They are also equipped with enable inputs with logic similar to the output enables of the 3-state devices.

With the exception of the quad buffers/drivers 125 and 126, which for convenience are included at the beginning of this chapter, all the parallel buffers/drivers listed here are enabled in groups rather than individually, with each output enable (OE) input serving as a common control for a group of two to ten driver outputs. The "active" level for these OE inputs is always the level that enables the outputs so that the device is put into its "transparent" mode and acts like a group of ordinary drivers. The "inactive" level, on the other hand, is the one that inhibits output or creates a high-impedance condition. Since most output enables are active-LOW, in the usual case these control inputs are externally designated

$\overline{OE}$ to indicate that it is the LOW level that enables the outputs and puts the device "on line" to the bus.

Two basic modes of organization are used with the common 6-line and 8-line buffers/drivers. In all cases, the arithmetic of package design (16 pins for the 6-line devices and 20 pins for the 8-line devices) allows six or eight pins for data input, six or eight for data output, one each for V_{CC} and GND, and two for output control; the basic difference is whether the two output enables each control one of two independent sections, or whether they are gated together to provide more sophisticated but also more general control over all the drivers as a unit. In the 6-bit devices with independent enables, one $\overline{OE}$ input controls a 4-line section, and the other $\overline{OE}$ controls a 2-line section. In the 8-bit devices with independent enables, each $\overline{OE}$ controls a 4-line section. In buffers/drivers with gated enables, on the other hand, both $\overline{OE}$ inputs must be LOW to enable all six or eight outputs, which are otherwise all disabled. Each mode of organization has its advantages for the logical demands of different applications.

While the corresponding types in each category are electronically very similar, they are distinctly different from a logical standpoint; the forms with dual independent enables are really two independent devices, like dual flip-flops or one-shots. To mark this difference, the dual devices are designed "4+2-line" or "4+4-line" and gathered adjacent to, but separately from, the NOR-enabled "6-line" and "8-line" devices. It should be noted that the 4+2-line and 4+4-line types can always be very easily converted to 6-line and 8-line devices by tying the two enable inputs together, while it is impossible to perform the reverse operation and restore independent subsection control in types with internally gated enables.

Within these larger categories, the buffers/drivers are functionally distinguished by the polarity of their outputs (inverting or noninverting) and, in the case of devices with independent enables, by the polarity of the two enable inputs (both active-LOW or one active-LOW and one active-HIGH). These independent distinctions combine to produce a broad array of types with slightly different logical functions, as will be seen in further detail below.

QUAD BUFFERS/DRIVERS

The four quad devices listed below are technically not "parallel circuits" in the sense used in this book, that is, devices that are enabled or activated in parallel groups by common enable or clock inputs. Rather, each buffer/driver in the quad package has its own output enable or $\overline{OE}$ input and can, in theory, be used as an independent one-line buffer like the hex buffers/drivers listed in Chapter 1. In practice, however, the 3-state outputs put these devices solidly in the "bus buffer" category, and for convenience they are included in this section with all the other 3-state buffers/drivers.

125   *Quad 3-state buffers/drivers with individual active-LOW enables.*
        Individually enabled 3-state bus buffers. Standard type, available

in TTL, LS, HC, HCT, and F versions from many different manufacturers.

425§   *Quad 3-state buffers/drivers with individual active-LOW enables.*
         TTL circuits that are identical to the TTL version of the 125 but
         omit an internal diode between each buffer output and V_{CC}.
         (Texas Instruments)

126   *Quad 3-state buffers/drivers with individual active-HIGH en-
         ables.* Same as the 125 but with individual active-HIGH (rather
         than active-LOW) 3-state output enables.

426§   *Quad 3-state buffers/drivers with individual active-HIGH en-
         ables.* TTL circuits that are identical to the TTL version of the
         126 but omit an internal diode between each buffer output and
         V_{CC}. (Texas Instruments)

FUNCTION TABLE 125, 425§		
INPUTS		OUTPUTS
$\overline{OE}$	A	Y
L	L	L
L	H	H
H	.	Z

. = Either LOW or HIGH level
Z = High impedance (disabled)

FUNCTION TABLE 126, 426§		
INPUTS		OUTPUTS
OE	A	Y
H	L	L
H	H	H
L	.	Z

. = Either LOW or HIGH level
Z = High impedance (disabled)

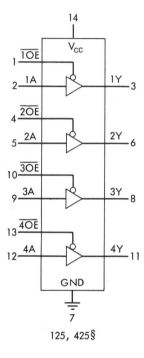

125, 425§

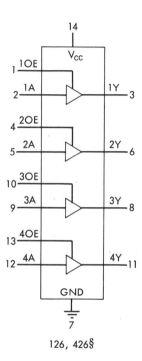

126, 426§

MANUFACTURERS	TTL	LS/A	HC	HCT	ALS	S	AS	F	AC	ACT
Fairchild	■	■								
Motorola			D							
National Semiconductor				■						
RCA				■						
SGS			▦							
Signetics	■	■	■	■				■		
Texas Instruments	■	■	■							
Toshiba			■							

Concurrent change control symbols: SGS makes both a 74LS125 and a 74LS125A

KEY PARAMETERS		TTL	LS/A	HC	HCT	ALS	S	AS	F	AC	ACT
I_{OH} (max.)	mA	-5.2	-2.6	-6	-6				-15		
I_{OL} (max.)	mA	16	24	6	6				64		
I_{IH} (max.)	mA	0.04	0.02						0.02		
I_{IL} (max.)	mA	-1.6	-0.4						-0.02		
I_{CC} (max.)	mA	54	20						40		
I_{CC} (quiesc.)	mA			0.08^g	0.08						
C_{pd} (typ.)	pF/gate			34^h	34^l						
t_{PLH} (max.)	ns	13^a	15	25^i	33^m				6.5		
t_{PHL} (max.)	ns	18	18	25^i	33^m				8		
t_{PZH} (max.)	ns	18^b	20^e	31^j	35^n				8.5		
t_{PZL} (max.)	ns	25	25	31^j	35^n				9		
t_{PHZ} (max.)	ns	16^c	20^f	31^k	35^o				6		
t_{PLZ} (max.)	ns	18^d	20^f	31^k	35^o				6		

[a] National 15 ns

[b] Fairchild and Signetics 17 ns

[c] National and Signetics 8 ns, Fairchild 12 ns

[d] Fairchild 8 ns, Signetics 12 ns, National 14 ns

[e] Fairchild 16 ns, National 25 ns

[f] Fairchild 25 ns

[g] SGS and Toshiba 0.04 mA

[h] SGS and Toshiba; Signetics 22 pF/gate, RCA 29 pF/gate,

TI and National 45 pF/gate

[i] SGS 26 ns, TI 30 ns

[j] SGS 22 ns, Toshiba 23 ns, TI 30 ns

[k] SGS 28 ns, TI and Toshiba 30 ns

[l] RCA; Signetics 24 ns

[m] Signetics; RCA 31 ns

[n] Signetics; RCA 31 ns

[o] RCA; Signetics 31 ns

MANUFACTURERS	TTL	LS/A	HC	HCT	ALS	S	AS	F	AC	ACT
Fairchild		■								
Motorola			D							
National Semiconductor	■			■						
RCA			■	■						
SGS		■	▨							
Signetics	■	■	■	■				■		
Texas Instruments	■	■		■						
Toshiba			■							

Missing change control symbol "A": Fairchild 74LS126. SGS makes both a 74LS126 and a 74LS126A

KEY PARAMETERS		TTL	LS/A	HC	HCT	ALS	S	AS	F	AC	ACT
I_{OH} (max.)	mA	-5.2	-2.6	-6	-6				-15		
I_{OL} (max.)	mA	16	24	6	6				64		
I_{IH} (max.)	mA	0.04	0.02						0.02		
I_{IL} (max.)	mA	-1.6	-0.4						0.02		
I_{CC} (max.)	mA	62	22^d						48		
I_{CC} (quiesc.)	mA			0.08^h	0.08						
C_{pd} (typ.)	pF/gate			34^i	36^m						
t_{PLH} (max.)	ns	13^a	15	25^j	30^n				7		
t_{PHL} (max.)	ns	18	18	25^j	30^n				8.5		
t_{PZH} (max.)	ns	18^b	25^e	31^k	31				8.5		
t_{PZL} (max.)	ns	25	35^f	31^k	31				8.5		
t_{PHZ} (max.)	ns	16	25^g	31^l	31				7.5		
t_{PLZ} (max.)	ns	18^c	25^g	31^l	35^o				8		

[a] National 15 ns
[b] National 19 ns
[c] National 20 ns
[d] SGS 24 mA
[e] Fairchild 20 ns, National 30 ns
[f] Fairchild and National 30 ns
[g] Fairchild 30 ns
[h] SGS 0.04 mA

[i] SGS and Toshiba; Signetics 23 pF/gate, RCA 30 pF/gate, TI and National 45 pF/gate
[j] SGS 26 ns, TI 30 ns
[k] SGS 22 ns, Toshiba 23 ns, TI 30 ns
[l] SGS 28 ns, TI and Toshiba 30 ns
[m] RCA; Signetics 24 pF/gate
[n] RCA; Signetics 25 ns
[o] RCA; Signetics 31 ns

4+2-LINE AND 6-LINE BUFFERS/DRIVERS

These components provide six buffer/driver circuits in each package. In the 4+2-line configuration, the two output enables independently control a 4-line section and a 2-line section, which can, if necessary, be used as two independent devices. In the 6-line configuration, the two active-LOW enables are ANDed together to provide the internal OE level that controls all six drivers, so that all six outputs present a high impedance to external devices unless both enables are LOW (this is identical to the NOR function, since the outputs are enabled only if neither of the enable inputs is HIGH).

4+2-line configuration:

368   *4+2-line 3-state inverting buffers/drivers with active-LOW enables.* Basic 6-bit bus buffers, widely supported and available in a number of technologies.

367   *4+2-line 3-state buffers/drivers with active-LOW enables.* Same as 368, but noninverting.

6-line configuration:

366   *6-line 3-state inverting buffers/drivers with NOR-gated enables.* 6-bit bus buffers with dual NOR-gated enables (that is, active-LOW AND-gated enables), widely supported and available in a number of technologies.

437§   *6-line inverting buffers/drivers with 200 mA totem-pole outputs.* S circuits specifically designed to drive MOS memories and other high-impedance loads. Pin-out is identical to that of the 366, but instead of causing the outputs to present a high-impedance state, either enable input, when taken HIGH, will set the outputs to HIGH for MOS RAM refresh applications. Key parameters are not given in the form used in this book; see manufacturer's data for details. (Texas Instruments)

436§   *6-line inverting buffers/drivers with 200 mA totem-pole outputs.* Same as the 437§, but includes a small series damping resistor (about 10 Ω) at each output to minimize output transients (overshoot). (Texas Instruments)

365   *6-line 3-state buffers/drivers with NOR-gated enables.* Same as the 366, but noninverting.

FUNCTION TABLE 368		
INPUTS		OUTPUTS
$\overline{OE}$	A	$\overline{Y}$
L	L	H
L	H	L
H	.	Z

. = Either LOW or HIGH level
Z = High impedance (disabled)

FUNCTION TABLE 367		
INPUTS		OUTPUTS
$\overline{OE}$	A	Y
L	L	L
L	H	H
H	.	Z

. = Either LOW or HIGH level
Z = High impedance (disabled)

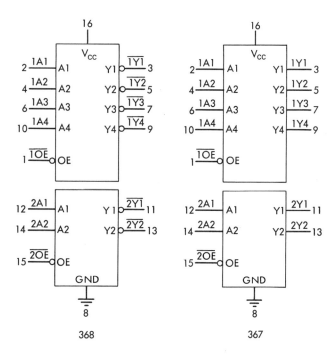

368

367

FUNCTION TABLE
366, 436§, 437§

INPUTS			OUTPUTS
$\overline{OE1}$	$\overline{OE2}$	A	$\overline{Y}$
L	L	L	H
L	L	H	L
H	.	.	Z*
.	H	.	Z*

. = Either LOW or HIGH level
Z = High impedance (disabled)
* = 436§, 437§ have H output
instead of high impedance

FUNCTION TABLE
365

INPUTS			OUTPUTS
$\overline{OE1}$	$\overline{OE2}$	A	Y
L	L	L	L
L	L	H	H
H	.	.	Z
.	H	.	Z

. = Either LOW or HIGH level
Z = High impedance (disabled)

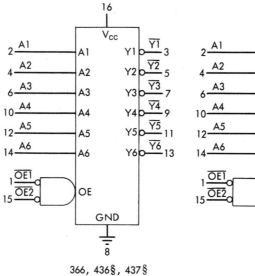

366, 436§, 437§

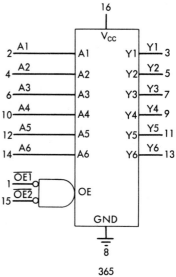

365

210

MANUFACTURERS	TTL/A	LS/A	HC	HCT	ALS	S	AS	F	AC	ACT
Fairchild		■						D		
Motorola		■	D							
National Semiconductor	■	■		■						
RCA		▒	▒	▒						
SGS		■	▒							
Signetics	■	■		■				■		
Texas Instruments	■	■	■		▒					
Toshiba		■	■						D	

Missing change control symbol "A": National 74368

KEY PARAMETERS		TTL/A	LS/A	HC	HCT	ALS	S	AS	F	AC	ACT
I_{OH} (max.)	mA	-5.2	-2.6[a]	-6	-6	-15			-15		
I_{OL} (max.)	mA	32	24	6	6	24[o]			64		
I_{IH} (max.)	mA	0.04	0.02			0.02			0.02		
I_{IL} (max.)	mA	-1.6	-0.4			-0.1			-0.02		
I_{CC} (max.)	mA	77	21			11			62		
I_{CC} (quiesc.)	mA			0.08[h]	0.08						
C_{pd} (typ.)	pF/gate			35[i]	42[m]						
t_{PLH} (max.)	ns	17	15[b]	24[j]	38[n]	n/a			7.5		
t_{PHL} (max.)	ns	16	18[c]	24[j]	38[n]	n/a			5.5		
t_{PZH} (max.)	ns	35	35[d]	38[k]	44	n/a			8.5		
t_{PZL} (max.)	ns	37	45[e]	38[k]	44	n/a			9		
t_{PHZ} (max.)	ns	11	32[f]	38[l]	44	n/a			7		
t_{PLZ} (max.)	ns	27	35[g]	38[l]	44	n/a			6.5		

[a] Fairchild -1.0 mA
[b] Fairchild 12 ns
[c] Fairchild 22 ns
[d] Fairchild 24 ns, National 30 ns
[e] Fairchild and National 30 ns
[f] National 20 ns, Fairchild 25 ns
[g] Fairchild and National 20 ns
[h] SGS and Toshiba 0.04 mA
[i] TI; Toshiba 29 pF/gate, Signetics 30 pF/gate, SGS 32 pF/gate, RCA 40 pF/gate, National 45 pF/gate
[j] SGS 22 ns, Toshiba 29 ns, RCA 31 ns
[k] SGS 24 ns, Toshiba 30 ns, National 47 ns, TI 48 ns
[l] SGS 34 ns, National 44 ns, TI 48 ns
[m] RCA; Signetics 30 pF/gate
[n] RCA; Signetics 30 ns
[o] Versions with -1 suffix have I_{OL} (max.) = 48 mA if V_{CC} is kept between 4.75 V and 5.25 V

MANUFACTURERS	TTL/A	LS/A	HC	HCT	ALS	S	AS	F	AC	ACT
Fairchild		■						D		
Motorola		■	D							
National Semiconductor	■	■		■						
RCA			▨	▨						
SGS		■	▨							
Signetics	■	■				■		■		
Texas Instruments	■	■		▨						
Toshiba		■	■						D	

Missing change control symbol "A": National 74367

KEY PARAMETERS		TTL/A	LS/A	HC	HCT	ALS	S	AS	F	AC	ACT
I_{OH} (max.)	mA	-5.2	-2.6^a	-6	-6	-15			-15		
I_{OL} (max.)	mA	32	24	6	6	24^n			64		
I_{IH} (max.)	mA	0.04	0.02			0.02			0.02		
I_{IL} (max.)	mA	-1.6	-0.4			-0.1			-0.02		
I_{CC} (max.)	mA	85	24			11			62		
I_{CC} (quiesc.)	mA			0.08^g	0.08						
C_{pd} (typ.)	pF/gate			35^h	42^l						
t_{PLH} (max.)	ns	16	16	24^i	38^m	n/a			7		
t_{PHL} (max.)	ns	22	22^b	24^i	38^m	n/a			7.5		
t_{PZH} (max.)	ns	35	35^c	38^j	44	n/a			8.5		
t_{PZL} (max.)	ns	37	40^d	38^j	44	n/a			9		
t_{PHZ} (max.)	ns	11	30^e	38^k	44	n/a			7		
t_{PLZ} (max.)	ns	27	35^f	38^k	44	n/a			6.5		

[a] Fairchild -1.0 mA
[b] National 16 ns
[c] Fairchild 24 ns, National 30 ns
[d] Fairchild and National 30 ns
[e] National 20 ns, Fairchild 25 ns
[f] Fairchild and National 20 ns
[g] SGS and Toshiba 0.04 mA
[h] TI; Signetics 30 pF/gate, Toshiba 32 pF/gate, SGS 34

pF/gate, RCA 40 pF/gate, National 45 pF/gate
[i] SGS 22 ns, Toshiba and National 30 ns, RCA 31 ns
[j] SGS 24 ns, Toshiba 30 ns, National 47 ns, TI 48 ns
[k] SGS 34 ns, National 44 ns, TI 48 ns
[l] RCA; Signetics 32 pF/gate
[m] RCA; Signetics 31 ns
[n] Versions with -1 suffix have I_{OL} (max.) = 48 mA if V_{CC} is kept between 4.75 V and 5.25 V

MANUFACTURERS	TTL/A	LS/A	HC	HCT	ALS	S	AS	F	AC	ACT
Fairchild		■						D		
Motorola		■	D							
National Semiconductor		■	■							
RCA		■	■		■					
SGS		■	▨							
Signetics	■	■						■		
Texas Instruments	■		■		▨					
Toshiba	■	■	■							

Missing change control symbol "A": National 74366

KEY PARAMETERS		TTL/A	LS/A	HC	HCT	ALS	S	AS	F	AC	ACT
I_{OH} (max.)	mA	-5.2	-2.6^a	-6	-6	-15			-15		
I_{OL} (max.)	mA	32	24	6	6	24^o			64		
I_{IH} (max.)	mA	0.04	0.02			0.02			0.02		
I_{IL} (max.)	mA	-1.6	-0.4			-0.1			-0.02		
I_{CC} (max.)	mA	77	21			11			62		
I_{CC} (quiesc.)	mA			0.08^h	0.08						
C_{pd} (typ.)	pF/gate			35^i	42^m						
t_{PLH} (max.)	ns	17	15^b	24^j	38^n	n/a			7.5		
t_{PHL} (max.)	ns	16	18^c	24^j	38^n	n/a			5.5		
t_{PZH} (max.)	ns	35	35^d	38^k	44	n/a			10		
t_{PZL} (max.)	ns	37	45^e	38^k	44	n/a			9.5		
t_{PHZ} (max.)	ns	11	32^f	38^l	44	n/a			7		
t_{PLZ} (max.)	ns	27	35^g	38^l	44	n/a			6.5		

[a] Fairchild -1.0 mA
[b] Fairchild 12 ns
[c] National 16 ns, Fairchild 22 ns
[d] Fairchild 24 ns, National 30 ns
[e] Fairchild and National 30 ns
[f] National 20 ns, Fairchild 25 ns
[g] Fairchild and National 20 ns
[h] SGS and Toshiba 0.04 mA
[i] TI; Signetics 30 pF/gate, Toshiba 31 pF/gate, SGS 32 pF/gate, RCA 40 pF/gate, National 45 pF/gate
[j] SGS 22 ns, Signetics 25 ns, Toshiba 29 ns, RCA 31 ns
[k] SGS 28 ns, TI 48 ns, National 55 ns
[l] Toshiba 44 ns, TI 48 ns, National 55 ns
[m] RCA; Signetics 30 pF/gate
[n] RCA; Signetics 30 ns
[o] Versions with -1 suffix have I_{OL} (max.) = 48 mA if V_{CC} is kept between 4.75 V and 5.25 V

365   6-line 3-state buffers/drivers with NOR-gated enables

MANUFACTURERS	TTL/A	LS/A	HC	HCT	ALS	S	AS	F	AC	ACT
Fairchild								D		
Motorola			D							
National Semiconductor										
RCA										
SGS										
Signetics										
Texas Instruments										
Toshiba										

Missing change control symbol "A": National 74365

KEY PARAMETERS		TTL/A	LS/A	HC	HCT	ALS	S	AS	F	AC	ACT
I_{OH} (max.)	mA	-5.2	-2.6[a]	-6	-6	-15			-15		
I_{OL} (max.)	mA	32	24	6	6	24[n]			64		
I_{IH} (max.)	mA	0.04	0.02			0.02			0.02		
I_{IL} (max.)	mA	-1.6	-0.4			-0.1			-0.02		
I_{CC} (max.)	mA	85	24			11			62		
I_{CC} (quiesc.)	mA			0.08[g]	0.08						
C_{pd} (typ.)	pF/gate			35[h]	42[l]						
t_{PLH} (max.)	ns	16	16	24[i]	38[m]	n/a			7		
t_{PHL} (max.)	ns	22	22[b]	24[i]	38[m]	n/a			7.5		
t_{PZH} (max.)	ns	35	35[c]	38[j]	44	n/a			10		
t_{PZL} (max.)	ns	37	40[d]	38[j]	44	n/a			9.5		
t_{PHZ} (max.)	ns	11	30[e]	38[k]	44	n/a			7		
t_{PLZ} (max.)	ns	27	35[f]	38[k]	44	n/a			6.5		

[a] Fairchild -1.0 mA

[b] National 16 ns

[c] Fairchild 24 ns, National 30 ns

[d] Fairchild and National 30 ns

[e] National 20 ns, Fairchild 25 ns

[f] Fairchild and National 20 ns

[g] SGS and Toshiba 0.04 mA

[h] TI; Toshiba 33 pF/gate, SGS 34 pF/gate, RCA and Signetics 40 pF/gate, National 45 pF/gate

[i] SGS 22 ns, Toshiba and National 30 ns, RCA 31 ns

[j] SGS 28 ns, TI 48 ns, National 55 ns

[k] Toshiba 44 ns, TI 48 ns, National 55 ns

[l] RCA; Signetics 40 pF/gate

[m] RCA; Signetics 31 ns

[n] Versions with -1 suffix have I_{OL} (max.) = 48 mA if V_{CC} is kept between 4.75 V and 5.25 V

4+4-LINE BUFFERS/DRIVERS

Like their 6-bit counterparts, the 8-bit buffers/drivers are configured in two basic ways; in this case either as dual, independently enabled 4-line devices or as singly enabled 8-line devices. Those consisting of two separately enabled 4-line sections are listed below, and those with gated enables controlling eight lines in common are listed separately further on.

The 4+4-line buffers/drivers are, as usual, subdivided according to their output polarity, the subcategories in this case including not just the ordinary inverting and noninverting varieties but also some types having four inverting outputs and four noninverting outputs. In addition, however, the 4+4-line buffers/drivers are also distinguished from one another by the polarity of their output enables, some types featuring enables that are both active-LOW, and some with complementary enables, one active-LOW and the other active-HIGH. (Complementary enables allow the circuits to combine multiplexer and driver functions.) The combination of these two independent sets of features provides the designer with a variety of logical configurations for meeting the needs of different applications.

As noted below, some of these circuits have Schmitt-trigger action (TTL hysteresis) at the data and enable inputs, which is particularly helpful when they are used as bus receivers in a noisy environment. All 4+4-line buffers/drivers are supplied in 20-pin DIPs.

4+4-line inverting buffers/drivers with active-LOW enables:

240 *4+4-line 3-state inverting buffers/drivers with active-LOW enables*. The basic 4+4-line inverting bus driver, universally supported and available in nearly all technologies. LS and S versions feature a moderate amount of Schmitt-trigger input hysteresis (min. 0.2 V).

1240 *4+4-line 3-state inverting buffers/drivers with active-LOW enables*. Low-power versions of ALS240.

340§ *4+4-line 3-state inverting buffers/drivers with active-LOW enables*. LS and S versions of the 240 that feature increased Schmitt-trigger input hysteresis (min. 0.4 V in LS and 0.5 V in S versions) for improved noise immunity. (Monolithic Memories)

	LS340	S340		LS340	S340
I_{OH} (max.)	-15 mA	-15 mA	t_{PLH} (max.)	25 ns	15 ns
I_{OL} (max.)	24 mA	64 mA	t_{PHL} (max.)	25 ns	22 ns
I_{IH} (max.)	0.02 mA	0.05 mA	t_{PZH} (max.)	35 ns	12 ns
I_{IL} (max.)	-0.2 mA	-0.25 mA	t_{PZL} (max.)	40 ns	15 ns
I_{CC} (max.)	50 mA	180 mA	t_{PHZ} (max.)	25 ns	12 ns
			t_{PLZ} (max.)	30 ns	15 ns

2240§ *4+4-line 3-state inverting MOS drivers with active-LOW enables*. ALS versions of the 240 specifically designed for driving MOS devices. The outputs are provided with 25 Ω series resistors so that no external resistors are required. (Texas Instruments)

I_{OH} (max.)	-15	mA	t_{PLH} (max.)	10 ns
I_{OL} (max.)	15	mA	t_{PHL} (max.)	10 ns
I_{IH} (max.)	0.02	mA	t_{PZH} (max.)	17 ns
I_{IL} (max.)	-0.1	mA	t_{PZL} (max.)	20 ns
I_{CC} (max.)	23	mA	t_{PHZ} (max.)	10 ns
			t_{PLZ} (max.)	15 ns

730§   *4+4-line 3-state inverting MOS drivers with active-LOW enables.* S version of the 240 provided with additional internal output resistors for controlling undershoot when driving highly capacitive loads, specifically the address and control lines of MOS dynamic RAMs. Also available in a −1 version with larger output resistors to control undershoot in more lightly capacitance-loaded circuits. (Monolithic Memories)

I_{OH} (max.)	-35	mA	t_{PLH} (max.)	17 ns
I_{OL} (max.)	40*	mA	t_{PHL} (max.)	17 ns
I_{IH} (max.)	0.02	mA	t_{PZH} (max.)	28 ns
I_{IL} (max.)	-0.2	mA	t_{PZL} (max.)	28 ns
I_{CC} (max.)	125	mA	t_{PHZ} (max.)	16 ns
			t_{PLZ} (max.)	24 ns

* −1 version has I_{OL} (max.) of 50 mA

468   *4+4-line 3-state inverting buffers/drivers with active-LOW enables.* These 8-line LS and ALS versions of the 6-line 368 are logically identical to the 240 but have a different pinout.

798§   *4+4-line 3-state inverting buffers/drivers with active-LOW enables.* A variant LS version of the 468. (Motorola)

I_{OH} (max.)	-5	mA	t_{PLH} (max.)	10 ns
I_{OL} (max.)	16	mA	t_{PHL} (max.)	17 ns
I_{IH} (max.)	0.02	mA	t_{PZH} (max.)	27 ns
I_{IL} (max.)	-0.4	mA	t_{PZL} (max.)	25 ns
I_{CC} (max.)	21	mA	t_{PHZ} (max.)	20 ns
			t_{PLZ} (max.)	27 ns

756§   *4+4-line inverting buffers/drivers with open-collector outputs and active-LOW enables.* Open-collector ALS and AS versions of the 240. (Texas Instruments)

	ALS756	AS756		ALS756	AS756
I_{OH} (max.), mA	0.1	0.1	t_{PLH} from A (max.), ns	24	19
I_{OL} (max.), mA	24*	64	t_{PHL} from A (max.), ns	10	6
I_{IH} (max.), mA	0.02	0.02	t_{PLH} from $\overline{OE}$ (max.), ns	24	19.5
I_{IL} (max.), mA	−0.1	−0.5	t_{PHL} from $\overline{OE}$ (max.), ns	20	7.5
I_{CC} (max.), mA	22	80			

* −1 version has I_{OL} (max.) = 48 mA if V_{CC} is kept between 4.75 V and 5.25 V

4+4-line noninverting buffers/drivers with active-LOW enables:

244   *4+4-line 3-state buffers/drivers with active-LOW enables.* The basic 4+4 noninverting bus driver, universally supported and available in nearly all technologies. LS and S versions feature a moderate amount of Schmitt-trigger input hysteresis (min. 0.2 V).

1244   *4+4-line 3-state buffers/drivers with active-LOW enables.* Low power versions of the ALS244.

344§   *4+4-line 3-state buffers/drivers with active-LOW enables.* LS and S versions of the 244 that feature increased Schmitt-trigger input hysteresis (min. 0.4 V in LS and 0.5 V in S versions) for improved noise immunity. (Monolithic Memories)

	LS344		S344			LS344	S344
I_{OH} (max.)	−15	mA	−15	mA	t_{PLH} (max.)	25 ns	22 ns
I_{OL} (max.)	24	mA	64	mA	t_{PHL} (max.)	25 ns	15 ns
I_{IH} (max.)	0.02	mA	0.05	mA	t_{PZH} (max.)	35 ns	12 ns
I_{IL} (max.)	−0.2	mA	−0.25	mA	t_{PZL} (max.)	40 ns	15 ns
I_{CC} (max.)	54	mA	200	mA	t_{PHZ} (max.)	25 ns	12 ns
					t_{PLZ} (max.)	30 ns	15 ns

734§   *4+4-line 3-state MOS drivers with active-LOW enables.* S version of the 244 provided with additional internal output resistors for controlling undershoot when driving highly capacitive loads, specifically the address and control lines of MOS dynamic RAMs. Also available in a −1 version with larger output resistors to control undershoot in more lightly capacitance-loaded circuits. (Monolithic Memories)

I_{OH} (max.)	−35	mA	t_{PLH} (max.)	17 ns	
I_{OL} (max.)	40*	mA	t_{PHL} (max.)	17 ns	
I_{IH} (max.)	0.02	mA	t_{PZH} (max.)	28 ns	
I_{IL} (max.)	−0.2	mA	t_{PZL} (max.)	28 ns	
I_{CC} (max.)	150	mA	t_{PHZ} (max.)	16 ns	
			t_{PLZ} (max.)	24 ns	

* −1 version has I_{OL} (max.) of 50 mA

467   *4+4-line 3-state buffers/drivers with active-LOW enables.* These 8-line LS and ALS versions of the 6-line 367 are logically identical to the 244 but have a different pinout.

797§   *4+4-line 3-state buffers/drivers with active-LOW enables.* A variant LS version of the 467. (Motorola)

I_{OH} (max.)	−5	mA	t_{PLH} (max.)	16 ns	
I_{OL} (max.)	16	mA	t_{PHL} (max.)	22 ns	
I_{IH} (max.)	0.02	mA	t_{PZH} (max.)	25 ns	
I_{IL} (max.)	−0.4	mA	t_{PZL} (max.)	20 ns	
I_{CC} (max.)	26	mA	t_{PHZ} (max.)	20 ns	
			t_{PLZ} (max.)	27 ns	

760§   *4+4-line buffers/drivers with open-collector outputs and active-LOW enables.* Open-collector AS version of the 244; ALS version under development. (Texas Instruments)

I_{OH} (max.)	0.1	mA	t_{PLH} from A (max.)	18.5 ns
I_{OL} (max.)	64	mA	t_{PHL} from A (max.)	6 ns
I_{IH} (max.)	0.02	mA	t_{PLH} from $\overline{OE}$ (max.)	18.5 ns
I_{IL} (max.)	−1	mA	t_{PHL} from $\overline{OE}$ (max.)	7 ns
I_{CC} (max.)	94	mA		

4+4-line inverting/noninverting buffers/drivers with active-LOW enables:

230   *4+4-line 3-state inverting/noninverting buffers/drivers with active-LOW enables.* These AS components are similar to the 240 and 244, but one of the two independent 4-line buffers/drivers in each package has inverted outputs and the other has noninverted outputs.

762§   *4+4-line inverting/noninverting buffers/drivers with open-collector outputs and active-LOW enables.* Open-collector AS version of the 230; ALS version under development. (Texas Instruments)

I_{OH} (max.)	0.1	mA	t_{PLH} from A (max.)	19	ns
I_{OL} (max.)	64	mA	t_{PHL} from A (max.)	6	ns
I_{IH} (max.)	0.02	mA	t_{PLH} from $\overline{OE}$ (max.)	19.5	ns
I_{IL} (max.)	−1	mA	t_{PHL} from $\overline{OE}$ (max.)	7.5	ns
I_{CC} (max.)	87	mA			

4+4-line inverting buffers/drivers with complementary enables:

231   *4+4-line 3-state inverting buffers/drivers with complementary enables.* ALS and AS components similar to the 240 but with one active-LOW and one active-HIGH output enable rather than two active-LOW enables.

210§   *4+4-line 3-state inverting buffers/drivers with complementary enables.* LS and S versions of the 231. Inputs have the same moderate Schmitt-trigger hysteresis (min. 0.2 V) as the common LS and S versions of 240, 244, and 241. (Monolithic Memories)

	LS210		S210			LS210	S210
I_{OH} (max.)	−15	mA	−15	mA	t_{PLH} (max.)	14 ns	7 ns
I_{OL} (max.)	24	mA	64	mA	t_{PHL} (max.)	18 ns	7 ns
I_{IH} (max.)	0.02	mA	0.05	mA	t_{PZH} (max.)	23 ns	12 ns
I_{IL} (max.)	−0.2	mA	−0.4	mA	t_{PZL} (max.)	30 ns	15 ns
I_{CC} (max.)	50	mA	150	mA	t_{PHZ} (max.)	18 ns	9 ns
					t_{PLZ} (max.)	25 ns	15 ns

310§   *4+4-line 3-state inverting buffers/drivers with complementary enables.* LS and S versions of the 210§ that feature increased Schmitt-trigger input hysteresis (min. 0.4 V in LS and 0.5 V in S versions) for improved noise immunity. (Monolithic Memories)

	LS310		S310			LS310	S310
I_{OH} (max.)	−15	mA	−15	mA	t_{PLH} (max.)	25 ns	15 ns
I_{OH} (max.)	24	mA	64	mA	t_{PHL} (max.)	25 ns	22 ns
I_{IH} (max.)	0.02	mA	0.05	mA	t_{PZH} (max.)	35 ns	12 ns
I_{IL} (max.)	−0.2	mA	−0.25	mA	t_{PZL} (max.)	40 ns	15 ns
I_{CC} (max.)	50	mA	180	mA	t_{PHZ} (max.)	25 ns	12 ns
					t_{PLZ} (max.)	30 ns	15 ns

700§   *4+4-line 3-state inverting MOS drivers with active-LOW enables.* S version of the 210§ (231) provided with additional internal output resistors for controlling undershoot when driving highly capacitive loads, specifically the address and control lines of MOS dynamic RAMs. Also available in a −1 version with larger output resistors to control undershoot in more lightly capacitance-loaded circuits. (Monolithic Memories)

I_{OH} (max.)	−35	mA	t_{PLH} (max.)	17 ns
I_{OL} (max.)	40*	mA	t_{PHL} (max.)	17 ns
I_{IH} (max.)	0.02	mA	t_{PZH} (max.)	28 ns
I_{IL} (max.)	−0.2	mA	t_{PZL} (max.)	28 ns
I_{CC} (max.)	125	mA	t_{PHZ} (max.)	16 ns
			t_{PLZ} (max.)	24 ns

* −1 version has I_{OL} (max.) of 50 mA

763§   *4+4-line inverting buffers/drivers with open-collector outputs and complementary enables.* Open-collector ALS and AS versions of the 231. (Texas Instruments)

	ALS763	AS763		ALS763	AS763
I_{OH} (max.), mA	0.1	0.1	t_{PLH} from A (max.), ns	25	19
I_{OL} (max.), mA	24*	64	t_{PHL} from A (max.), ns	9	6
I_{IH} (max.), mA	0.02	0.02	t_{PLH} from $\overline{OE}$ (max.), ns	25	20
I_{IL} (max.), mA	−0.1	−0.5	t_{PHL} from $\overline{OE}$ (max.), ns	21	8
I_{CC} (max.). mA	22	82			

* −1 version has I_{OL} (max.) = 48 mA if V_{CC} is kept between 4.75 and 5.25 V

4+4-line noninverting buffers/drivers with complementary enables:

241   *4+4-line 3-state buffers/drivers with complementary enables.* Standard components similar to the 244 but with one active-HIGH and one active-LOW output enable rather than two active-LOW enables. Universally supported and available in nearly all technologies. LS and S versions feature a moderate amount of Schmitt-trigger input hysteresis (min. 0.2 V).

1241 *4+4-line 3-state buffers/drivers with complementary enables.* Low-power versions of the ALS241 and F241.

341§ *4+4-line 3-state buffers/drivers with active-LOW enables.* LS and S versions of the 241 that feature increased Schmitt-trigger input hysteresis (min. 0.4 V in LS and 0.5 V in S versions) for improved noise immunity. (Monolithic Memories)

	LS341		S341			LS341	S341
I_{OH} (max.)	−15	mA	−15	mA	t_{PLH} (max.)	25 ns	22 ns
I_{OL} (max.)	24	mA	64	mA	t_{PHL} (max.)	25 ns	15 ns
I_{IH} (max.)	0.02	mA	0.05	mA	t_{PZH} (max.)	35 ns	12 ns
I_{IL} (max.)	−0.2	mA	−0.25	mA	t_{PZL} (max.)	40 ns	15 ns
I_{CC} (max.)	54	mA	200	mA	t_{PHZ} (max.)	25 ns	12 ns
					t_{PLZ} (max.)	30 ns	15 ns

731§   *4+4-line 3-state MOS drivers with complementary enables.* S version of the 241 provided with additional internal output resistors for controlling undershoot when driving highly capacitive loads, specifically the address and control lines of MOS dynamic RAMs. Also available in a −1 version with larger output resistors to control undershoot in more lightly capacitance-loaded circuits. (Monolithic Memories)

I_{OH} (max.)	−35	mA	t_{PLH} (max.)	17 ns
I_{OL} (max.)	40*	mA	t_{PHL} (max.)	17 ns
I_{IH} (max.)	0.02	mA	t_{PZH} (max.)	28 ns
I_{IL} (max.)	−0.2	mA	t_{PZL} (max.)	28 ns
I_{CC} (max.)	150	mA	t_{PHZ} (max.)	16 ns
			t_{PLZ} (max.)	24 ns

* −1 version has I_{OL} (max.) of 50 mA

757§   *4+4-line buffers/drivers with open-collector outputs and complementary enables.* Open-collector AS version of the 241. (Texas Instruments)

I_{OH} (max.)	0.1	mA	t_{PLH} from A (max.)	18.5 ns
I_{OL} (max.)	64	mA	t_{PHL} from A (max.)	6 ns
I_{IH} (max.)	0.02	mA	t_{PLH} from $\overline{OE}$ (max.)	21 ns
I_{IL} (max.)	−1	mA	t_{PHL} from $\overline{OE}$ (max.)	7.5 ns
I_{CC} (max.)	95	mA		

FUNCTION TABLE
240, 1240, 340§, 2240§, 730§, 468§, 798§, 756§

INPUTS		OUTPUTS
$\overline{OE}$	A	$\overline{Y}$
L	L	H
L	H	L
H	.	Z

. = Either LOW or HIGH level
Z = High impedance (disabled)

FUNCTION TABLE
244, 1244, 344§, 734§, 467, 797§, 760§

INPUTS		OUTPUTS
$\overline{OE}$	A	Y
L	L	L
L	H	H
H	.	Z

. = Either LOW or HIGH level
Z = High impedance (disabled)

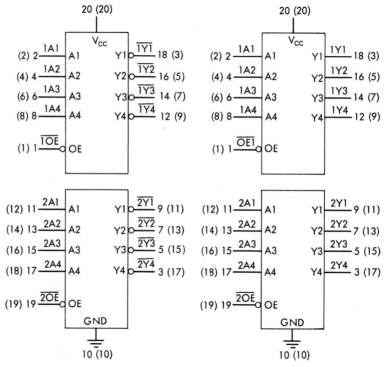

3-state outputs: 240, 1240, 340§, 2240§, 730§, (468), (798§)
OC outputs: 756§

3-state outputs: 244, 1244, 344§, 734§, (467), (797§)
OC outputs: 760§

FUNCTION TABLE
230, 762§
Buffer/driver 1

INPUTS		OUTPUTS
$\overline{OE}$	A	$\overline{Y}$
L	L	H
L	H	L
H	.	Z

Buffer/driver 2

INPUTS		OUTPUTS
$\overline{OE}$	A	Y
L	L	L
L	H	H
H	.	Z

. = Either LOW or HIGH level
Z = High impedance (disabled)

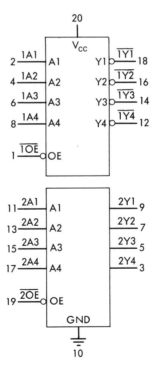

3-state outputs: 230
OC outputs: 762§

FUNCTION TABLE
231, 210§, 310§, 700§, 763§
Buffer/driver 1

INPUTS		OUTPUTS
$\overline{OE}$	A	$\overline{Y}$
L	L	H
L	H	L
H	.	Z

Buffer/driver 2

INPUTS		OUTPUTS
OE	A	$\overline{Y}$
H	L	H
H	H	L
L	.	Z

. = Either LOW or High level
Z = High impedance (disabled)

FUNCTION TABLE
241, 1241, 341§, 731§, 757§
Buffer/driver 1

INPUTS		OUTPUTS
$\overline{OE}$	A	Y
L	L	L
L	H	H
H	.	Z

Buffer/driver 2

INPUTS		OUTPUTS
OE	A	Y
H	L	L
H	H	H
L	.	Z

. = Either LOW or HIGH level
Z = High impedance (disabled)

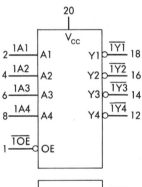

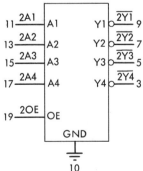

3- state outputs: 231, 210§,
310§, 700§
OC outputs 763§

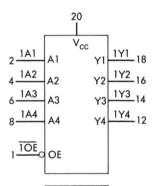

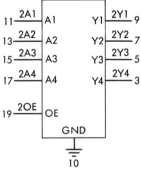

3-state outputs: 241, 1241,
341§, 731§
OC outputs: 757§

240   4+4-line 3-state inverting buffers/drivers with active-LOW enables

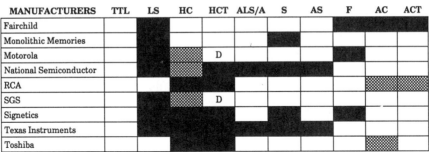

MANUFACTURERS	TTL	LS	HC	HCT	ALS/A	S	AS	F	AC	ACT
Fairchild		■							■	■
Monolithic Memories		■				■				
Motorola		■	▒	D						
National Semiconductor		■	▒							
RCA			■	■					▒	▒
SGS		■	▒	D						
Signetics		■				■		■		
Texas Instruments		■								
Toshiba			■	■					▒	

ALS -1 versions available from National and TI

KEY PARAMETERS		TTL	LS	HC	HCT	ALS/A	S	AS	F	AC	ACT
I_{OH} (max.)	mA		-15	-6	-6	-15	-15	-15	-15	-24	-24
I_{OL} (max.)	mA		24	6	6	24^m	64	64	64	24	24
I_{IH} (max.)	mA		0.02			0.02	0.05	0.02	0.02		
I_{IL} (max.)	mA		-0.2			-0.1	-2	-0.5	-1		
I_{CC} (max.)	mA		50			25	150	75	75		
I_{CC} (quiesc.)	mA			0.08^c	0.08^h					0.08	0.08
C_{pd} (typ.)	pF/gate			40^d	40^i					45^r	45^u
t_{PLH} (max.)	ns		14	25^e	28^j	9	7	6.5	8^n	7^s	9.5^v
t_{PHL} (max.)	ns		18	25^e	28^j	9	7	5.7	5.7^o	6.5^s	8.5^v
t_{PZH} (max.)	ns		23	38^f	38^k	13	10	6.4	5.7^p	8^t	9.5^w
t_{PZL} (max.)	ns		30	38^f	38^k	18	15	9	10	8.5^t	10.5^w
t_{PHZ} (max.)	ns		18^a	38^g	38^l	10	9	5	6.3^q	9.5^t	10.5^w
t_{PLZ} (max.)	ns		25^b	38^g	38^l	12	15	9.5	9.5^q	9.5^t	10.5^w

[a] TI 25 ns
[b] TI 20 ns
[c] SGS and Toshiba 0.04 mA
[d] Signetics 30 pF/gate, TI 35 pF/gate, RCA 38 pF/gate, Motorola and National 50 pF/gate
[e] SGS 27 ns
[f] SGS 24 ns, Toshiba 25 ns
[g] SGS 40 ns
[h] Toshiba 0.04 mA
[i] Signetics 30 pF/gate, Toshiba 42 pF/gate, National 90 pF/gate
[j] National and Signetics 25 ns, TI 32 ns, Toshiba 42 ns
[k] TI and Toshiba 44 ns

[l] RCA; Signetics 31 ns, National 32 ns, TI 44 ns, Toshiba 57 ns
[m] Versions with -1 suffix have I_{OL} (max.) = 48 mA if V_{CC} is kept between 4.75 V and 5.25 V
[n] Signetics 7.5 ns
[o] Signetics 5 ns
[p] Signetics 4 ns
[q] Signetics 7.5 ns
[r] RCA 95 pF/gate
[s] RCA 6.8 ns
[t] RCA 12 ns
[u] Toshiba 19 pF/gate, RCA 115 pF/gate
[v] RCA 7.8 ns; Toshiba data not available
[w] RCA 13 ns; Toshiba data not available

1240   4+4-line 3-state inverting buffers/drivers with active-LOW enables

MANUFACTURERS	TTL	LS	HC	HCT	ALS	S	AS	F	AC	ACT
National Semiconductor					■					
Signetics								■		
Texas Instruments					■					

ALS -1 versions available from National and TI

KEY PARAMETERS		TTL	LS	HC	HCT	ALS	S	AS	F	AC	ACT
I_{OH} (max.)	mA					-15			-15		
I_{OL} (max.)	mA					16^a			64		
I_{IH} (max.)	mA					0.02			0.07^c		
I_{IL} (max.)	mA					-0.1			-0.07^d		
I_{CC} (max.)	mA					14^b			75		
t_{PLH} (max.)	ns					13			7.5		
t_{PHL} (max.)	ns					13			5		
t_{PZH} (max.)	ns					20			8		
t_{PZL} (max.)	ns					22			9.5		
t_{PHZ} (max.)	ns					10			6.5		
t_{PLZ} (max.)	ns					13			6		

[a] Versions with -1 suffix have I_{OL} (max.) = 24 mA if V_{CC} is kept between 4.75 V and 5.25 V
[b] National 13 mA
[c] Including 0.05 mA off-state output current
[d] Including -0.05 mA off-state output current

468   4+4-line 3-state inverting buffers/drivers with active-LOW enables

MANUFACTURERS	TTL	LS	HC	HCT	ALS/A	S	AS	F	AC	ACT
National Semiconductor		■	■		■					
Texas Instruments		■			■					

KEY PARAMETERS		TTL	LS	HC	HCT	ALS/A	S	AS	F	AC	ACT
I_{OH} (max.)	mA		-2.6^a			-15					
I_{OL} (max.)	mA		24			24					
I_{IH} (max.)	mA		0.02			0.02					
I_{IL} (max.)	mA		-0.2^b			-0.1					
I_{CC} (max.)	mA		28^c			27^j					
t_{PLH} (max.)	ns		12^d			12					
t_{PHL} (max.)	ns		15^e			9					
t_{PZH} (max.)	ns		40^f			16					
t_{PZL} (max.)	ns		45^g			23					
t_{PHZ} (max.)	ns		40^h			10					
t_{PLZ} (max.)	ns		45^i			17					

[a] National -5.2 mA
[b] National -0.05 mA
[c] National 21 mA
[d] National 10 ns
[e] National 17 ns
[f] National 15 ns
[g] National 35 ns
[h] National 20 ns
[i] National 27 ns
[j] National 10 mA

MANUFACTURERS	TTL	LS	HC	HCT	ALS/A	S	AS	F	AC	ACT
Fairchild		■						■	■	
Monolithic Memories		■				■				
Motorola			▨	D					■	
National Semiconductor		■						■		
RCA		■							▨	▨
SGS		■	▨	D						
Signetics		■					■			
Texas Instruments		■						■		
Toshiba									▨	▨

ALS -1 versions available from National and TI

KEY PARAMETERS		TTL	LS	HC	HCT	ALS/A	S	AS	F	AC	ACT
I_{OH} (max.)	mA		-15	-6	-6	-15	-15	-15	-15	-24	-24
I_{OL} (max.)	mA		24	6	6	24[m]	64	64	64	24	24
I_{IH} (max.)	mA		0.02			0.02	0.05	0.02	0.02		
I_{IL} (max.)	mA		-0.2			-0.1	-2	-1	-1.6		
I_{CC} (max.)	mA		54			27	180	90	90		
I_{CC} (quiesc.)	mA			0.08[c]	0.08[h]					0.08	0.08
C_{pd} (typ.)	pF/gate			40[d]	40[i]					45[n]	45[q]
t_{PLH} (max.)	ns		18	29[e]	33[j]	10	9	6.2	6.2	7.5[o]	10[r]
t_{PHL} (max.)	ns		18	29[e]	33[j]	10	9	6.2	6.5	7.5[o]	10[r]
t_{PZH} (max.)	ns		23	38[f]	38[k]	20	12	9	6.7	8[p]	9.5[s]
t_{PZL} (max.)	ns		30	38[f]	38[k]	20	15	7.5	8	8.5[p]	10.5[s]
t_{PHZ} (max.)	ns		18[a]	38[g]	38[l]	10	9	6	7	9.5[p]	10.5[s]
t_{PLZ} (max.)	ns		25[b]	38[g]	38[l]	13	15	9	7	9.5[p]	10.5[s]

[a] TI 25 ns

[b] TI 20 ns

[c] SGS and Toshiba 0.04 mA

[d] Signetics and TI 35 pF/gate, RCA 46 pF/gate, Motorola and National 50 pF/gate

[e] Signetics 25 ns, SGS 27 ns, RCA and Toshiba 28 ns

[f] SGS 24 ns, Toshiba 25 ns

[g] SGS 40 ns

[h] Toshiba 0.04 mA

[i] Signetics 35 pF/gate, Toshiba 46 pF/gate, National 90 pF/gate

[j] RCA; National 25 ns, Signetics 28 ns, TI 35 ns, Toshiba 44 ns

[k] TI and Toshiba 44 ns

[l] RCA; Signetics 31 ns, National 32 ns, TI 44 ns, Toshiba 57 ns

[m] Versions with -1 suffix have I_{OL} (max.) = 48 mA if V_{CC} is kept between 4.75 V and 5.25 V

[n] Toshiba 19 pF/gate, RCA 80 pF/gate

[o] RCA 8.6 ns; Toshiba data not available

[p] RCA 12 ns; Toshiba data not available

[q] RCA 95 pF/gate

[r] RCA 9.6 ns

[s] RCA 13 ns

1244   4+4-line 3-state buffers/drivers with active-LOW enables

MANUFACTURERS	TTL	LS	HC	HCT	ALS/A	S	AS	F	AC	ACT
National Semiconductor					■					
Signetics					■			■		
Texas Instruments					■					

ALS -1 versions available from National and TI

KEY PARAMETERS		TTL	LS	HC	HCT	ALS/A	S	AS	F	AC	ACT
I_{OH} (max.)	mA					-15			-15		
I_{OL} (max.)	mA					16[a]			64		
I_{IH} (max.)	mA					0.02			0.07[b]		
I_{IL} (max.)	mA					-0.1			-0.07[c]		
I_{CC} (max.)	mA					17			75		
t_{PLH} (max.)	ns					14			7.5		
t_{PHL} (max.)	ns					14			6		
t_{PZH} (max.)	ns					22			8.5		
t_{PZL} (max.)	ns					22			8.5		
t_{PHZ} (max.)	ns					10			6		
t_{PLZ} (max.)	ns					13			6		

[a] Versions with -1 suffix have I_{OL} (max.) = 24 mA if V_{CC} is kept between 4.75 V and 5.25 V

[b] Including 0.05 mA off-state output current

[c] Including -0.05 mA off-state output current

467   4+4-line 3-state buffers/drivers with active-LOW enables

MANUFACTURERS	TTL	LS	HC	HCT	ALS/A	S	AS	F	AC	ACT
National Semiconductor		■			■					
Texas Instruments		■			■					

KEY PARAMETERS		TTL	LS	HC	HCT	ALS/A	S	AS	F	AC	ACT
I_{OH} (max.)	mA		-2.6[a]			-15					
I_{OL} (max.)	mA		24			24					
I_{IH} (max.)	mA		0.02			0.02					
I_{IL} (max.)	mA		-0.2[b]			-0.1					
I_{CC} (max.)	mA		37[c]			33					
t_{PLH} (max.)	ns		15[d]			13					
t_{PHL} (max.)	ns		18[e]			12					
t_{PZH} (max.)	ns		40[f]			23					
t_{PZL} (max.)	ns		45[g]			25					
t_{PHZ} (max.)	ns		40[h]			10					
t_{PLZ} (max.)	ns		45[i]			18					

[a] National -5.2 mA

[b] National -0.05 mA

[c] National 26 mA

[d] National 16 ns

[e] National 28 ns

[f] National 25 ns

[g] National 30 ns

[h] National 20 ns

[i] National 27 ns

230   4+4-line 3-state inverting/non-inverting buffers/drivers

MANUFACTURERS	TTL	LS	HC	HCT	ALS	S	AS	F	AC	ACT
National Semiconductor							■			
Texas Instruments					D		■			

KEY PARAMETERS		TTL	LS	HC	HCT	ALS	S	AS	F	AC	ACT
I_{OH} (max.)	mA							-15			
I_{OL} (max.)	mA							64			
I_{IH} (max.)	mA							0.02			
I_{IL} (max.)	mA							-0.5[a]			
I_{CC} (max.)	mA							87			
t_{PLH} (max.)	ns							6.5			
t_{PHL} (max.)	ns							6.2			
t_{PZH} (max.)	ns							9			
t_{PZL} (max.)	ns							8.5			
t_{PHZ} (max.)	ns							6			
t_{PLZ} (max.)	ns							9.5			

[a] Input 2A -1 mA

231   4+4-line 3-state inverting buffers/drivers with complem. enables

MANUFACTURERS	TTL	LS	HC	HCT	ALS	S	AS	F	AC	ACT
National Semiconductor							■			
Texas Instruments					■		■			

ALS -1 version available from TI

KEY PARAMETERS		TTL	LS	HC	HCT	ALS	S	AS	F	AC	ACT
I_{OH} (max.)	mA					-15		-15			
I_{OL} (max.)	mA					24[a]		64			
I_{IH} (max.)	mA					0.02		0.02			
I_{IL} (max.)	mA					-0.1		-0.5			
I_{CC} (max.)	mA					22		82			
t_{PLH} (max.)	ns					9		6.5			
t_{PHL} (max.)	ns					9		5.7			
t_{PZH} (max.)	ns					16		6.4			
t_{PZL} (max.)	ns					19		9			
t_{PHZ} (max.)	ns					10		6			
t_{PLZ} (max.)	ns					13		9.5			

[a] Version with -1 suffix has I_{OL} (max.) = 48 mA if V_{CC} is
   kept between 4.75 V and 5.25 V

MANUFACTURERS	TTL	LS	HC	HCT	ALS/A	S	AS	F	AC	ACT
Fairchild		■						■		
Monolithic Memories							■			
Motorola			▨	D				■		
National Semiconductor			▨							
RCA		▨	■						▨	▨
SGS		■	▨	D						
Signetics		■								
Texas Instruments		■					■			
Toshiba									D	

ALS -1 versions available from National and TI

KEY PARAMETERS		TTL	LS	HC	HCT	ALS/A	S	AS	F	AC	ACT
I_{OH} (max.)	mA		-15	-6	-6	-15	-15	-15	-15	-24	-24
I_{OL} (max.)	mA		24	6	6	24^m	64	64	64	24	24
I_{IH} (max.)	mA		0.02			0.02	0.05	0.02	0.02		
I_{IL} (max.)	mA		-0.2			-0.1	-2	-1	-1.6		
I_{CC} (max.)	mA		54			30	180	90	90		
I_{CC} (quiesc.)	mA			0.08^c	0.08^h					0.08	0.08
C_{pd} (typ.)	pF/gate			40^d	40^i					45^o	45^r
t_{PLH} (max.)	ns		18	28^e	28^j	11	9	6.2	6.2	7.5^p	10^s
t_{PHL} (max.)	ns		18	28^e	28^j	10	9	6.2	6.5	7.5^p	10^s
t_{PZH} (max.)	ns		23	38^f	38^k	21	12	10.5	6.7	9.5^q	10^t
t_{PZL} (max.)	ns		30	38^f	38^k	21	15	8.5	8	9.5^q	11^t
t_{PHZ} (max.)	ns		18^a	38^g	38^l	10	9	7	7	10.5^q	11.5^t
t_{PLZ} (max.)	ns		25^b	38^g	38^l	15	15	12	7^n	10.5^q	11.5^t

[a] TI 25 ns
[b] TI 20 ns
[c] SGS and Toshiba 0.04 mA
[d] Signetics 30 pF/gate, RCA 34 pF/gate, TI 35 pF/gate, Motorola and National 50 pF/gate
[e] Signetics and Motorola 25 ns, SGS 27 ns, National and TI 29 ns
[f] SGS 24 ns, Toshiba 25 ns
[g] SGS 40 ns
[h] Toshiba 0.04 mA
[i] Texas Instruments; Signetics 30 pF/gate, RCA 38 pF/gate, Toshiba 49 pF/gate, National 90 pF/gate

[j] National 25 ns, TI 32 ns, RCA 33 ns, Toshiba 44 ns
[k] TI and Toshiba 44 ns
[l] RCA; National 32 ns, TI 44 ns, Toshiba 57 ns
[m] Versions with -1 suffix have I_{OL} (max.) = 48 mA if V_{CC} is kept between 4.75 V and 5.25 V
[n] Motorola 7.5 ns
[o] RCA 95 pF/gate
[p] RCA 8.6 ns
[q] RCA 12 ns
[r] RCA 115 pF/gate
[s] RCA 9.6 ns
[t] RCA 13 ns

1241   4+4-line 3-state buffers/drivers with dual 4-bit enables

MANUFACTURERS	TTL	LS	HC	HCT	ALS	S	AS	F	AC	ACT
National Semiconductor					■					
Signetics								■		
Texas Instruments					D					

ALS -1 version available from National

KEY PARAMETERS		TTL	LS	HC	HCT	ALS	S	AS	F	AC	ACT
I_{OH} (max.)	mA					-15			-15		
I_{OL} (max.)	mA					16^a			64		
I_{IH} (max.)	mA					0.02			0.07^b		
I_{IL} (max.)	mA					-0.1			-0.07^c		
I_{CC} (max.)	mA					17			80		
t_{PLH} (max.)	ns					11			6		
t_{PHL} (max.)	ns					12			7		
t_{PZH} (max.)	ns					22			7.5		
t_{PZL} (max.)	ns					21			8.5		
t_{PHZ} (max.)	ns					11			8.5		
t_{PLZ} (max.)	ns					16			8.5		

[a] Versions with -1 suffix have I_{OL} (max.) = 24 mA if V_{CC} is kept between 4.75 V and 5.25 V

[b] Including 0.05 mA off-state output current
[c] Including -0.05 mA off-state output current

8-LINE BUFFERS/DRIVERS

The 8-line bus drivers are enabled as a single unit rather than as two independent units like the 4+4-line buffers/drivers listed above. The common output enable is controlled by two active-LOW enable inputs that are ANDed together to produce the internal OE signal to the 3-state outputs. AND-gating two active-LOW inputs is logically identical to NOR-gating two active-HIGH ones, so for brevity in the following descriptions these circuits are said to have "NOR-gated enables" rather than "AND-gated active-LOW enables." Both expressions mean that the eight output lines are in the high-impedance or Z (off) state if either of the two enable inputs is HIGH; they are enabled to transmit data to the data bus only if both inputs are LOW.

The categorization of the 8-line buffers/drivers is much simpler than that of the 4+4 types; the same NOR-gated enables are used for all of them, so the only logical distinction is between inverting and noninverting outputs. All are supplied in 20-pin DIPs.

8-line inverting buffers/drivers with NOR-gated enables:

466   *8-line 3-state inverting buffers/drivers with NOR-gated enables.* 8-line LS and ALS versions of the 4+4-line 468 and the 6-line 366.

796§ *8-line 3-state inverting buffers/drivers with NOR-gated enables.*
A variant LS version of the 466. (Motorola)

I_{OH} (max.)	−5	mA	t_{PLH} (max.)	10 ns
I_{OL} (max.)	16	mA	t_{PHL} (max.)	17 ns
I_{IH} (max.)	0.02	mA	t_{PZH} (max.)	27 ns
I_{IL} (max.)	−0.4	mA	t_{PZL} (max.)	25 ns
I_{CC} (max.)	21	mA	t_{PHZ} (max.)	20 ns
			t_{PLZ} (max.)	27 ns

540   *8-line 3-state inverting buffers/drivers with NOR-gated enables.*
Widely supported LS, HC, HCT, and ALS versions of the 466
with the input and output pins arranged on opposite sides of the
package for easier layout and greater PC board density. This
pinout makes the 540 especially useful as an output port for microprocessors and other bus-oriented systems. LS versions feature a moderate amount of Schmitt-trigger input hysteresis (min.
0.2 V).

940§ *8-line 3-state inverting buffers/drivers with NOR-gated enables.*
S version of the 540 with a somewhat different pinout. Inputs are
provided with a moderate amount of Schmitt-trigger hysteresis
(min. 0.2 V). (National Semiconductor)

I_{OH} (max.)	−15	mA	t_{PLH} (max.)	7 ns
I_{OL} (max.)	64	mA	t_{PHL} (max.)	7 ns
I_{IH} (max.)	0.05	mA	t_{PZH} (max.)	10 ns
I_{IL} (max.)	−0.4	mA*	t_{PZL} (max.)	15 ns
I_{CC} (max.)	150	mA	t_{PHZ} (max.)	9 ns
			t_{PLZ} (max.)	15 ns

* Enable inputs −2 mA

2540§ *8-line 3-state inverting MOS drivers with NOR-gated enables.*
ALS version of the 540 specifically designed for driving MOS
devices. The outputs are provided with 25 Ω series resistors so
that no external resistors are required. (Texas Instruments)

I_{OH} (max.)	−15	mA	t_{PLH} (max.)	12 ns
I_{OL} (max.)	30	mA	t_{PHL} (max.)	11 ns
I_{IH} (max.)	0.02	mA	t_{PZH} (max.)	15 ns
I_{IL} (max.)	−0.1	mA	t_{PZL} (max.)	20 ns
I_{CC} (max.)	22	mA	t_{PHZ} (max.)	10 ns
			t_{PLZ} (max.)	12 ns

8-line noninverting buffers/drivers with NOR-gated enables:

465   *8-line 3-state buffers/drivers with NOR-gated enables.* 8-line LS
            and ALS versions of the 4+4-line 467 and the 6-line 367.

795§   *8-line 3-state buffers/drivers with NOR-gated enables.* A variant
            LS version of the 465. (Motorola)

I_{OH} (max.)	−5	mA	t_{PLH} (max.)	16 ns	
I_{OL} (max.)	16	mA	t_{PHL} (max.)	22 ns	
I_{IH} (max.)	0.02	mA	t_{PZH} (max.)	25 ns	
I_{IL} (max.)	−0.4	mA	t_{PZL} (max.)	20 ns	
I_{CC} (max.)	26	mA	t_{PHZ} (max.)	20 ns	
			t_{PLZ} (max.)	27 ns	

541   *8-line 3-state buffers/drivers with NOR-gated enables.* Widely
            supported versions of the 465 with the input and output pins
            arranged on opposite sides of the package for easier layout and
            greater PC board density. This pinout makes the 541 especially
            useful as an output port for microprocessors and other bus-ori-
            ented systems. LS versions feature a moderate amount of
            Schmitt-trigger input hysteresis (min. 0.2 V).

941§   *8-line 3-state buffers/drivers with NOR-gated enables.* S version
            of the 541 with a somewhat different pinout. Inputs are provided
            with a moderate amount of Schmitt-trigger hysteresis (min. 0.2
            V). (National Semiconductor)

I_{OH} (max.)	−15	mA	t_{PLH} (max.)	9 ns	
I_{OL} (max.)	64	mA	t_{PHL} (max.)	9 ns	
I_{IH} (max.)	0.05	mA	t_{PZH} (max.)	12 ns	
I_{IL} (max.)	−0.4	mA*	t_{PZL} (max.)	15 ns	
I_{CC} (max.)	180	mA	t_{PHZ} (max.)	9 ns	
			t_{PLZ} (max.)	15 ns	

* Enable inputs −2 mA

2541§   *8-line 3-state MOS drivers with NOR-gated enables.* ALS ver-
            sion of the 541 specifically designed for driving MOS devices.
            The outputs are provided with 25 Ω series resistors so that no
            external resistors are required. (Texas Instruments)

I_{OH} (max.)	-15	mA	t_{PLH} (max.)	15 ns
I_{OL} (max.)	30	mA	t_{PHL} (max.)	12 ns
I_{IH} (max.)	0.02	mA	t_{PZH} (max.)	15 ns
I_{IL} (max.)	-0.1	mA	t_{PZL} (max.)	20 ns
I_{CC} (max.)	25	mA	t_{PHZ} (max.)	10 ns
			t_{PLZ} (max.)	12 ns

FUNCTION TABLE
466, 796§, 540, 940§, 2540§

INPUTS			OUTPUTS
$\overline{OE1}$	$\overline{OE2}$	A	$\overline{Y}$
L	L	L	H
L	L	H	L
.	H	.	Z
H	.	.	Z

. = Either LOW or HIGH level
Z = High impedance (disabled)

FUNCTION TABLE
465, 795§, 541, 941§, 2541§

INPUTS			OUTPUTS
$\overline{OE1}$	$\overline{OE2}$	A	Y
L	L	L	L
L	L	H	H
.	H	.	Z
H	.	.	Z

. = Either LOW or HIGH level
Z = High impedance (disabled)

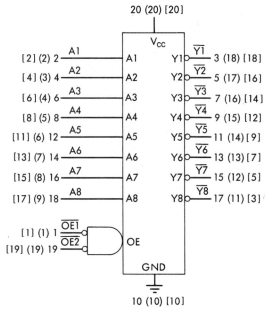

466, 796§, (540), [940§], (2540§)

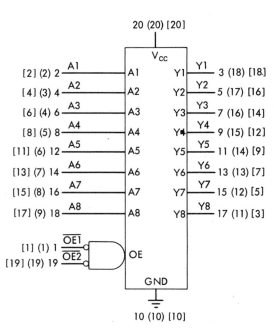

465, 795§, (541), [941§], (2541§)

MANUFACTURERS	TTL	LS	HC	HCT	ALS/A	S	AS	F	AC	ACT
National Semiconductor		■			■					
Texas Instruments		■			■					

KEY PARAMETERS		TTL	LS	HC	HCT	ALS/A	S	AS	F	AC	ACT
I_{OH} (max.)	mA		-2.6[a]			-15					
I_{OL} (max.)	mA		24			24					
I_{IH} (max.)	mA		0.02			0.02					
I_{IL} (max.)	mA		-0.2[b]			-0.1					
I_{CC} (max.)	mA		28[c]			27[j]					
t_{PLH} (max.)	ns		12[d]			12					
t_{PHL} (max.)	ns		15[e]			9					
t_{PZH} (max.)	ns		40[f]			16					
t_{PZL} (max.)	ns		45[g]			23					
t_{PHZ} (max.)	ns		40[h]			10					
t_{PLZ} (max.)	ns		45[i]			17					

[a] National -5.2 mA
[b] National -0.05 mA
[c] National 21 mA
[d] National 10 ns
[e] National 17 ns

[f] National 15 ns
[g] National 35 ns
[h] National 20 ns
[i] National 27 ns
[j] National 10 mA

8-line 3-state inverting buffers/drivers with NOR-gated enables

MANUFACTURERS	TTL	LS	HC	HCT	ALS	S	AS	F	AC	ACT
Fairchild		■						D	■	■
Motorola			D							
National Semiconductor			■	■						
RCA			■						■	■
SGS		■	■	D						
Signetics			D	D				■		
Texas Instruments			■	■	■					
Toshiba			■	■					D	

ALS -1 version available from Texas Instruments

KEY PARAMETERS		TTL	LS	HC	HCT	ALS	S	AS	F	AC	ACT
I_{OH} (max.)	mA		-15^a	-6	-6	-15			-15	-24	-24
I_{OL} (max.)	mA		24	6	6	24^s			64	24	24
I_{IH} (max.)	mA		0.02			0.02			0.02		
I_{IL} (max.)	mA		-0.2			-0.1			-0.02		
I_{CC} (max.)	mA		52^b			29			75		
I_{CC} (quiesc.)	mA			0.08^i	0.08^n					0.08	0.08
C_{pd} (typ.)	pF/gate			50^j	55^o					30^t	30^w
t_{PLH} (max.)	ns		15^c	28^k	30^p	12			7.5	6.5^u	8^x
t_{PHL} (max.)	ns		15^d	28^k	30^p	9			5	6^u	8^x
t_{PZH} (max.)	ns		25^e	40^l	44^q	15			8	9.5^v	13.2^x
t_{PZL} (max.)	ns		38^f	40^l	44^q	20			10	8.5^v	13.2^x
t_{PHZ} (max.)	ns		18^g	40^m	44^r	10			6.5	11^v	13.2^x
t_{PLZ} (max.)	ns		25^h	40^m	44^r	12			6	9^v	13.2^x

[a] Fairchild -3 mA
[b] Fairchild 50 mA
[c] Fairchild 14 ns
[d] Fairchild 18 ns
[e] Fairchild 23 ns
[f] Fairchild 30 ns
[g] TI 25 ns
[h] TI 18 ns
[i] SGS and Toshiba 0.04 mA
[j] Toshiba 33 pF/gate, TI 35 pF/gate
[k] RCA; National 25 ns, SGS and Toshiba 26 ns. Max. values for TI not available
[l] RCA; SGS 35 ns, Toshiba 36 ns, National 38 ns. Max. values for TI not available
[m] RCA and Toshiba; National and SGS 38 ns. Max. values for TI not available

[n] Toshiba 0.04 mA
[o] RCA; TI 35 pF/gate, Toshiba 37 pF/gate, National 50 pF/gate. SGS data not available
[p] RCA; National 25 ns, Toshiba 32 ns. Max. values for TI not available
[q] RCA and Toshiba; National 38 ns. Max. values for TI not available
[r] RCA; National 38 ns, Toshiba 39 ns. Max. values for TI not available
[s] Version with -1 suffix has I_{OL} (max.) = 48 mA if V_{CC} is kept between 4.75 V and 5.25 V
[t] RCA 95 pF/gate
[u] RCA 6.6 ns
[v] RCA 12 ns
[w] RCA 115 pF/gate
[x] RCA; comparable Fairchild data not available

MANUFACTURERS	TTL	LS	HC	HCT	ALS/A	S	AS	F	AC	ACT
National Semiconductor		■			■					
SGS		■								
Texas Instruments		■			■					

KEY PARAMETERS		TTL	LS	HC	HCT	ALS/A	S	AS	F	AC	ACT
I_{OH} (max.)	mA		-2.6^a			-15					
I_{OL} (max.)	mA		24			24					
I_{IH} (max.)	mA		0.02			0.02					
I_{IL} (max.)	mA		-0.2^b			-0.1					
I_{CC} (max.)	mA		37^c			33					
t_{PLH} (max.)	ns		15^d			13					
t_{PHL} (max.)	ns		18^e			12					
t_{PZH} (max.)	ns		40^f			23					
t_{PZL} (max.)	ns		45^g			25					
t_{PHZ} (max.)	ns		40^h			10					
t_{PLZ} (max.)	ns		45^i			18					

[a] National -5.2 mA
[b] National -0.05 mA
[c] National 26 mA
[d] National 16 ns
[e] National 28 ns

[f] National 25 ns
[g] National 30 ns
[h] National 20 ns
[i] National 27 ns

541   8-line 3-state inverting buffers/drivers with NOR-gated enables

MANUFACTURERS	TTL	LS	HC	HCT	ALS	S	AS	F	AC	ACT
Fairchild								D	▓	░░
Motorola		▓	D							
National Semiconductor			░░	░░						
RCA			▓	▓					░░	░░
SGS		░░		D						
Signetics		▓	D	D					▓	
Texas Instruments			░░	░░	▓					
Toshiba			▓						D	

ALS -1 version available from Texas Instruments

KEY PARAMETERS		TTL	LS	HC	HCT	ALS	S	AS	F	AC	ACT
I_{OH} (max.)	mA		-15	-6	-6	-15			-15	-24	-24
I_{OL} (max.)	mA		24	6	6	24^n			64	24	24
I_{IH} (max.)	mA		0.02			0.02			0.02		
I_{IL} (max.)	mA		-0.2			-0.1			-0.02		
I_{CC} (max.)	mA		55^a			19			72		
I_{CC} (quiesc.)	mA			0.08^d	0.08^i					0.08	0.08
C_{pd} (typ.)	pF/gate			50^e	55^j					30^o	30^r
t_{PLH} (max.)	ns		15	29^f	35^k	14			7	6.5^p	9.8^s
t_{PHL} (max.)	ns		18	29^f	35^k	10			7.5	6.5^p	9.8^s
t_{PZH} (max.)	ns		32	44^g	44^l	15			7.5	9.5^q	13.2^s
t_{PZL} (max.)	ns		38	44^g	44^l	20			9.5	8.5^q	13.2^s
t_{PHZ} (max.)	ns		18^b	44^h	44^m	10			7.5	10.5^q	13.2^s
t_{PLZ} (max.)	ns		29^c	44^h	44^m	12			7.5	8.5^q	13.2^s

[a] Signetics 52 mA
[b] TI 29 ns
[c] TI 18 ns
[d] SGS and Toshiba 0.04 mA
[e] TI 35 pF/gate, Toshiba 36 pF/gate, RCA 48 pF/gate
[f] RCA and Toshiba; SGS 26 ns. Max. values for TI not available
[g] RCA; SGS 35 ns, Toshiba 36 ns, National 38 ns. Max. values for TI not available
[h] RCA; National and SGS 38 ns, Toshiba 40 ns. Max. values for TI not available
[i] Toshiba 0.04 mA
[j] RCA; TI 35 pF/gate, Toshiba 39 pF/gate, National 45 pF/gate. SGS data not available

[k] RCA; National 29 ns, Toshiba 36 ns. Max. values for TI not available
[l] RCA and Toshiba; National 38 ns. Max. values for TI not available
[m] RCA; National 38 ns, Toshiba 39 ns. Max. values for TI not available
[n] Version with -1 suffix has I_{OL} (max.) = 48 mA if V_{CC} is kept between 4.75 V and 5.25 V
[o] RCA 80 pF/gate
[p] RCA 8.6 ns
[q] RCA 12 ns
[r] RCA 100 pF/gate
[s] RCA; comparable Fairchild data not available

8-LINE BUFFERS/DRIVERS WITH AUXILIARY DEVICES

In this category are three relatively complex single-sourced devices that combine ordinary 3-state buffer/driver functions with additional capabilities useful in certain common applications. Only brief descriptions are provided here; refer to manufacturer's data for additional details.

7340§ *8-line 3-state buffers/drivers with bidirectional storage register.* An ordinary 8-line 3-state bus driver is combined with an 8-bit D register (array of eight D flip-flops) and provided with a common clock and two control inputs. The output bus (B bus) can be switched to transmit data from the A bus directly or to read data previously loaded into the register from the A bus. The A bus itself is bidirectional and can be used to read the internal register as well as to write to it. (HC from Texas Instruments)

655§ *8-line 3-state inverting buffers/drivers with parity generator/ checker.* An 8-line buffer/driver is combined with a parity generator/checker for automatic parity checking on the data being transmitted. In addition to the eight data outputs, the 655 provides a SUM EVEN output that is HIGH when an even number of data inputs are HIGH and a SUM ODD output that is HIGH when an odd number of data inputs are HIGH. (F from Signetics; F under development by Fairchild)

656§ *8-line 3-state buffers/drivers with parity generator/checker.* Noninverting versions of the 655. (F from Signetics; F under development by Fairchild)

10-LINE BUFFERS/DRIVERS

Most bus-oriented systems handle data in multiples of eight bits and consequently are well served by the 8-line buffers/drivers described above. There are applications, however, which require bus lines in multiples of nine or ten. For example, systems that transmit parity bits for error-checking need one or two extra lines for each group of eight to carry the parity bits, and some microcomputer systems achieve memory addressing beyond the usual 16 bits by using address buses with 18 or 20 lines (that is, 2×9 or 2×10).

A small number of 9- and 10-line parallel circuits have been introduced in the last few years to support these applications. There are no 7400-series 9-line buffers/drivers in production at the time of this writing, but the following two 10-line circuits will meet the basic requirements of most applications that demand one or two extra channels. Both are provided in 24-pin packages that allow for ten input and output lines plus a pair of NOR-gated 3-state output enables.

These two designs were originally manufactured by Advanced Micro Devices, which continues to provide them in its 29800, 29800A, and 29C800 series of bus interface components. While not identical to the normal 7400-series technologies, the AMD versions are fully TTL-compatible and provide an important source for these types. In the data pages that follow, the 29827 and 29828 (comparable to S or AS) are described in the columns usually reserved for S components;

the improved 29827A and 29828A are included with the 74AS827 and 74AS828, which they strongly resemble; and the 29C827 and 29C828 (high-speed TTL-compatible CMOS) are specified in the ACT column.

Both of these 10-bit drivers are 3-state devices supplied in 24-pin DIPs.

828   *10-line 3-state inverting buffers/drivers with NOR-gated enables*. 10-line inverting bus drivers with pinout and characteristics similar to those of the 8-line 540. All the AMD versions feature a moderate amount of Schmitt-trigger hysteresis (typically 0.2 V) at the data inputs to improve noise immunity.

827   *10-line 3-state buffers/drivers with NOR-gated enables*. Same as the 828, but noninverting.

FUNCTION TABLE
828

INPUTS			OUTPUTS
$\overline{OE1}$	$\overline{OE2}$	A	$\overline{Y}$
L	L	L	H
L	L	H	L
.	H	.	Z
H	.	.	Z

. = Either LOW or HIGH level
Z = High impedance (disabled)

FUNCTION TABLE
827

INPUTS			OUTPUTS
$\overline{OE1}$	$\overline{OE2}$	A	Y
L	L	L	L
L	L	H	H
.	H	.	Z
H	.	.	Z

. = Either LOW or HIGH level
Z = High impedance (disabled)

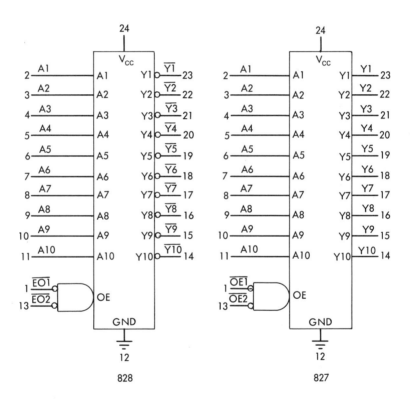

828   827

828   10-line 3-state inverting buffers/drivers with NOR-gated enables

MANUFACTURERS	TTL	LS	HC	HCT	ALS	*	**	F	AC	***
Advanced Micro Devices						███				███
Fairchild								D		
Signetics								▒▒▒		

The * column refers to the Am29828 (comparable to S and AS)
The ** column refers to the Am29828A (comparable to AS)
The *** column refers to the Am29C828 (comparable to ACT)

KEY PARAMETERS		TTL	LS	HC	HCT	ALS	*	**	F	AC	***
I_{OH} (max.)	mA						-24	-24	-3		-15
I_{OL} (max.)	mA						48	48	48		24
I_{IH} (max.)	mA						0.05	0.05	0.02		
I_{IL} (max.)	mA						-1	-0.5	-0.6		
I_{CC} (max.)	mA						80	80	90		
I_{CC} (quiesc.)	mA										0.12
C_{pd} (typ.)	pF/gate										a
t_{PLH} (max.)	ns						9.0	7	n/a		10
t_{PHL} (max.)	ns						9.5	9	n/a		10
t_{PZH} (max.)	ns						17	11	n/a		13
t_{PZL} (max.)	ns						17	12	n/a		13
t_{PHZ} (max.)	ns						19	10	n/a		13
t_{PLZ} (max.)	ns						12	10	n/a		13

[a] C_{pd} not given; instead, AMD specifies a dynamic supply
current I_{CCD} (max.) of 0.275 mA/MHz/bit

827　10-line 3-state buffers/drivers with NOR-gated enables

MANUFACTURERS	TTL	LS	HC	HCT	ALS	*	**	F	AC	***
Advanced Micro Devices						■				■
Fairchild								D		
Signetics							▨			

The * column refers to the Am29827 (comparable to S and AS)
The ** column refers to the Am29827A (comparable to AS)
The *** column refers to the Am29C827 (comparable to ACT)

KEY PARAMETERS		TTL	LS	HC	HCT	ALS	*	**	F	AC	***
I_{OH} (max.)	mA						-24	-24	-3		-15
I_{OL} (max.)	mA						48	48	48		24
I_{IH} (max.)	mA						0.05	0.05	0.02		
I_{IL} (max.)	mA						-1	-0.5	-0.6		
I_{CC} (max.)	mA						80	80	90		
I_{CC} (quiesc.)	mA										0.12
C_{pd} (typ.)	pF/gate										a
t_{PLH} (max.)	ns						10	8	n/a		10
t_{PHL} (max.)	ns						10	8	n/a		10
t_{PZH} (max.)	ns						17	11	n/a		13
t_{PZL} (max.)	ns						17	12	n/a		13
t_{PHZ} (max.)	ns						19	10	n/a		13
t_{PLZ} (max.)	ns						12	10	n/a		13

^{a} C_{pd} not given; instead, AMD specifies a dynamic supply
current I_{CCD} (max.) of 0.275 mA/MHz/bit

5

Parallel Latches

In keeping with the scheme of organization used in this book, the category of parallel latches includes all packages of multiple "D-type" latches (transparent latches) in which groups of latches are enabled simultaneously through common latch enable inputs. These configurations put the maximum number of latches in each package by sacrificing individual latching control over each bit of data (though the data themselves remain accessible through individual input and output lines).

Virtually all parallel latches are 3-state devices intended to interface with a bus, or set of parallel lines that serve a number of devices in common. The third state (the high-impedance or "Z" state) puts the latch outputs into a condition where they neither load nor drive the common bus, effectively disconnecting the latch from the larger system. The outputs of these 3-state latches are similar to 3-state driver outputs and are designed for driving bus lines or other relatively high-capacitance or low-impedance loads without additional interface or pull-up circuitry.

Two parallel latch types, the 100§ and the 116, do not have 3-state bus driver outputs, only the ordinary 2-state kind. They are included in this section because the individual D latches in each package are controlled in parallel through a common latch enable, as in the 3-state types.

Like the one-bit transparent latches discussed in Chapter 2, the parallel latches have two basic modes of operation. When enabled, each parallel latch acts exactly like a parallel line driver, simply transmitting the data at inputs D1, D2, etc. to the outputs Q1, Q2, etc. In this mode the 3-state parallel latches are functionally identical to the 3-state parallel buffers/drivers of Chapter 4 and can be similarly switched on or off the bus using the active-LOW output enables, $\overline{\text{OE}}$.

The latches, however, have an additional capability: at any moment, the latch inputs can be disabled by changing the level at the latch enable, E. When E is taken LOW, the data at the D inputs are "latched" and will thereafter continue to be steadily present at the Q outputs regardless of any further input changes. In this latched mode the 3-state output controls serve to place the latched data on, or remove them from, the common bus.

The 3-state output enables do not affect the internal functioning of the latches. Old data will be retained, or new data entered (depending on the state of the E input), even when the outputs are in the "off" or high-impedance condition.

Some latches are provided with direct clear (reset) inputs, direct set (preset) inputs, or both. As a general rule these inputs are asynchronous and override all other inputs to directly load a LOW or HIGH level in every bit position controlled by the $\overline{\text{R}}$ or $\overline{\text{S}}$ input.

NOTE ON INVERTING AND NONINVERTING VERSIONS

In previous sections, inverting versions of each functional type have been placed before noninverting versions of that type because the inverting versions are electrically more primitive than the corresponding noninverting types. They tend to be faster and consume less power, making them the more often used of the two forms. In storage devices such as parallel latches and registers, however, the noninverting forms are felt strongly to be the more "natural" of the two, and the timing advantage enjoyed by simple inverting gates disappears in the more complex bus-oriented devices. Consequently, the noninverting types are more widely used and supported. The noninverting latches and registers are therefore considered the primary forms and have been placed before the corresponding inverting types in the following listings.

DUAL 4-BIT LATCHES

These circuits provide two sets of transparent D latches, four latches in each set, with each set of four enabled or disabled by a common E input. Unlike the other devices in this chapter, the outputs of these circuits are the ordinary 2-state variety and are not suited to driving bus lines or other heavy loads. In other words, these latches are just collections of the ordinary transparent latches of Chapter 2, grouped together for convenience and controlled by common enable inputs to minimize the pin count. Their switching characteristics are specified for a load capacitance of 15 pF, like the ordinary TTL and LS circuits of Part I.

116   *Dual 4-bit latches with direct clears and NOR-gated latch enables.* TTL latches from Signetics and TI. Each set of four D latches is controlled by a common internal latch enable input, E, produced by gating together two active-LOW external inputs $\overline{E1}$ and $\overline{E2}$. The four latches in each group are "transparent" and act like line drivers as long as both enable inputs stay LOW; when either goes HIGH, the latches refuse further input and retain the last valid inputs, D, at the four outputs, Q. The outputs of each group of four latches can be directly set to LOW by a common active-LOW clear (reset) input, $\overline{R}$. Supplied in 24-pin DIP.

100§   *Dual 4-bit latches.* Cut-down version of the 116 (TTL) that eliminates the clear inputs and omits one of the dual latch enables in each group, leaving only one enable for each group of four. Supplied in 24-pin DIP similar to that of the 116, but with four pins left unused. (Texas Instruments)

I_{OH} (max.)	−0.4	mA	t_W	(min.)	20 ns
I_{OL} (max.)	16	mA	t_{SU}	(min.)	20 ns
I_{IH} (max.)	0.08*	mA	t_H	(min.)	5 ns
I_{IL} (max.)	−3.2†	mA	t_{pd} $\underline{D}$ to Q (max.)		30 ns
I_{CC} (max.)	21	mA	t_{pd} $\underline{E}$ to Q (max.)		25 ns

* D inputs; max. at E 0.32 mA

† D inputs; max. at E −12.8 mA

INPUTS				OUTPUTS
$\overline{R}$	$\overline{E1}$	$\overline{E2}$	D	Q
H	L	L	L	L
H	L	L	H	H
H	.	H	.	Q_0
H	H	.	.	Q_0
L	.	.	.	L

. = Either LOW or HIGH level

INPUTS		OUTPUTS
$\overline{E}$	D	Q
L	L	L
L	H	H
H	.	Q_0

. = Either LOW or HIGH level

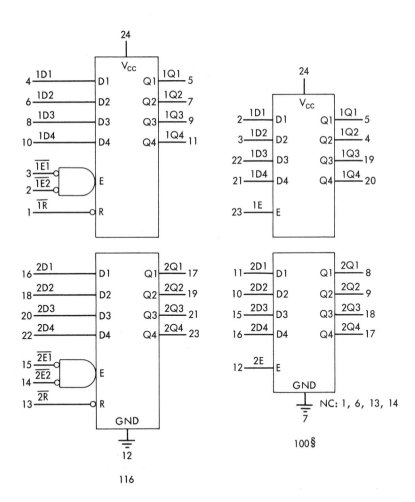

116

100§

116   **Dual 4-bit latches with direct clears and NOR-gated latch enables**

MANUFACTURERS	TTL	LS	HC	HCT	ALS	S	AS	F	AC	ACT
Signetics										
Texas Instruments										

KEY PARAMETERS		TTL	LS	HC	HCT	ALS	S	AS	F	AC	ACT
I_{OH} (max.)	mA	-0.8									
I_{OL} (max.)	mA	16									
I_{IH} (max.)	mA	0.06^a									
I_{IL} (max.)	mA	-1.6^b									
I_{CC} (max.)	mA	100									
t_W (min.)	ns	18									
t_{SU} (min.)	ns	14									
t_H (min.)	ns	8									
t_{pd} D to Q (max.)	ns	18									
t_{pd} $\overline{E}$ to Q (max.)	ns	30									
t_{pd} $\overline{R}$ to Q (max.)	ns	22									

^{a} D inputs; $\overline{E}$ and $\overline{R}$ inputs 0.04 mA ^{b} D inputs; $\overline{E}$ and $\overline{R}$ inputs -1.6 mA. Initial peak (max.) at D inputs -2.4 mA.

6-BIT LATCHES

Only one 7400-series device is available in this category.

268§   *6-bit 3-state latches.* The basic 6-bit transparent latch, provided in S technology. A moderate amount of Schmitt-trigger hysteresis (typ. 0.4 V) at the $\overline{E}$ input improves noise rejection. 16-pin DIP. (Texas Instruments)

I_{OH} (max.)	−6.5 mA	t_W	(min.)	7.3 ns
I_{OL} (max.)	20 mA	t_{SU}	(min.)	0 ns
I_{IH} (max.)	0.05 mA	t_H	(min.)	10 ns
I_{IL} (max.)	−0.25 mA	t_{pd} D to Q	(max.)	12 ns
I_{CC} (max.)	136 mA	t_{pd} $\overline{E}$ to Q	(max.)	18 ns
		t_{en}	(max.)	18 ns
		t_{dis}	(max.)	12 ns

FUNCTION TABLE
268§

INPUTS			OUTPUTS
$\overline{OE}$	$\overline{E}$	D	Q
L	H	L	L
L	H	H	H
L	L	.	Q_0
H	.	.	Z

. = Either LOW or HIGH level
Z = High impedance (disabled)

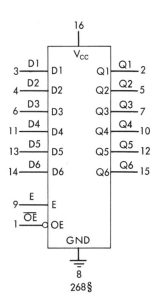

268§

4+4-BIT LATCHES

8-bit latches, like 8-bit buffers/drivers, are supplied in two basic configurations: a true 8-bit arrangement in which all eight constituent latches are controlled by a common latch enable and a common output enable, and a "4+4" arrangement in which half the constituent latches are controlled by one set of common control inputs and the other half by a second, independent set of latch and output enables. As shown below, there are two types of commercially available 4+4-bit latches, one noninverting and the other inverting; both types are provided in 24-pin DIPs.

873   *4+4-bit 3-state latches with direct clears.* Basic 4+4-bit bus latches in ALS and AS implementations. Since each 4-bit section has its own latch enable, E, active-LOW output enable, $\overline{OE}$, and active-LOW direct clear (reset) input, $\overline{R}$, each section can, if necessary, function as an independent 4-bit latch. On the other hand, if the appropriate controls are wired together to form single E, $\overline{OE}$, and $\overline{R}$ inputs, the device will function as a single 8-bit parallel bus latch. Corresponding bus register: 874.

880   *4+4-bit 3-state inverting latches with direct presets.* Inverting versions of the ALS873 and AS873. The input labeled $\overline{R}$ that sets all the outputs to LOW in the 873 also sets them all to LOW on the 880, but since the 880's inverted outputs are normally HIGH, this input is referred to as "set" or "preset" and labeled $\overline{S}$. Corresponding bus register: 876.

FUNCTION TABLE 873				
INPUTS				OUTPUTS
$\overline{OE}$	$\overline{R}$	E	D	Q
L	H	H	L	L
L	H	H	H	H
L	H	L	.	Q_0
L	L	.	.	L
H	.	.	.	Z

. = Either LOW or HIGH level
Z = High impedance (disabled)

FUNCTION TABLE 880				
INPUTS				OUTPUTS
$\overline{OE}$	$\overline{S}$	E	D	$\overline{Q}$
L	H	H	L	H
L	H	H	H	L
L	H	L	.	$\overline{Q}_0$
L	L	.	.	L
H	.	.	.	Z

. = Either LOW or HIGH level
Z = High impedance (disabled)

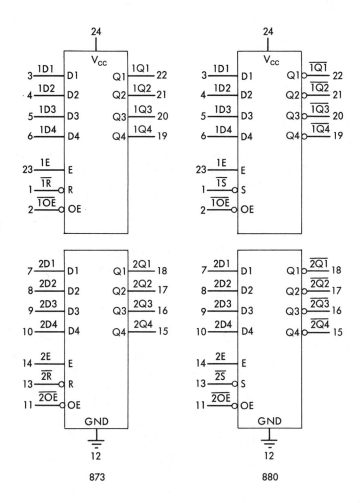

873

880

873   4+4-bit 3-state latches with direct clears

MANUFACTURERS	TTL	LS	HC	HCT	ALS	S	AS	F	AC	ACT
National Semiconductor					■		■			
Texas Instruments					■		■			

Added change control symbol: TI 74ALS873B

KEY PARAMETERS		TTL	LS	HC	HCT	ALS	S	AS	F	AC	ACT
I_{OH} (max.)	mA					-2.6		-15			
I_{OL} (max.)	mA					24		48			
I_{IH} (max.)	mA					0.02		0.02			
I_{IL} (max.)	mA					-0.2[a]		-0.5			
I_{CC} (max.)	mA					31		129			
t_W (min.)	ns					15		4.5			
t_{SU} (min.)	ns					10		2			
t_H (min.)	ns					7		3			
t_{pd} D to Q (max.)	ns					14		6			
t_{pd} E to Q (max.)	ns					22[b]		11.5			
t_{pd} $\overline{R}$ to Q (max.)	ns					20		7.5			
t_{en} (max.)	ns					18		9.5			
t_{dis} (max.)	ns					15[c]		7.5			

[a] National -0.1 mA   [c] National 12 ns
[b] National 21 ns

880   4+4-bit 3-state inverting latches with direct presets

MANUFACTURERS	TTL	LS	HC	HCT	ALS	S	AS	F	AC	ACT
National Semiconductor					■		■			
Texas Instruments					■		■			

Added change control symbol: TI 74ALS880A

KEY PARAMETERS		TTL	LS	HC	HCT	ALS	S	AS	F	AC	ACT
I_{OH} (max.)	mA					-2.6		-15			
I_{OL} (max.)	mA					24		48			
I_{IH} (max.)	mA					0.02		0.02			
I_{IL} (max.)	mA					-0.2[a]		-0.5			
I_{CC} (max.)	mA					31		137			
t_W (min.)	ns					15		3.5			
t_{SU} (min.)	ns					10		2			
t_H (min.)	ns					10		1			
t_{pd} D to $\overline{Q}$ (max.)	ns					20		9.5			
t_{pd} E to $\overline{Q}$ (max.)	ns					24		11.5			
t_{pd} $\overline{S}$ to $\overline{Q}$ (max.)	ns					21		10			
t_{en} (max.)	ns					18[b]		10			
t_{dis} (max.)	ns					17		8			

[a] National -0.1 mA
[b] National 13 ns

8-BIT LATCHES

Together with the 8-bit bus registers (Chapter 7), the 8-bit latches are among the most important and commonly found interface circuits in all kinds of byte-oriented devices, especially in communications equipment and at the input and output ports of digital computers.

In perhaps the most typical arrangement, the D inputs to an 8-bit latch are connected to a set of internal lines on which data are placed asynchronously by a sending device. At some point, the 3-state outputs of this latch are enabled, placing these bits on a common data bus external to the sending device. Meanwhile, an input register, typically edge-triggered, in the receiving device is monitoring the data through input lines connected to the same bus. At a particular transition (typically the negative edge) of a shared clock or write strobe signal, the data from the sending device are latched into its output latch, holding the voltages on the external bus steady to provide valid setup to the receiving device. Simultaneously, the same strobe edge triggers the input register in the receiving device to read and store the valid data from the data bus, effecting a secure transfer of data between devices that are otherwise operating asynchronously.

There are a number of other possible arrangements, however, and in many applications it is the latch that serves as the input circuit. Like registers, latches can also serve as temporary storage areas for data being worked on in digital processing.

Most 7400-series 8-bit latches are supplied in 20-pin DIPs, which allow for just a single latch enable and a single output enable to control the eight input and output lines. Some other 8-bit latches that have more extensive control circuitry and require larger packages are described in a separate section.

The basic 8-bit latches listed here come in two logical varieties, noninverting and inverting, and in two pin configurations, a "neighboring" arrangement in which each data input pin is placed adjacent to the corresponding output pin and a symmetrical or "bus-structured" arrangement in which each data input pin is located directly opposite its output pin on the other side of the package. Some single-sourced variations on these basic types provide improved or specialized output characteristics.

Noninverting 8-bit latches:

373   *8-bit 3-state latches.* Standard 8-bit bus latches, universally supported and available in nearly all technologies. Some LS and S versions feature a moderate amount of Schmitt-trigger hysteresis (typically about 0.4 V) at the latch enable input for improved noise rejection. Corresponding bus register: 374.

573   *8-bit 3-state latches.* Same as 373 but with symmetrical placement of I/O pins to simplify PC board layout in bus-oriented systems.

Widely supported in a variety of implementations. Corresponding bus register: 574.

531§   *8-bit 3-state latches.* S version of the 373 with I_{OL} (max.) increased to 32 mA from the usual maximum of 20 mA for the S373. Corresponding bus register: 532§. (Monolithic Memories)

I_{OH} (max.)	−6.5 mA	t_W	(min.)	7.3 ns
I_{OL} (max.)	32 mA	t_{SU}	(min.)	0 ns
I_{IH} (max.)	0.05 mA	t_H	(min.)	10 ns
I_{IL} (max.)	−0.25 mA	t_{pd} D to Q (max.)		12* ns
I_{CC} (max.)	160 mA	t_{pd} E to Q (max.)		18* ns
		t_{en}	(max.)	18* ns
		t_{dis}	(max.)	12 ns

* At C_L = 15 pF instead of the usual 45 pF or 50 pF

363§   *8-bit 3-state latches with MOS driver outputs.* LS version of the 373 that drives its outputs about 1 V closer to V_{CC} than the usual 3-state buffer, over 3.5 V at minimum V_{CC}. The circuit is intended for driving MOS memories, microprocessors, and other devices with thresholds of 2.4 V to 3.5 V. Corresponding bus register: 364§. (Signetics)

I_{OH} (max.)	−2.6 mA	t_W	(min.)	15 ns
I_{OL} (max.)	24 mA	t_{SU}	(min.)	0 ns
I_{IH} (max.)	0.02 mA	t_H	(min.)	10 ns
I_{IL} (max.)	−0.4 mA	t_{pd} D to Q (max.)		27 ns
I_{CC} (max.)	70 mA	t_{pd} E to Q (max.)		36 ns
		t_{en}	(max.)	36 ns
		t_{dis}	(max.)	25 ns

Inverting 8-bit latches:

533   *8-bit 3-state inverting latches.* Standard inverting versions of the 373 ("neighboring" or nonsymmetrical pinout), universally supported and available in most technologies. Corresponding bus register: 534.

563   *8-bit 3-state inverting latches.* Inverting versions of the 573 (bus-structured pinout), supported by a number of manufacturers. Corresponding bus register: 564.

580   *8-bit 3-state inverting latches.* ALS and AS versions of the 563 (bus-structured pinout). Corresponding bus register: 576.

535§   *8-bit 3-state inverting latches.* S version of the 533 ("neighboring" or nonsymmetrical pinout) with I_{OL} (max.) increased to 32 mA from the usual maximum of 20 mA for the S533. Corresponding bus register: 536§. (Monolithic Memories)

I_{OH} (max.)	−6.5 mA	t_W	(min.)	7.3 ns
I_{OL} (max.)	32 mA	t_{SU}	(min.)	0 ns
I_{IH} (max.)	0.05 mA	t_H	(min.)	10 ns
I_{IL} (max.)	−0.25 mA	t_{pd} D to Q (max.)		18* ns
I_{CC} (max.)	160 mA	t_{pd} E to Q (max.)		22* ns
		t_{en}	(max.)	20* ns
		t_{dis}	(max.)	16 ns

* At C_L = 15 pF instead of the usual 45 pF or 50 pF

FUNCTION TABLE
373, 573, 531§ , 363§

INPUTS			OUTPUTS
$\overline{OE}$	E	D	Q
L	H	L	L
L	H	H	H
L	L	.	Q_0
H	.	.	Z

. = Either LOW or HIGH level
Z = High impedance (disabled)

FUNCTION TABLE
533, 563, 580, 535§

INPUTS			OUTPUTS
$\overline{OE}$	E	D	$\overline{Q}$
L	H	L	H
L	H	H	L
L	L	.	$\overline{Q}_0$
H	.	.	Z

. = Either LOW or HIGH level
Z = High impedance (disabled)

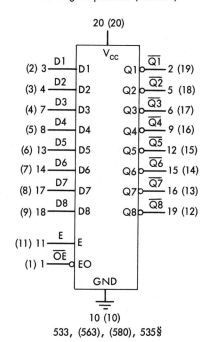

373, (573), 531§ , 363§   533, (563), (580), 535§

MANUFACTURERS	TTL	LS	HC	HCT	ALS	S	AS	F	AC	ACT
Fairchild		■							■	■
Monolithic Memories						■				
Motorola			▦	D				■		
National Semiconductor				▦				■		
RCA			■						▦	▦
SGS			▦	D						
Signetics		■				■	■	■		
Texas Instruments		■		▦		■	■	■		
Toshiba		■	■							

KEY PARAMETERS		TTL	LS	HC	HCT	ALS	S	AS	F	AC	ACT
I_{OH} (max.)	mA		-2.6	-6	-6	-2.6	-6.5	-15	-3	-24	-24
I_{OL} (max.)	mA		24	6	6	24	20	48	24	24	24
I_{IH} (max.)	mA		0.02			0.02	0.05	0.02	0.02		
I_{IL} (max.)	mA		-0.4			-0.1	-0.25	-0.5	-0.6		
I_{CC} (max.)	mA		40			27	160[v]	100	55		
I_{CC} (quiesc.)	mA			0.08[d]	0.08[m]					0.08	0.08
C_{pd} (typ.)	pF/channel			50[e]	53[n]					40[aa]	40[hh]
t_W (min.)	ns		15	20[f]	20[o]	10	7.3[w]	4.5	6	4.5[bb]	8[bb]
t_{SU} (min.)	ns		5[a]	13[g]	13[p]	10	0	2	2	4.5[cc]	8[cc]
t_H (min.)	ns		20[b]	13[h]	13[q]	7	10	3	3	0[dd]	1[dd]
t_{pd} D to Q (max.)	ns		18[c]	38[i]	44[r]	16	12[x]	6	8	10.5[ee]	11.5[ii]
t_{pd} $\overline{E}$ to Q (max.)	ns		30	44[j]	44[s]	23	18[y]	11.5	13	10.5[ff]	11.5[jj]
t_{en} (max.)	ns		36	38[k]	44[t]	20	18[z]	9.5	12	9.5[gg]	10.5[gg]
t_{dis} (max.)	ns		25	38[l]	44[u]	12	12	7	7.5	12.5[gg]	12.5[gg]

[a] Fairchild 0 ns, SGS 20 ns
[b] Fairchild 10 ns
[c] Fairchild 20 ns
[d] SGS and Toshiba 0.04 mA
[e] SGS and Toshiba 41 pF/channel, Signetics 45 pF/channel, RCA 51 pF/channel, TI 100 pF/channel
[f] SGS 18 ns, Toshiba 19 ns
[g] Motorola 5 ns, National 6 ns, SGS 18 ns
[h] RCA, Signetics, and TI 5 ns, SGS 12 ns
[i] Motorola 25 ns, SGS 35 ns, Toshiba 36 ns, National 37 ns
[j] Motorola 30 ns, SGS 42 ns
[k] Motorola 28 ns, SGS 35 ns, National 37 ns
[l] Motorola 25 ns, National 37 ns, SGS 42 ns
[m] Toshiba 0.04 mA
[n] RCA; Signetics 41 pF/channel, TI 50 pF/channel, Toshiba 55 pF/channel. National data not available
[o] National 16 ns, Toshiba 19 ns, TI 25 ns
[p] National 5 ns, Toshiba 6 ns, Signetics 15 ns
[q] Signetics 4 ns, TI 5 ns, National 10 ns

[r] National 37 ns, Signetics 38 ns, RCA 40 ns
[s] Signetics 40 ns
[t] National 37 ns, Signetics 40 ns
[u] National 37 ns, Signetics and Toshiba 38 ns
[v] TI 190 mA
[w] National 15 ns
[x] National 17 ns
[y] National 21 ns
[z] National 23 ns
[aa] RCA 90 pF/channel
[bb] RCA 6 ns
[cc] RCA 2 ns
[dd] RCA 3 ns
[ee] RCA 8.2 ns
[ff] RCA 11.8 ns
[gg] RCA 13 ns
[hh] RCA 108 pF/channel
[ii] RCA 9.4 ns
[jj] RCA 13 ns

MANUFACTURERS	TTL	LS	HC	HCT	ALS	S	AS	F	AC	ACT
Fairchild		■						■	▨	■
Motorola			D							
National Semiconductor			■	▨			■			
RCA			■	■					▨	▨
SGS		▨	▨							
Signetics			■	■				▨		
Texas Instruments			■	■	■		■			
Toshiba			■		■				D	

Added change control symbol: TI 74ALS573B

KEY PARAMETERS		TTL	LS	HC	HCT	ALS	S	AS	F	AC	ACT
I_{OH} (max.)	mA		-2.6	-6	-6	-2.6		-15	-3	-24	-24
I_{OL} (max.)	mA		24	6	6	24		48	24	24	24
I_{IH} (max.)	mA		0.02			0.02		0.02	0.02		
I_{IL} (max.)	mA		-0.4			-0.1		-0.5	-0.6		
I_{CC} (max.)	mA		40			27		106	55		
I_{CC} (quiesc.)	mA			0.08^d	0.08^d					0.08	0.08
C_{pd} (typ.)	pF/channel			50^e	50^m					25^w	25^y
t_W (min.)	ns		15	20^f	20^n	10^u		4.5	6	6^x	4^z
t_{SU} (min.)	ns		20^a	13^g	13^o	10		2	2	2^x	3.5^{aa}
t_H (min.)	ns		20^b	10^h	11^p	7		3	3	3^x	0^{bb}
t_{pd} D to Q (max.)	ns		20^c	35^i	44^q	14		6	8	8.2^x	12^{cc}
t_{pd} E to Q (max.)	ns		30	38^j	44^r	20		11.5	13	11.8^x	12^{dd}
t_{en} (max.)	ns		36	35^k	44^s	18		9.5	12	13^x	11^{dd}
t_{dis} (max.)	ns		25	38^l	44^t	15^v		7	7.5	13^x	12.5^{dd}

[a] Fairchild 0 ns

[b] Fairchild 10 ns

[c] SGS 18 ns

[d] SGS and Toshiba 0.04 mA

[e] Signetics 26 pF/channel, SGS and Toshiba 41 pF/channel, RCA 51 pF/channel

[f] SGS 18 ns, Toshiba 19 ns

[g] SGS 18 ns, National 19 ns

[h] RCA; Signetics and TI 5 ns, National 6 ns, SGS 12 ns, Toshiba 13 ns

[i] SGS; National 28 ns, Toshiba 33 ns, Signetics 38 ns, RCA and TI 44 ns

[j] National 29 ns, SGS 42 ns, RCA and TI 44 ns

[k] RCA and TI 38 ns

[l] National 31 ns, SGS 42 ns

[m] Signetics 26 pF/channel, SGS and Toshiba 41 pF/channel, RCA 53 pF/channel

[n] SGS 18 ns, Toshiba 19 ns, TI 25 ns. Max. data for National not available

[o] RCA and Signetics 16 ns, SGS 18 ns. Max. data for National not available

[p] Signetics; TI and Toshiba 5 ns, SGS 6 ns, RCA 13 ns. National data not available

[q] SGS 42 ns, RCA 50 ns. Max. data for National not available

[r] SGS 46 ns. Max. data for National not available

[s] Signetics 38 ns, SGS 42 ns. Max. data for National not available

[t] Signetics 38 ns, SGS 40 ns, Toshiba 46 ns. Max. data for National not available

[u] National 15 ns

[v] National 12 ns

[w] RCA 90 pF/channel

[x] RCA; comparable Fairchild data not available

[y] RCA 108 pF/channel

[z] RCA 6 ns

[aa] RCA 2 ns

[bb] RCA 3 ns

[cc] RCA 9.4 ns

[dd] RCA 13 ns

533 8-bit 3-state inverting latches

MANUFACTURERS	TTL	LS	HC	HCT	ALS	S	AS	F	AC	ACT
Fairchild		■						■	▦	▦
Monolithic Memories		■				■				
Motorola			▦					■		
National Semiconductor			▦	▦	■		■			
RCA									▦	▦
SGS		▦	▦							
Signetics			▦				■			
Texas Instruments			▦	▦	■	■	■	■		
Toshiba			■						D	

KEY PARAMETERS		TTL	LS	HC	HCT	ALS	S	AS	F	AC	ACT
I_{OH} (max.)	mA		-2.6^a	-6	-6	-2.6	-6.5	-15	-3	-24	-24
I_{OL} (max.)	mA		24	6	6	24	20	48	24	24	24
I_{IH} (max.)	mA		0.02			0.02	0.05	0.02	0.02		
I_{IL} (max.)	mA		-0.4			-0.1	-0.25	-0.5	-0.6		
I_{CC} (max.)	mA		40^b			28	160	110	61		
I_{CC} (quiesc.)	mA			0.08^g	0.08					0.08	0.08
C_{pd} (typ.)	pF/channel			50^h	50^p					90^y	108^y
t_W (min.)	ns		15^c	20^i	20^q	15	7.3	2^x	6	6^y	6^y
t_{SU} (min.)	ns		0^d	13^j	13^r	15	0	2	2	2^y	2^y
t_H (min.)	ns		10^c	13^k	10^s	7	10	3	3	3^y	3^y
t_{pd} D to $\overline{Q}$ (max.)	ns		25^c	38^l	43^t	19	18	7.5	10	10.2^y	11.4^y
t_{pd} E to $\overline{Q}$ (max.)	ns		30^e	44^m	48^u	23	22	9	13	11.8^y	13^y
t_{en} (max.)	ns		36^c	38^n	44^v	18	20	9.5	11	13^y	13^y
t_{dis} (max.)	ns		25^f	38°	38^w	16	16	7	7	13^y	13^y

[a] SGS -0.4 mA
[b] Monolithic 48 mA
[c] SGS data not available
[d] Monolithic 3 ns; SGS data not available
[e] Monolithic 35 ns; SGS data not available
[f] Monolithic 29 ns; SGS data not available
[g] SGS and Toshiba 0.04 mA
[h] Signetics 34 pF/channel, SGS 41 pF/channel, RCA 42 pF/channel
[i] SGS 18 ns, Toshiba 19 ns
[j] National 6 ns, SGS 18 ns
[k] TI 5 ns, RCA and Signetics 9 ns, SGS 12 ns
[l] SGS 35 ns, Toshiba 36 ns, National 37 ns, RCA 41 ns, TI 44 ns

[m] SGS 42 ns
[n] National 37 ns
[o] National 37 ns
[p] TI; Signetics 34 pF/channel, RCA 42 pF/channel. National data not available
[q] TI 25 ns
[r] National 6 ns
[s] RCA and Signetics; TI 5 ns, National 13 ns
[t] National 37 ns, TI 44 ns
[u] National and TI 44 ns
[v] National 37 ns
[w] National 37 ns, TI 44 ns
[x] National 4.5 ns
[y] RCA; comparable Fairchild data not available

MANUFACTURERS	TTL	LS	HC	HCT	ALS	S	AS	F	AC	ACT
Fairchild		■						■	▒	▒
Motorola			D							
National Semiconductor				▒	▒					
RCA									▒	▒
SGS			▒	▒						
Signetics								▒		
Texas Instruments			■	■	■					
Toshiba			■	■					D	

Added change control symbol: TI 74ALS563A

KEY PARAMETERS		TTL	LS	HC	HCT	ALS	S	AS	F	AC	ACT
I_{OH} (max.)	mA		-2.6	-6	-6	-2.6			-3	-24	-24
I_{OL} (max.)	mA		24	6	6	24			24	24	24
I_{IH} (max.)	mA		0.02			0.02			0.02		
I_{IL} (max.)	mA		-0.4			-0.1			-0.6		
I_{CC} (max.)	mA		40			29^r			55^t		
I_{CC} (quiesc.)	mA			0.08^a	0.08^a					0.08	0.08
C_{pd} (typ.)	pF/channel			41^b	41^j					50^z	50^{bb}
t_W (min.)	ns		15	20^c	20^k	15			7^u	6^{aa}	3^{cc}
t_{SU} (min.)	ns		0	13^d	13^l	10			2.5^v	2^{aa}	4.5^{dd}
t_H (min.)	ns		10	5^e	5^m	10			3.5^w	3^{aa}	0^{ee}
t_{pd} D to $\overline{Q}$ (max.)	ns		25	36^f	44^n	18			10	10.2^{aa}	12.5^{ff}
t_{pd} E to $\overline{Q}$ (max.)	ns		30	41^g	44^o	22			12.5^x	11.8^{aa}	11.5^{gg}
t_{en} (max.)	ns		36	38^h	44^p	18			11.5^y	13^{aa}	10^{gg}
t_{dis} (max.)	ns		25	38^i	44^q	15^s			7	13^{aa}	11.5^{gg}

[a] SGS and Toshiba 0.04 mA

[b] Signetics 30 pF/channel, RCA 42 pF/channel, National and TI 50 pF/channel

[c] SGS 18 ns, Toshiba 19 ns

[d] SGS 18 ns, National 19 ns

[e] TI; RCA and Signetics 4 ns, National 6 ns, SGS 12 ns, Toshiba 13 ns

[f] Signetics; National 28 ns, Toshiba 33 ns, SGS 35 ns, RCA 38 ns, TI 44 ns

[g] RCA; National 29 ns, Signetics 36 ns, Toshiba 38 ns, SGS 42 ns, TI 44 ns

[h] National, SGS, and Toshiba 35 ns

[i] National 31 ns, SGS 42 ns

[j] Signetics 30 pF/channel, RCA 42 pF/channel, TI 50 pF/channel. National data not available

[k] SGS 18 ns, Toshiba 19 ns, TI 25 ns. National data not available

[l] SGS 18 ns. National data not available

[m] SGS 6 ns. National data not available

[n] RCA and Signetics 38 ns, SGS 42 ns. Max. data for National not available

[o] SGS 46 ns. Max. data for National not available

[p] SGS 42 ns. Max. data for National not available

[q] SGS 40 ns, Toshiba 46 ns. Max. data for National not available

[r] National 27 mA

[s] National 13 ns

[t] Signetics 61 mA

[u] Signetics 6 ns

[v] Signetics 2 ns

[w] Signetics 3 ns

[x] Signetics 13 ns

[y] Signetics 11 ns

[z] RCA 90 pF/channel

[aa] RCA; comparable Fairchild data not available

[bb] RCA 108 pF/channel

[cc] RCA 6 ns

[dd] RCA 2 ns

[ee] RCA 3 ns

[ff] RCA 11.4 ns

[gg] RCA 13 ns

580 8-bit 3-state inverting latches

MANUFACTURERS	TTL	LS	HC	HCT	ALS	S	AS	F	AC	ACT
National Semiconductor					■		■			
Texas Instruments					■		■			

KEY PARAMETERS		TTL	LS	HC	HCT	ALS	S	AS	F	AC	ACT
I_{OH} (max.)	mA					-2.6		-15			
I_{OL} (max.)	mA					24		48			
I_{IH} (max.)	mA					0.02		0.02			
I_{IL} (max.)	mA					-0.1		-0.5			
I_{CC} (max.)	mA					29^a		115			
t_W (min.)	ns					15		2			
t_{SU} (min.)	ns					10		2			
t_H (min.)	ns					10		3			
t_{pd} D to $\overline{Q}$ (max.)	ns					18		7.5			
t_{pd} E to $\overline{Q}$ (max.)	ns					22		9			
t_{en} (max.)	ns					18		9.5			
t_{dis} (max.)	ns					15^b		7			

a National 27 mA
b National 13 ns

8-BIT LATCHES WITH MULTIPLE CONTROLS

For applications in which a latch must be controlled by more complex logic than that afforded by a single latch enable and output enable, the two types described below offer a more complete set of control inputs at the price of four additional pins, which enlarge the 20-pin complement of the ordinary 8-bit bus latch to 24.

Both of these types were originally designed by Advanced Micro Devices, which continues to supply them in its 29800 and 29800A series of bus interface components. While not identical to the usual 7400-series technology categories, these families are fully TTL-compatible and offer an important alternative to other sources. In the data pages that follow, the 29845 and 29846 (which have performance characteristics that fall roughly between those of S and AS devices) are listed in the columns usually reserved for S components, while the improved 29845A is described along with the 74AS845, which it strongly resembles.

845   *8-bit 3-state latches with preset, clear, and multiuser controls.* Parallel latches with bus-structured pinout and a full set of control inputs, consisting of latch enable, E; active-LOW set or preset, $\overline{S}$; active-LOW reset or clear, $\overline{R}$; and three AND-gated active-LOW output enables, $\overline{OE1}$, $\overline{OE2}$, and $\overline{OE3}$. The 3-state outputs are in the high-impedance condition (effectively disconnected from the bus) whenever any of the three output enables is taken HIGH, and en-

abled (put on the bus) only when all three output enables are held
LOW. Thus, any one (or more) of up to three "users" can switch
the latch off the bus on its own initiative, without the permission of
any of the other users. The preset (set) control, when taken LOW,
will override any other input condition and cause all the Q outputs
to go HIGH; the clear (reset) control, when taken LOW, will cause
all the outputs to go LOW. Taking both $\overline{S}$ and $\overline{R}$ LOW at the same
time will result in the preset ($\overline{S}$) condition. TI's ALS and AS versions of the 845 feature "power-up high impedance state." Corresponding bus register: 825.

846   *8-bit 3-state inverting latches with preset, clear, and multiuser controls.* Same as 845, but inverting. NOTE: The outputs of the 846
are conventionally considered to be noninverting and the act of
logical inversion is considered to take place at the inputs instead.
This is logically the same as calling the data inputs normal and the
outputs inverting, but it has the advantage of keeping the $\overline{S}$ and $\overline{R}$
pin assignments and functional descriptions the same as those of
the 845, making the rule easy to remember: in both the 845 and the
846, a LOW at $\overline{S}$ (pin 14) will set all Q HIGH and a LOW at $\overline{R}$ (pin
11) will set all Q LOW. (This is true regardless of what the inputs
and outputs are called, but giving them the usual labels D and $\overline{Q}$
would necessitate switching the names of the $\overline{S}$ and $\overline{R}$ pins and
make the rule governing them more difficult to remember.) Corresponding bus register: 826.

FUNCTION TABLE
845

INPUTS							OUTPUTS	
$\overline{OE1}$	$\overline{OE2}$	$\overline{OE3}$	$\overline{S}$	$\overline{R}$	E	D	Q	
L	L	L	H	H	H	L	L	Driver mode
L	L	L	H	H	H	H	H	
L	L	L	H	H	L	.	Q_0	Latched (E is LOW)
L	L	L	L	.	.	.	H	Direct set
L	L	L	H	L	.	.	L	Direct reset
H	.	.	.	.	.	.	Z	
.	H	.	.	.	.	.	Z	Outputs off
.	.	H	.	.	.	.	Z	

. = Either LOW or HIGH level
Z = High impedance (disabled)

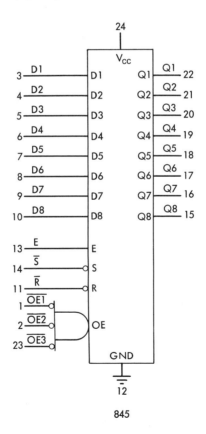

845

FUNCTION TABLE
846

INPUTS							OUTPUTS	
$\overline{OE1}$	$\overline{OE2}$	$\overline{OE3}$	$\overline{S}$	$\overline{R}$	E	$\overline{D}$	Q	
L	L	L	H	H	H	H	L	Driver mode
L	L	L	H	H	H	L	H	
L	L	L	H	H	L	.	Q_0	Latched (E is LOW)
L	L	L	L	.	.	.	H	Direct set
L	L	L	H	L	.	.	L	Direct reset
H	.	.	.	.	.	.	Z	Outputs off
.	H	.	.	.	.	.	Z	
.	.	H	.	.	.	.	Z	

. = Either LOW or HIGH level
Z = High impedance (disabled)

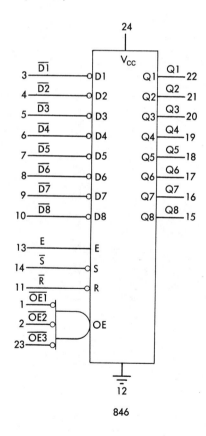

846

MANUFACTURERS	TTL	LS	HC	HCT	ALS	*	AS	F	AC	ACT
Advanced Micro Devices					███	███				
Fairchild								D	▒▒	▒▒
National Semiconductor							▒▒			
Signetics							▒▒			
Texas Instruments					███					

ALS -1 version available from TI
The * column refers to the Am29845 (comparable to S and AS)
The AS column includes the Am29845A, which is similar to the 74AS845

KEY PARAMETERS		TTL	LS	HC	HCT	ALS	*	AS	F	AC	ACT
I_{OH} (max.)	mA					-2.6	-24	-24	-3	-24	-24
I_{OL} (max.)	mA					24[a]	48	48	24	24	24
I_{IH} (max.)	mA					0.02	0.05	0.02[d]	0.07[i]		
I_{IL} (max.)	mA					-0.1	-1	-0.5	-0.07[j]		
I_{CC} (max.)	mA					72	120	85[e]	75		
I_{CC} (quiesc.)	mA									0.08	0.08
C_{pd} (typ.)	pF/gate									n/a	n/a
t_W E HIGH (min.)	ns					20[b]	6[c]	4[f]	n/a	n/a	n/a
t_{SU} (min.)	ns					10	2.5	2.5	n/a	n/a	n/a
t_H (min.)	ns					5	2.5	2.5	n/a	n/a	n/a
t_{pd} D to Q (max.)	ns					20	10	9	n/a	n/a	n/a
t_{pd} E to Q (max.)	ns					27	12	12	n/a	n/a	n/a
t_{pd} $\overline{S}$ to Q (max.)	ns					26	12	10[g]	n/a	n/a	n/a
t_{pd} $\overline{R}$ to Q (max.)	ns					23	21	13	n/a	n/a	n/a
t_{en} (max.)	ns					18	14	13.5[h]	n/a	n/a	n/a
t_{dis} (max.)	ns					12	15	8	n/a	n/a	n/a

[a] Versions with -1 suffix have I_{OL} (max.) = 48 mA if V_{CC} is kept between 4.75 V and 5.25 V
[b] $\overline{S}$ or $\overline{R}$ LOW (min.) 35 ns
[c] $\overline{S}$ or $\overline{R}$ LOW (min.) 6 ns
[d] AMD 0.05 mA
[e] AMD 97 mA

[f] $\overline{R}$ LOW (min.) also 4 ns; $\overline{S}$ LOW (min.) 4 ns (National, TI) or 5 ns (AMD)
[g] AMD 12 ns
[h] National and AMD 11.5 ns
[i] Including 0.05 mA off-state output current
[j] Including -0.05 mA off-state output current

846   8-bit 3-state inverting latches with preset, clear, and mult. controls

MANUFACTURERS	TTL	LS	HC	HCT	ALS	*	AS	F	AC	ACT
Advanced Micro Devices						███				
Fairchild								D	▒▒▒	
National Semiconductor							▒▒▒			
Signetics								▒▒▒		
Texas Instruments					███		███			

ALS -1 version available from TI
The * column refers to the Am29846 (comparable to S and AS)

KEY PARAMETERS		TTL	LS	HC	HCT	ALS	*	AS	F	AC	ACT
I_{OH} (max.)	mA					-2.6	-24	-24	-3	-24	-24
I_{OL} (max.)	mA					24[a]	48	48	48	24	24
I_{IH} (max.)	mA					0.02	0.05	0.02	0.07[h]		
I_{IL} (max.)	mA					-0.1	-1	-0.5	-0.07[i]		
I_{CC} (max.)	mA					72	120	87[d]	75		
I_{CC} (quiesc.)	mA									0.08	0.08
C_{pd} (typ.)	pF/gate									n/a	n/a
t_W E HIGH (min.)	ns					20[b]	6[c]	4[e]	n/a	n/a	n/a
t_{SU} (min.)	ns					10	2.5	2.5	n/a	n/a	n/a
t_H (min.)	ns					5	2.5	2.5	n/a	n/a	n/a
t_{pd} $\overline{D}$ to Q (max.)	ns					20	10	10[f]	n/a	n/a	n/a
t_{pd} E to Q (max.)	ns					27	12	13[g]	n/a	n/a	n/a
t_{pd} $\overline{S}$ to Q (max.)	ns					26	12	10[f]	n/a	n/a	n/a
t_{pd} $\overline{R}$ to Q (max.)	ns					23	21	13.5[f]	n/a	n/a	n/a
t_{en} (max.)	ns					18	14	13.5[f]	n/a	n/a	n/a
t_{dis} (max.)	ns					12	15	8[f]	n/a	n/a	n/a

[a] Versions with -1 suffix have I_{OL} (max.) = 48 mA if V_{CC} is kept between 4.75 V and 5.25 V
[b] $\overline{S}$ or $\overline{R}$ LOW (min.) 35 ns
[c] $\overline{S}$ or $\overline{R}$ LOW (min.) 8 ns
[d] National 53 mA

[e] $\overline{S}$ or $\overline{R}$ LOW (min.) also 4 ns
[f] Max. data for National not available
[g] National data not available
[h] Including 0.05 mA off-state output current
[i] Including -0.05 mA off-state output current

8-BIT SPECIAL-FUNCTION LATCHES

Advances in circuit integration have made possible several types of "educated" 8-bit latches (most of them single-sourced) that add other useful functions to the basic parallel latch. In certain applications these devices can make possible a substantially decreased component count by incorporating functions that would otherwise have to be implemented by added circuitry.

412   *8-bit 3-state multi-mode latches.* S and F versions of the popular Intel 8212, a type specifically designed for implementing bus-orga-

nized I/O ports. Each 24-pin DIP contains an 8-bit 3-state latch controlled by mutually gated mode (M), device select ($\overline{SE1}$, SE2), and strobe (STB) inputs. By using an appropriate combination of these controls, the latches can be made to receive data at the D inputs without transmitting it, transmit data at the Q outputs without receiving it, operate in a fully transparent mode, or operate in the basic enabled (strobed) mode. See function table and manufacturer's data (TI, Fairchild, Intel) for further details. Also included is a status flip-flop in which the output is LOW when the device is selected or when a strobe input is received. This output, $\overline{INT}$, can be used to generate an interrupt request or indicate a busy condition. The S version from Texas Instruments features high-level outputs (typically 4 V) for directly driving low-threshold MOS devices.

432   *8-bit 3-state inverting multi-mode latches.* Inverting version of the F412.

793§   *8-bit latches with readback.* LS and ACT components specifically designed for I/O operations on microprocessor data buses. The pinout is identical to the standard bus-structured 8-bit latches (573, 563, etc.). The general functioning of the latch is also very similar except that pin 1, the output enable ($\overline{OE}$), is not the usual 3-state enable, but rather a readback enable that puts the latched data from the Q outputs back out at the corresponding D inputs. Thus, in the latched mode (latch enable E in the LOW state), taking $\overline{OE}$ LOW will place the byte stored in the latch back onto the bus from which it came, allowing the microprocessor CPU to quickly read the latched data for verification or updating. See manufacturer's data for further information. Corresponding bus register: 794§. (Monolithic Memories)

LS793§:

I_{OH} (max.)	−2.6 mA	t_W	(min.)	15 ns
I_{OL} (max.)	24 mA	t_{SU}	(min.)	10 ns
I_{IH} (max.)	0.04 mA	t_H	(min.)	10 ns
I_{IL} (max.)	−0.25 mA	t_{pd} D to Q (max.)		18 ns
I_{CC} (max.)	120 mA	t_{pd} E to Q (max.)		25 ns
		t_{en}	(max.)	20 ns
		t_{dis}	(max.)	20 ns

ACT793§:

I_{OH} (max.)	−2.6 mA	t_W	(min.)	15 ns
I_{OL} (max.)	24 mA	t_{SU}	(min.)	8 ns
I_{CC} (max.)	0.08 mA	t_H	(min.)	8 ns
		t_{pd} D to Q	(max.)	40 ns
		t_{pd} E to Q	(max.)	40 ns
		t_{en}	(max.)	30 ns
		t_{dis}	(max.)	33 ns

549§   *8/8-bit 3-state pipelined latches.* LS components that provide two
8-bit latches in the same 24-pin DIP. Internal multiplexing circuitry allows the latches to be configured for nose-to-tail or side-
by-side operation. Applications include two-stage buffers for
pipelined I/O, automatic backup storage for diagnostic purposes,
etc. See manufacturer's data for further information. Corresponding bus register: 548§. (Monolithic Memories)

I_{OH} (max.)	−2.6 mA	t_W	(min.)	16 ns
I_{OL} (max.)	32 mA	t_{SU}	(min.)	6 ns
I_{IH} (max.)	0.02 mA	t_H	(min.)	10 ns
I_{IL} (max.)	−0.25* mA	t_{pd} D to Q	(max.)	18 ns
I_{CC} (max.)	160 mA	t_{pd} E† to Q	(max.)	25 ns
		t_{en}	(max.)	20 ns
		t_{dis}	(max.)	17 ns

* D or Y (Q); all others −0.4 mA

† Or G

INPUTS						OUTPUTS	
$\overline{SE1}$	SE2	$\overline{R}$	M	STB	D	Q	
L	H	H	H	.	L	L	Transparent driver
L	H	H	H	.	H	H	
L	H	H	L	H	L	L	Strobed driver
L	H	H	L	H	H	H	
H	L	H	H	.	.	Q_0	Hold data
L	H	H	L	L	.	Q_0	
H	.	L	H	.	.	L	Clear (reset)
L	H	L	L	L	.	L	
H	.	.	L	.	.	Z	Deselect (outputs off)
.	L	.	L	.	.	Z	

FUNCTION TABLE
412 status latch

INPUTS				OUTPUTS
$\overline{SE1}$	SE2	$\overline{R}$	STB	$\overline{INT}$
H	.	L	.	H
.	L	L	.	H
.	.	H	↟	L
L	H	H	.	L

. = Either LOW or HIGH level
Z = High impedance (disabled)

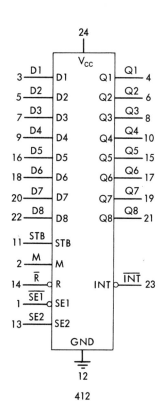

412

FUNCTION TABLE
432

INPUTS						OUTPUTS	
$\overline{SE1}$	SE2	$\overline{R}$	M	STB	D	$\overline{Q}$	
L	H	H	H	.	L	H	Transparent driver
L	H	H	H	.	H	L	
L	H	H	L	H	L	H	Strobed driver
L	H	H	L	H	H	L	
H	L	H	H	.	.	$\overline{Q}_0$	Hold data
L	H	H	L	L	.	$\overline{Q}_0$	
H	.	L	H	.	.	H	Clear (reset)
L	H	L	L	L	.	H	
H	.	.	L	.	.	Z	Deselect (outputs off)
.	L	.	L	.	.	Z	

FUNCTION TABLE
432 status latch

INPUTS				OUTPUTS
$\overline{SE1}$	SE2	$\overline{R}$	STB	$\overline{INT}$
H	.	L	.	H
.	L	L	.	H
.	.	H	↑	L
L	H	H	.	L

. = Either LOW or HIGH level
Z = High impedance (disabled)

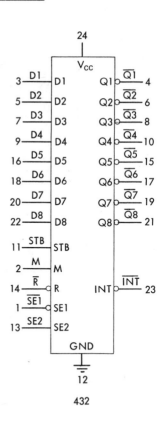

432

268

412   8-bit 3-state multi-mode latches

MANUFACTURERS	TTL	LS	HC	HCT	ALS	S	AS	F	AC	ACT
Fairchild								■		
Signetics								▦		
Texas Instruments						■				

KEY PARAMETERS		TTL	LS	HC	HCT	ALS	S	AS	F	AC	ACT
I_{OH} (max.)	mA						-1		-3		
I_{OL} (max.)	mA						20		24		
I_{IH} (max.)	mA						0.01		0.02		
I_{IL} (max.)	mA						-0.25[a]		-0.6		
I_{CC} (max.)	mA						130		60		
t_W (min.)	ns						25		9		
t_{SU} (min.)	ns						15		1		
t_H (min.)	ns						20		9		
t_{pd} (max.)	ns						b		9.5[d]		
t_{en} (max.)	ns						c		19[e]		
t_{dis} (max.)	ns						20		15[f]		

[a] Mode input -0.75 mA, $\overline{SE1}$ input -1 mA

[b] Data at C_L = 45 pF or 50 pF not available. At C_L = 30 pF, t_{pd} D to Q (max.) = 20 ns, t_{pd} any control input to Q (max.) = 27 ns

[c] Data at C_L = 45 pF or 50 pF not available. At C_L = 30 pF, t_{en} (max.) = 40 ns

[d] D to Q; max. 17.5 ns from $\overline{MR}$ to Q, 20.5 ns from control inputs to Q

[e] $\overline{SE1}$ to Q; max. 17.5 ns from SE2 to Q, 12 ns from M to Q

[f] $\overline{SE1}$ to Q; max. 13 ns from SE2 to Q, 12 ns from M to Q

432   8-bit 3-state inverting multi-mode latches

MANUFACTURERS	TTL	LS	HC	HCT	ALS	S	AS	F	AC	ACT
Fairchild								■		
Signetics								▦		

KEY PARAMETERS		TTL	LS	HC	HCT	ALS	S	AS	F	AC	ACT
I_{OH} (max.)	mA								-3		
I_{OL} (max.)	mA								24		
I_{IH} (max.)	mA								0.02		
I_{IL} (max.)	mA								-0.6		
I_{CC} (max.)	mA								65		
t_W (min.)	ns								9		
t_{SU} (min.)	ns								1[a]		
t_H (min.)	ns								9[b]		
t_{pd} (max.)	ns								12[c]		
t_{en} (max.)	ns								20		
t_{dis} (max.)	ns								17.5		

[a] Fairchild 0 ns

[b] Fairchild 12.5 ns

[c] D to Q; max. 20.5 ns from $\overline{MR}$ to Q, 23 ns from control inputs to Q

9-BIT LATCHES

Byte-oriented systems operate on data in 8-bit groups and are therefore usually well served by 8-bit components like the 8-bit latches listed above. In many applications, however, one or two extra bits per byte are needed for parity checking, expanded addressing, extra control lines, or other special requirements, resulting in buses whose lines are organized in groups of nine or ten. A minimum set of 9-bit and 10-bit latches is available for these applications. The 9-bit latches are listed below, and the 10-bit latches follow separately.

Both of the 9-bit designs described here were originally manufactured by Advanced Micro Devices, which continues to provide them in its 29800, 29800A, and 29C800 series of bus interface components. While not identical to the normal 7400-series technologies, the AMD versions are fully TTL-compatible and provide an important source for these types. In the data pages that follow, the 29843 and 29844 (comparable to S or AS) are described in the columns usually reserved for S components; the improved 29843A is included with the 74AS843, which it strongly resembles; and the 29C843 (high-speed TTL-compatible CMOS) is included in the ACT column.

All of these 9-bit latches are 3-state devices supplied in 24-pin DIPs.

843   *9-bit 3-state latches with preset and clear.* Implementations of the basic 9-bit bus latch. Data input (D) and output (Q) pins are assigned in a symmetrical configuration similar to that of the bus-structured 8-bit types (563, 573, 580). Assignment of latch enable (E), active-LOW output enable ($\overline{OE}$), V_{CC}, and ground pins leaves two remaining control inputs, which are assigned the sometimes very useful functions of direct preset (set), $\overline{S}$, and direct clear (reset), $\overline{R}$. TI's ALS and AS versions of the 843 feature "power-up high impedance state." Corresponding bus register: 823.

844   *9-bit 3-state inverting latches with preset and clear.* Same as 843, but inverting. NOTE: As in the case of 8-bit inverting latches with preset and clear, the 844 is considered to have noninverted outputs (Q), the inversion being considered to take place at the inputs ($\overline{D}$) instead. This convention has no effect on the basic data logic—in either case, an output always shows the inverse or logical complement of its input—but it keeps the pin labeling and logical descriptions of the 844's $\overline{S}$ and $\overline{R}$ inputs identical to those for the 843. Corresponding bus register: 824.

FUNCTION TABLE
843

INPUTS					OUTPUTS	
$\overline{OE}$	$\overline{S}$	$\overline{R}$	E	D	Q	
L	H	H	H	L	L	Driver Mode
L	H	H	H	H	H	
L	H	H	L	.	Q_0	Latched (E is LOW)
L	L	.	.	.	H	Direct set
L	H	L	.	.	L	Direct reset
H	.	.	.	.	Z	Outputs off

. = Either LOW or HIGH level
Z = High impedance (disabled)

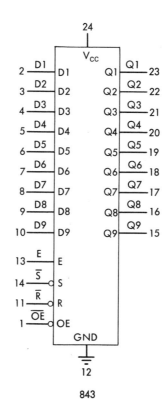

843

FUNCTION TABLE
844

INPUTS					OUTPUTS	
$\overline{OE}$	$\overline{S}$	$\overline{R}$	E	$\overline{D}$	Q	
L	H	H	H	H	L	Driver mode
L	H	H	H	L	H	
L	H	H	L	.	Q_0	Latched (E is LOW)
L	L	.	.	.	H	Direct set
L	H	L	.	.	L	Direct reset
H	.	.	.	.	Z	Outputs off

. = Either LOW or HIGH level
Z = High impedance (disabled)

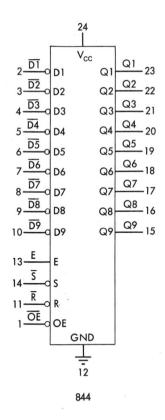

844

843   9-bit 3-state latches with preset and clear

MANUFACTURERS	TTL	LS	HC	HCT	ALS	*	AS	F	AC	ACT
Advanced Micro Devices						■				
Fairchild								D	■	■
National Semiconductor							■			
Signetics								■		
Texas Instruments					■					

ALS -1 version available from TI
The * column refers to the Am29843 (comparable to S and AS)
The AS column includes the Am29843A, which is similar to the 74AS843
The ACT column includes the Am29C843, which is comparable to the 74ACT843

KEY PARAMETERS		TTL	LS	HC	HCT	ALS	*	AS	F	AC	ACT
I_{OH} (max.)	mA					-2.6	-24	-24	-3	-24	-24^k
I_{OL} (max.)	mA					24^a	48	48	48	24	24
I_{IH} (max.)	mA					0.02	0.05	0.02^d	0.07^i		
I_{IL} (max.)	mA					-0.1	-1	-0.5	-0.07^j		
I_{CC} (max.)	mA					67	120	92^e	75		
I_{CC} (quiesc.)	mA									0.08	0.08^l
C_{pd} (typ.)	pF/gate									n/a	m
t_W E HIGH (min.)	ns					20^b	6^c	4^f	n/a	n/a	6^n
t_{SU} (min.)	ns					10	2.5	2.5	n/a	n/a	3^o
t_H (min.)	ns					5	2.5	2.5	n/a	n/a	4^o
t_{pd} D to Q (max.)	ns					18	10	9	n/a	n/a	11^o
t_{pd} E to Q (max.)	ns					26	12	12	n/a	n/a	12^o
t_{pd} $\overline{S}$ to Q (max.)	ns					30	12	10^g	n/a	n/a	13^o
t_{pd} $\overline{R}$ to Q (max.)	ns					30	21	13	n/a	n/a	12°
t_{en} (max.)	ns					14	14	13.5^h	n/a	n/a	12°
t_{dis} (max.)	ns					12	15	8	n/a	n/a	12°

[a] Version with -1 suffix has I_{OL} (max.) = 48 mA if V_{CC} is kept between 4.75 V and 5.25 V

[b] $\overline{S}$ or $\overline{R}$ LOW (min.) 35 ns

[c] $\overline{S}$ or $\overline{R}$ LOW (min.) 6 ns

[d] AMD 0.05 mA

[e] AMD 97 mA

[f] $\overline{R}$ LOW (min.) also 4 ns; $\overline{S}$ LOW (min.) 4 ns (National, TI) or 5 ns (AMD)

[g] AMD 12 ns

[h] National and AMD 11.5 ns

[i] Including 0.05 mA off-state output current

[j] Including -0.05 mA off-state output current

[k] AMD -15 mA

[l] AMD 0.12 mA

[m] AMD specifies dynamic supply current I_{CCD} (max.) of 0.275 mA/MHz/bit; Fairchild C_{pd} not available

[n] E input; t_W $\overline{S}$ or $\overline{R}$ LOW (min.) is 8 ns (AMD). Comparable Fairchild data not available

[o] AMD; comparable Fairchild data not available

844   9-bit 3-state inverting latches with preset and clear

MANUFACTURERS	TTL	LS	HC	HCT	ALS	*	AS	F	AC	ACT
Advanced Micro Devices						▮				
Fairchild									▨	▨
National Semiconductor							▨			
Signetics								▨		
Texas Instruments					▮		▮			

ALS -1 version available from TI
The * column refers to the Am29844 (comparable to S and AS)

KEY PARAMETERS		TTL	LS	HC	HCT	ALS	*	AS	F	AC	ACT
I_{OH} (max.)	mA					-2.6	-24	-24	-3	-24	-24
I_{OL} (max.)	mA					24[a]	48	48	48	24	24
I_{IH} (max.)	mA					0.02	0.05	0.02	0.07[e]		
I_{IL} (max.)	mA					-0.1	-1	-0.5	-0.07[f]		
I_{CC} (max.)	mA					72	120	95	75		
I_{CC} (quiesc.)	mA									0.08	0.08
C_{pd} (typ.)	pF/gate									n/a	n/a
t_W E HIGH (min.)	ns					20[b]	6[c]	4[d]	n/a	n/a	n/a
t_{SU} (min.)	ns					10	2.5	2.5	n/a	n/a	n/a
t_H (min.)	ns					5	2.5	2.5	n/a	n/a	n/a
t_{pd} $\overline{D}$ to Q (max.)	ns					20	10	10	n/a	n/a	n/a
t_{pd} E to Q (max.)	ns					29	12	13	n/a	n/a	n/a
t_{pd} $\overline{S}$ to Q (max.)	ns					30	12	10	n/a	n/a	n/a
t_{pd} $\overline{R}$ to Q (max.)	ns					30	21	13.5	n/a	n/a	n/a
t_{en} (max.)	ns					20	14	13.5	n/a	n/a	n/a
t_{dis} (max.)	ns					12	15	8	n/a	n/a	n/a

[a] Version with -1 suffix has I_{OL} (max.) = 48 mA if V_{CC} is kept between 4.75 V and 5.25 V
[b] $\overline{S}$ or $\overline{R}$ LOW (min.) 35 ns
[c] $\overline{S}$ or $\overline{R}$ LOW (min.) 8 ns
[d] $\overline{S}$ or $\overline{R}$ LOW (min.) also 4 ns
[e] Including 0.05 mA off-state output current
[f] Including -0.05 mA off-state output current

10-BIT LATCHES

Like the 9-bit latches, these 10-bit types are all 3-state bus-driving devices supplied in 24-pin DIPs. They provide the widest data path possible in a nonmultiplexed configuration limited to 24 pins. Since two additional pins are needed for the tenth input and output lines, the preset and clear functions provided in the 9-bit versions must be omitted to keep the 24-pin package, leaving the 10-bit types with just the basic 3-state latch functions.

Advanced Micro Devices originated and remains an important source for these components as parts of its 29800, 29800A, and 29C800 families of TTL-compatible bus interface circuits. As with the 9-bit latches, the 29841 and 29842 (comparable to S or AS devices) are listed in the columns usually reserved for S components, while the 29C841 (high-speed TTL-compatible CMOS) is included with the 74ACT841, and the 29841A is included with the 74AS841.

841   *10-bit 3-state latches.* Components with a simple bus-structured 10-line data pinout very similar to that of the 9-bit latches, plus the minimum latch enable (E) and active-LOW output enable ($\overline{\text{OE}}$) controls. Logic is identical to that of the basic 8-bit types (373, 573, etc.). TI's ALS and AS versions of the 841 feature "power-up high impedance state." Corresponding bus register: 821.

842   *10-bit 3-state inverting latches.* Same as 841, but inverting. Logic is identical to that of the basic 8-bit inverting types (533, 563, etc.). NOTE: The inverting function of the 842 is conventionally ascribed to the inputs ($\overline{\text{D}}$) rather than the outputs (Q) to keep the style identical to that adopted for the 9-bit latches (see above).

FUNCTION TABLE
841

INPUTS			OUTPUTS
$\overline{OE}$	E	D	Q
L	H	L	L
L	H	H	H
L	L	.	Q_0
H	.	.	Z

. = Either LOW or HIGH level
Z = High impedance (disabled)

FUNCTION TABLE
842

INPUTS			OUTPUTS
$\overline{OE}$	E	$\overline{D}$	Q
L	H	H	L
L	H	L	H
L	L	.	Q_0
H	.	.	Z

. = Either LOW or HIGH level
Z = High impedance (disabled)

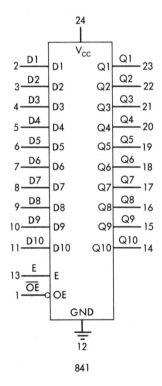

841

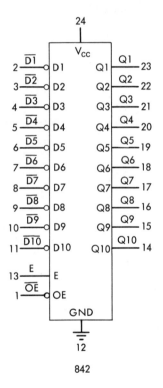

842

841 10-bit 3-state latches

MANUFACTURERS	TTL	LS	HC	HCT	ALS	*	AS	F	AC	ACT
Advanced Micro Devices						████				
Fairchild								D	▒▒▒▒	▒▒▒▒
National Semiconductor							▒▒▒▒			
Signetics								▒▒▒▒		
Texas Instruments					████					

ALS -1 version available from TI
The * column refers to the Am29841 (comparable to S and AS)
The AS column includes the Am29841A, which is similar to the 74AS841
The ACT column includes the Am29C841, which is comparable to the 74ACT841

KEY PARAMETERS		TTL	LS	HC	HCT	ALS	*	AS	F	AC	ACT
I_{OH} (max.)	mA					-2.6	-24	-24	-3	-24	-24^g
I_{OL} (max.)	mA					24^a	48	48	48	24	24
I_{IH} (max.)	mA					0.02	0.05	0.02^b	0.07^e		
I_{IL} (max.)	mA					-0.1	-1	-0.5	-0.07^f		
I_{CC} (max.)	mA					62	120	94^c	75		
I_{CC} (quiesc.)	mA									0.08	0.08^h
C_{pd} (typ.)	pF/gate									n/a	i
t_W (min.)	ns					20	6	4	n/a	n/a	6^j
t_{SU} (min.)	ns					10	2.5	2.5	n/a	n/a	3^k
t_H (min.)	ns					5	2.5	2.5	n/a	n/a	4^k
t_{pd} D to Q (max.)	ns					13	10	9	n/a	n/a	11^k
t_{pd} E to Q (max.)	ns					26	12	12	n/a	n/a	12^k
t_{en} (max.)	ns					12	14	13.5^d	n/a	n/a	12^k
t_{dis} (max.)	ns					12	15	8	n/a	n/a	12^k

[a] Version with -1 suffix has I_{OL} (max.) = 48 mA if V_{CC} is kept between 4.75 V and 5.25 V
[b] AMD 0.05 mA
[c] AMD 97 mA
[d] National and AMD 11.5 ns
[e] Including 0.05 mA off-state output current
[f] Including -0.05 mA off-state output current

[g] AMD -15 mA
[h] AMD 0.12 mA
[i] AMD specifies dynamic supply current I_{CCD} (max.) of 0.275 mA/MHz/bit; Fairchild C_{pd} not available
[j] E input; t_W $\overline{S}$ or $\overline{R}$ LOW (min.) is 8 ns (AMD). Comparable Fairchild data not available
[k] AMD; comparable Fairchild data not available

MANUFACTURERS	TTL	LS	HC	HCT	ALS	*	AS	F	AC	ACT
Advanced Micro Devices						■				
Fairchild									▨	▨
National Semiconductor							▨			
Signetics								▨		
Texas Instruments					■		■			

ALS -1 version available from TI
The * column refers to the Am29842 (comparable to S and AS)

KEY PARAMETERS		TTL	LS	HC	HCT	ALS	*	AS	F	AC	ACT
I_{OH} (max.)	mA					-2.6	-24	-24	-3	-24	-24
I_{OL} (max.)	mA					24^a	48	48	48	24	24
I_{IH} (max.)	mA					0.02	0.05	0.02	0.07^b		
I_{IL} (max.)	mA					-0.1	-1	-0.5	-0.07^c		
I_{CC} (max.)	mA					74	120	97	75		
I_{CC} (quiesc.)	mA									0.08	0.08
C_{pd} (typ.)	pF/gate									n/a	n/a
t_W (min.)	ns					20	6	4	n/a	n/a	n/a
t_{SU} (min.)	ns					10	2.5	2.5	n/a	n/a	n/a
t_H (min.)	ns					5	2.5	2.5	n/a	n/a	n/a
t_{pd} $\overline{D}$ to Q (max.)	ns					18	10	9	n/a	n/a	n/a
t_{pd} E to Q (max.)	ns					27	12	12	n/a	n/a	n/a
t_{en} (max.)	ns					12	14	12.5	n/a	n/a	n/a
t_{dis} (max.)	ns					12	15	8	n/a	n/a	n/a

a Version with -1 suffix has I_{OL} (max.) = 48 mA if V_{CC} is kept between 4.75 V and 5.25 V

b Including 0.05 mA off-state output current
c Including -0.05 mA off-state output current

6

Registers

The word "register" is used in this book to refer to all packages of edge-triggered D-type flip-flops in which a group of flip-flops is triggered in parallel by a common clock pulse (CP). Structurally, registers strongly resemble the parallel latches of Chapter 5, and in many cases the only external difference between a given register and the corresponding latch is the appearance of an edge-triggered CP input in place of the level-triggered latch enable.

Functionally, however, registers differ from parallel latches in the way they read input data. Unlike latches, whose outputs reflect changing input data as long as the latch enable remains at its active level, registers read their inputs only at the instant that the clock pulse undergoes a triggering transition, which in all the 7400-series registers is the positive (LOW-to-HIGH) edge of the clock pulse. For proper operation, valid data must be present at the D inputs at least one setup time t_{SU} before the positive edge of the clock pulse and remain there at least one hold time t_H thereafter. (See data sheets.)

As with latches, the category of registers includes ordinary 2-state types, as well as the open-collector types and the more common 3-state bus-oriented devices. In the case of registers, however, the 2-state standard-output types are important enough to be gathered here in a separate section of their own. The 3-state and open-collector types, which are all specifically designed to drive bus lines and other highly capacitive loads, have been separately categorized as "bus registers" and can be found in Chapter 7.

The 2-state registers in this section are used where multiple D flip-flops are triggered by a common clock pulse. This includes applications where a number of individual flip-flops are needed to perform one-line functions, but can be actuated by a single system clock. In these instances, a substantial reduction in component count can be achieved by using one of these registers in place of a collection of separately housed flip-flops of the kind described in Chapter 2. More commonly, however, these devices are used to implement word-oriented working registers of various kinds (instruction registers, storage registers, shift registers, pattern generators, etc.) within communications and information-processing equipment.

Since the outputs of the devices listed in this section are the ordinary 2-state "totem-pole" kind, the timing data presented here are valid under test conditions associated with the ordinary gates and flip-flops of Part I. In particular, it should be remembered that TTL, LS, and S performance is specified using a load capacitance of 15 pF at 25°C, while all other versions are tested at $C_L = 50$ pF over the entire operating temperature range.

The registers in this section are provided in three basic configurations: a 4-bit arrangement with complementary outputs housed in a 16-pin package; a 6-bit form with noninverting outputs, also in a 16-pin package; and an 8-bit configuration with noninverting outputs, housed in a 20-pin package to accommodate the two extra input and output lines. In each case two pins are available for control functions, one of which is used for the clock input. Each of the three main configurations is available in two varieties, depending on the function assigned to the remaining control input. In one variety, this pin is given an asynchronous clear (reset) function; in the other, it performs as a clock enable.

The various kinds of registers resulting from these different options are summed up in Table 6.1.

TABLE 6.1

	Asynchronous Clear ($\overline{R}$) Input	Clock Enable ($\overline{CE}$) Input
4-bit registers with complementary outputs	175, 171§	379
6-bit registers with noninverting outputs	174	378
8-bit registers with noninverting outputs	273	377

4-BIT REGISTERS

These 2-state edge-triggered registers provide four D flip-flops activated by a common clock pulse (CP) input. Each flip-flop has both a noninverting output, Q, and an inverting output, $\overline{Q}$. As with the simpler gates and flip-flops, there is a slightly less propagation delay from the D inputs to the inverting $\overline{Q}$ outputs than to the noninverting Q outputs, presumably due to an extra inverting stage at each Q output that produces the externally noninverted signal. All versions of these 4-bit registers are available in 16-pin DIPs.

175   *4-bit registers with asynchronous clear.* The standard 4-bit register, available in most technologies and supported by most manufacturers. When taken LOW, the asynchronous clear (reset) input, $\overline{R}$, will override all other inputs and cause all four flip-flops to be directly loaded with a logic 0, storing a LOW at each Q output and a HIGH at each $\overline{Q}$ output.

171§   *4-bit registers with asynchronous clear.* An LS version of the 175 with a different pinout and slightly different input structure and switching characteristics. (Texas Instruments)

I_{OH} (max.)	−0.4	mA	t_W	(min.)	20 ns
I_{OL} (max.)	8	mA	t_{SU}	(min.)	20† ns
I_{IH} (max.)	0.02	mA	t_H	(min.)	5 ns
I_{IL} (max.)	−0.4*	mA	f_{MAX}	(min.)	20 MHz
I_{CC} (max.)	25	mA	t_{pd} D to Q (max.)		30 ns
			t_{pd} $\overline{R}$ to Q (max.)		40 ns

* D inputs; all others −0.2 mA
† D to CP; t_{SU} $\overline{R}$ LOW (min.) is 25 ns

379   *4-bit registers with clock enable.* Same as the 175, but with a clock enable ($\overline{CE}$) input in place of the asynchronous clear. If $\overline{CE}$ is taken HIGH, the clock (CP) input will be inhibited and the register will continue to maintain its previous outputs regardless of changes at the D inputs, just as if the clock were held LOW. Available in LS, HC, and F versions.

FUNCTION TABLE
175, 171§

INPUTS			OUTPUTS	
$\overline{R}$	CP	D	Q	$\overline{Q}$
H	↑	L	L	H
H	↑	H	H	L
H	–	•	NO CHANGE	
L	•	•	L	H

. = Either LOW or HIGH level
– = Clock inactive (L, H, or ↑)

FUNCTION TABLE
379

INPUTS			OUTPUTS	
$\overline{CE}$	CP	D	Q	$\overline{Q}$
L	↑	L	L	H
L	↑	H	H	L
L	–	•	NO CHANGE	
H	•	•	NO CHANGE	

. = Either LOW or HIGH level
– = Clock inactive (L, H, or ↑)

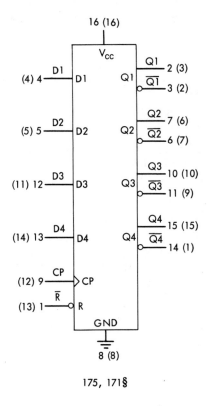

175, 171§

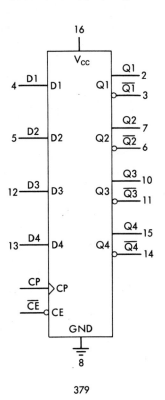

379

282

MANUFACTURERS	TTL	LS	HC	HCT	ALS	S	AS	F	AC	ACT
Fairchild										
Motorola										
National Semiconductor										
RCA									D	D
SGS										
Signetics										
Texas Instruments										
Toshiba									D	

Added change control symbol: TI 74AS175A

KEY PARAMETERS		TTL	LS	HC	HCT	ALS	S	AS	F	AC	ACT
I_{OH} (max.)	mA	-0.8	-0.4	-4	-4	-0.4	-1	-2	-1	-24	-24
I_{OL} (max.)	mA	16	8	4	4	8	20	20	20	24	24
I_{IH} (max.)	mA	0.04	0.02			0.02	0.05	0.02	0.02		
I_{IL} (max.)	mA	-1.6	-0.4[g]			-0.1	-2	-0.5	-0.6		
I_{CC} (max.)	mA	45	18			14	96	34	34		
I_{CC} (quiesc.)	mA			0.08[i]	0.08					0.08	0.08
C_{pd} (typ.)	pF/channel			65[m]	67[t]					45	45
t_W (min.)	ns	20[a]	20[h]	20[n]	25	10	7[x]	4[y]	5	n/a	4
t_{SU} (min.)	ns	20[b]	20[i]	25[o]	25[u]	10[w]	5	3[w]	3	n/a	2.5
t_H (min.)	ns	5[c]	5[c]	0[p]	5	0	3	1	1	n/a	1
f_{MAX} (min.)	MHz	25[d]	30	24[q]	20	50	75	100[z]	100	n/a	145
t_{pd} CP to Q (max.)	ns	35[e]	25[j]	41[r]	41	17	17	10	9.5	n/a	12
t_{pd} $\overline{R}$ to Q (max.)	ns	35[f]	30[k]	44[s]	50[v]	23	22	13[aa]	13	n/a	10.5

[a] National gives 10 ns for CP HIGH, 25 ns for CP LOW, and 20 ns for $\overline{R}$ LOW, Fairchild 20 ns for CP HIGH and 20 ns for $\overline{R}$ LOW

[b] D to CP. TI further states that t_{SU} $\overline{R}$ inactive (min.) is 25 ns

[c] National 0 ns

[d] National 30 MHz

[e] National 25 ns

[f] National 40 ns

[g] National -0.36 mA at D inputs, SGS -0.36 mA at all inputs

[h] Fairchild 15 ns for CP HIGH and 18 ns for $\overline{R}$ LOW

[i] D to CP (Fairchild 10 ns). TI further states that t_{SU} $\overline{R}$ inactive (min.) is 25 ns

[j] National 30 ns

[k] Fairchild 33 ns, National 35 ns

[l] SGS and Toshiba 0.04 mA

[m] RCA; TI 30 pF/channel, Signetics 32 pF/channel, Motorola 60 pF/channel, SGS and Toshiba 71 pF/channel.

National has 150 pF/package

[n] SGS 18 ns, Toshiba 19 ns

[o] SGS 12 ns, Toshiba 13 ns, RCA and Signetics 20 ns

[p] RCA and Signetics 5 ns

[q] SGS 20 ns, RCA and TI 25 ns

[r] Toshiba; Motorola, National, and TI 38 ns, RCA, Signetics, and SGS 44 ns

[s] RCA; Motorola and National 32 ns, Signetics and TI 38 ns, Toshiba 46 ns, SGS 47 ns

[t] RCA; Signetics 34 pF/channel

[u] RCA; Signetics 20 ns

[v] RCA; Signetics 48 ns

[w] D to CP; t_{SU} $\overline{R}$ inactive (min.) is 6 ns

[x] CP; t_W $\overline{R}$ (min.) is 10 ns (Fairchild 7 ns)

[y] CP HIGH; t_W CP LOW (min.) is 3 ns (National) or 5 ns (TI), and t_W $\overline{R}$ (min.) is 5 ns

[z] National 105 MHz

[aa] National 14 ns

379   4-bit registers with clock enable

MANUFACTURERS	TTL	LS	HC	HCT	ALS	S	AS	F	AC	ACT
Fairchild		■						■	▒	▒
Motorola		■								
SGS		■						■		
Signetics								■		
Texas Instruments		■								

KEY PARAMETERS		TTL	LS	HC	HCT	ALS	S	AS	F	AC	ACT
I_{OH} (max.)	mA		-0.4	-4					-1	-24	-24
I_{OL} (max.)	mA		8	4					20	24	24
I_{IH} (max.)	mA		0.02						0.02		
I_{IL} (max.)	mA		-0.4						-0.6		
I_{CC} (max.)	mA		15[a]						40		
I_{CC} (quiesc.)	mA			0.04						0.08	0.08
C_{pd} (typ.) pF/channel				30						n/a	n/a
t_W (min.)	ns		20[b]	25					5	n/a	n/a
t_{SU} (min.)	ns		20[c]	25					3[d]	n/a	n/a
t_H (min.)	ns		5	0					1[e]	n/a	n/a
f_{MAX} (min.)	MHz		30	20					100[f]	n/a	n/a
t_{pd} CP to Q (max.)	ns		27	40					9.5	n/a	n/a

[a] SGS 16 mA, Fairchild 18 mA

[b] Fairchild and SGS 17 ns

[c] D to CP; t_{SU} $\overline{CE}$ inactive (min.) is 10 ns, t_{SU} $\overline{CE}$ active (min.) 25 ns, except for Fairchild, which has t_{SU} $\overline{CE}$ HIGH or LOW of 25 ns, and SGS, which has t_{SU} $\overline{CE}$ HIGH or LOW of 30 ns

[d] D to CP; t_{SU} $\overline{CE}$ HIGH or LOW (min.) is 6 ns

[e] D inputs; t_H $\overline{CE}$ HIGH or LOW (min.) is 0 ns (Fairchild, Signetics) or 2 ns (Motorola)

[f] Signetics 90 MHz

6-BIT REGISTERS

These 2-state edge-triggered registers provide six flip-flops activated by a common clock pulse (CP) input. Like the 4-bit registers, these devices are supplied in 16-pin DIPs; the four pins used to provide inverted $\overline{Q}$ outputs in the 4-bit designs are here used to accommodate the extra pairs of input and output lines, so only a single noninverted Q output is externally available from each of the six flip-flops.

174   *6-bit registers with asynchronous clear.* The standard 6-bit register, available in nearly all technologies and supported by most manufacturers. When taken LOW, the asynchronous clear (reset) input, $\overline{R}$, will override all other inputs and cause all six flip-flops to be directly loaded with a logic 0, storing a LOW at each Q output.

378   *6-bit registers with clock enable.* Same as the 174, but with a clock enable ($\overline{CE}$) input in place of the asynchronous clear. If $\overline{CE}$

is taken HIGH, the clock (CP) input will be inhibited and the register will continue to maintain its previous outputs regardless of changes at the D inputs, just as if the clock were held LOW. Available in LS, HC, and F versions.

FUNCTION TABLE
174

INPUTS			OUTPUTS
$\overline{R}$	CP	D	Q
H	↑	L	L
H	↑	H	H
H	–	.	NO CHANGE
L	.	.	L

. = Either LOW or HIGH level
– = Clock inactive (L, H, or ↓)

FUNCTION TABLE
378

INPUTS			OUTPUTS
$\overline{CE}$	CP	D	Q
L	↑	L	L
L	↑	H	H
L	–	.	NO CHANGE
H	.	.	NO CHANGE

. = Either Low or HIGH level
– = Clock inactive (L, H, or ↓)

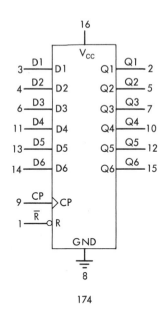

174

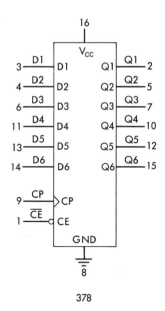

378

174   6-bit registers with asynchronous clear

MANUFACTURERS	TTL	LS	HC	HCT	ALS	S	AS	F	AC	ACT
Fairchild	■	□	■	□	□	□	□	■	■	□
Motorola	□	■	▨	□	□	□	□	□	■	■
National Semiconductor	■	■	□	□	■	□	▨	□	□	□
RCA	□	■	□	□	□	□	□	□	D	D
SGS	□	■	▨	□	□	□	□	□	□	□
Signetics	■	■	□	□	■	■	■	□	□	□
Texas Instruments	■	■	□	■	■	□	□	□	□	□
Toshiba	□	□	■	■	□	□	□	□	D	□

KEY PARAMETERS		TTL	LS	HC	HCT	ALS	S	AS	F	AC	ACT
I_{OH} (max.)	mA	-0.8	-0.4	-4	-4	-0.4	-1	-2	-1	-24	-24
I_{OL} (max.)	mA	16	8	4	4	8	20	20	20	24	24
I_{IH} (max.)	mA	0.04	0.02			0.02	0.05	0.02	0.02		
I_{IL} (max.)	mA	-1.6	-0.4[g]			-0.1	-2	-0.5	-0.6		
I_{CC} (max.)	mA	65	26			19	144	45	45		
I_{CC} (quiesc.)	mA			0.08[l]	0.08					0.08	0.08
C_{pd} (typ.)	pF/channel			38[m]	44[l]					85	85
t_W (min.)	ns	20[a]	20[h]	20[n]	25[u]	10	7[aa]	4[bb]	4[bb]	5	3.5
t_{SU} (min.)	ns	20[b]	20[i]	25[o]	25[v]	10[z]	5	3[z]	4	5.5	1.5
t_H (min.)	ns	5[c]	5[c]	5[p]	5	0	3	1	0	3	2
f_{MAX} (min.)	MHz	25[d]	30	22[q]	20[w]	50	75	100[cc]	80	100	140
t_{pd} CP to Q (max.)	ns	35[e]	30[j]	41[r]	44[x]	17	17	10	11	9.5	11.5
t_{pd} $\overline{R}$ to Q (max.)	ns	35[f]	35[k]	41[s]	44[y]	23	22	14	15	10.5	11

[a] National gives 10 ns for CP HIGH, 25 ns for CP LOW, and 20 ns for $\overline{R}$ LOW, Fairchild 20 ns for CP HIGH and 20 ns for $\overline{R}$ LOW

[b] D to CP. TI further states that t_{SU} $\overline{R}$ inactive (min.) is 25 ns

[c] National 0 ns

[d] National 30 MHz

[e] National 25 ns

[f] National 40 ns

[g] National -0.36 mA at D inputs, SGS -0.36 mA at all inputs

[h] Fairchild 18 ns for CP HIGH or $\overline{R}$ LOW

[i] D to CP (Fairchild 10 ns). TI further states that t_{SU} $\overline{R}$ inactive (min.) is 25 ns

[j] Fairchild and TI 25 ns

[k] TI 30 ns

[l] SGS and Toshiba 0.04 mA

[m] RCA; Signetics 17 pF/channel, TI 27 pF/channel, SGS and Toshiba 53 pF/channel. Motorola and National have 136 pF/package

[n] SGS 18 ns, Toshiba 19 ns

[o] Signetics 15 ns, SGS 18 ns, Toshiba 19 ns, RCA 20 ns

[p] TI and Toshiba 0 ns, Signetics 3 ns, SGS 6 ns

[q] Toshiba; SGS 20 MHz, Motorola and National 21 MHz, RCA and Signetics 24 MHz, TI 25 MHz

[r] TI 40 ns, Toshiba 45 ns, SGS 47 ns

[s] Motorola and National; RCA and Signetics 38 ns, TI 40 ns, Toshiba 45 ns, SGS 47 ns

[t] RCA; Signetics 17 pF/channel

[u] CP (RCA); Signetics 20 ns. t_W $\overline{R}$ (min.) is 25 ns (Signetics) or 31 ns (RCA)

[v] RCA; Signetics 20 ns

[w] RCA; Signetics 24 MHz

[x] Signetics; RCA 50 ns

[y] Signetics; RCA 55 ns

[z] D to CP; t_{SU} $\overline{R}$ inactive (min.) is 6 ns

[aa] CP; t_W $\overline{R}$ (min.) is 10 ns (Fairchild 7 ns)

[bb] CP HIGH; t_W CP LOW (min.) is 6 ns, and t_W $\overline{R}$ (min.) is 5 ns

[cc] National 105 MHz

378 6-bit registers with clock enable

MANUFACTURERS	TTL	LS	HC	HCT	ALS	S	AS	F	AC	ACT
Fairchild		■	■					■	▓	▓
Motorola		■						■		
SGS		■								
Signetics		■						■		
Texas Instruments		■	■							

KEY PARAMETERS		TTL	LS	HC	HCT	ALS	S	AS	F	AC	ACT
I_{OH} (max.)	mA		-0.4	-4					-1	-24	-24
I_{OL} (max.)	mA		8	4					20	24	24
I_{IH} (max.)	mA		0.02						0.02		
I_{IL} (max.)	mA		-0.4						-0.6		
I_{CC} (max.)	mA		22[a]						45		
I_{CC} (quiesc.)	mA			0.08						0.08	0.08
C_{pd} (typ.)	pF/channel			30						n/a	n/a
t_W (min.)	ns		20	25					6	n/a	n/a
t_{SU} (min.)	ns		20[b]	25					4[c]	n/a	n/a
t_H (min.)	ns		5	0					0[d]	n/a	n/a
f_{MAX} (min.)	MHz		30	20					80	n/a	n/a
t_{pd} CP to Q (max.)	ns		27	40					9.5	n/a	n/a

[a] Signetics 24 mA, SGS 27 mA

[b] D to CP; t_{SU} $\overline{CE}$ inactive (min.) is 10 ns, t_{SU} $\overline{CE}$ active (min.) is 25 ns, except Fairchild and SGS, which have t_{SU} $\overline{CE}$ HIGH or LOW (min.) of 30 ns

[c] D to CP; t_{SU} $\overline{CE}$ (min.) is 6 ns (Motorola) or 4 ns for $\overline{CE}$ HIGH and 10 ns for $\overline{CE}$ LOW (Fairchild, Signetics)

[d] D inputs; t_H $\overline{CE}$ (min.) is 0 ns (Fairchild, Signetics) or 2 ns (Motorola)

8-BIT REGISTERS

These 2-state edge-triggered registers provide eight D flip-flops activated by a common clock pulse (CP) input. As with the 6-bit designs, each flip-flop provides a single noninverted external Q output; to accommodate the extra pairs of input and output lines, each register is housed in a 20-pin DIP. Note that three 8-bit 273s can replace four 6-bit 174s, and three 8-bit 377s can replace four 6-bit 378s.

273 *8-bit registers with asynchronous clear.* The standard 8-bit register, available in most technologies and supported by most manufacturers. When taken LOW, the asynchronous clear (reset) input, $\overline{R}$, will override all other inputs and cause all eight flip-flops to be directly loaded with a logic 0, storing a LOW at each Q output.

377 *8-bit registers with clock enable.* Same as the 273, but with a clock enable ($\overline{CE}$) input in place of the asynchronous clear. If $\overline{CE}$

is taken HIGH, the clock (CP) input will be inhibited and the register will continue to maintain its previous outputs regardless of changes at the D inputs, just as if the clock were held LOW. Available in LS, HC, S, and F versions.

FUNCTION TABLE
273

INPUTS			OUTPUTS
$\overline{R}$	CP	D	Q
H	↑	L	L
H	↑	H	H
H	–	.	NO CHANGE
L	.	.	L

. = Either LOW or HIGH level
- = Clock inactive (L, H, or ↑)

FUNCTION TABLE
377

INPUTS			OUTPUTS
$\overline{CE}$	CP	D	Q
L	↑	L	L
L	↑	H	H
L	–	.	NO CHANGE
H	.	.	NO CHANGE

. = Either LOW or HIGH level
- = Clock inactive (L, H, or ↑)

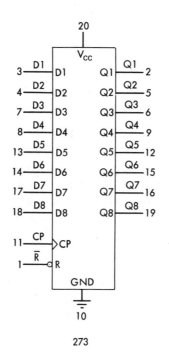

273

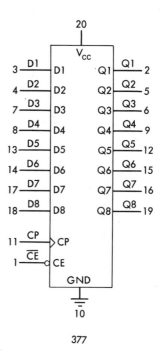

377

MANUFACTURERS	TTL	LS	HC	HCT	ALS	S	AS	F	AC	ACT
Fairchild		■						D		▨
Monolithic Memories		■								
Motorola			▨							
National Semiconductor			▨	▨	■					
RCA			■						D	D
SGS		■	▨							
Signetics		■						■		
Texas Instruments	■	■			■	■				
Toshiba		■	■						D	

KEY PARAMETERS		TTL	LS	HC	HCT	ALS	S	AS	F	AC	ACT
I_{OH} (max.)	mA	-0.8	-0.4	-4	-4	-2.6	-1		-1	-24	-24
I_{OL} (max.)	mA	16	8	4	4	24	20		20	24	24
I_{IH} (max.)	mA	0.04[a]	0.02			0.02	0.05		0.02		
I_{IL} (max.)	mA	-1.6[b]	-0.4			-0.2	-2[w]		-0.02		
I_{CC} (max.)	mA	94	27			29	150		88		
I_{CC} (quiesc.)	mA			0.08[f]	0.08					0.08	0.08
C_{pd} (typ.)	pF/channel			35[g]	25[n]					50	50
t_W (min.)	ns	16.5	20	20[h]	25[o]	14[u]	7[x]		4	4.5	n/a
t_{SU} (min.)	ns	20[c]	20[d]	25[i]	25[p]	10[v]	5		1.5	4.5	n/a
t_H (min.)	ns	5	5	3[j]	3[q]	0	3		0	1	n/a
f_{MAX} (min.)	MHz	30	30	21[k]	20[r]	35	75		120	125	n/a
t_{pd} CP to Q (max.)	ns	27	27[e]	40[l]	44[s]	15	15		12	11	n/a
t_{pd} $\overline{R}$ to Q (max.)	ns	27	27	40[m]	49[t]	18	15		10.5	10.5	n/a

[a] CP and D inputs; 0.08 mA at $\overline{R}$

[b] CP and D inputs; -3.2 mA at $\overline{R}$

[c] D to CP; t_{SU} $\overline{R}$ inactive (min.) is 25 ns

[d] D to CP (Fairchild 15 ns). TI further states that t_{SU} $\overline{R}$ inactive (min.) is 25 ns

[e] Fairchild 24 ns

[f] SGS and Toshiba 0.04 mA

[g] TI; Signetics 20 pF/channel, RCA 25 pF/channel, SGS and Toshiba 53 pF/channel, Motorola 60 pF/channel. National has 175 pF/package

[h] CP and $\overline{R}$ (SGS 18 ns, Toshiba 19 ns) for all except RCA, which gives 20 ns for t_W CP (min.) but 15 ns for t_W $\overline{R}$ (min.)

[i] RCA and Signetics 15 ns, SGS 18 ns, Toshiba 19 ns

[j] RCA and Signetics; Motorola, National, TI, and Toshiba 0 ns, SGS 6 ns

[k] SGS 20 MHz, Toshiba 22 MHz, Signetics 24 MHz, RCA 25 MHz

[l] RCA 38 ns, Toshiba 45 ns, SGS 47 ns

[m] RCA 38 ns, Toshiba 44 ns, SGS 47 ns

[n] RCA; Signetics 23 pF/channel. National data not available

[o] CP (RCA); National 20 ns, Signetics 33 ns. For t_W $\overline{R}$ (min.) RCA gives 15 ns, National 20 ns, Signetics 21 ns

[p] National; RCA 15 ns, Signetics 20 ns

[q] National 5 ns

[r] RCA; Signetics 15 MHz, National 22 MHz

[s] RCA 38 ns

[t] National; RCA 40 ns, Signetics 49 ns

[u] CP HIGH or LOW; t_W $\overline{R}$ LOW (min.) is 10 ns

[v] D to CP; t_{SU} $\overline{R}$ inactive (min.) is 15 ns

[w] Monolithic Memories -0.25 mA

[x] CP; t_W $\overline{R}$ (min.) is 10 ns

377   8-bit registers with clock enable

MANUFACTURERS	TTL	LS	HC	HCT	ALS	S	AS	F	AC	ACT
Fairchild		■						D	■	
Monolithic Memories		■				■				
Motorola			■							
RCA			▨	■						
SGS		■	▨							
Signetics		■							■	
Texas Instruments		■								
Toshiba		■	■							

KEY PARAMETERS		TTL	LS	HC	HCT	ALS	S	AS	F	AC	ACT
I_{OH} (max.)	mA		-0.4	-4	-4		-1		-1	-24	-24
I_{OL} (max.)	mA		8	4	4		20		20	24	24
I_{IH} (max.)	mA		0.02				0.05		0.02		
I_{IL} (max.)	mA		-0.4				-0.25		-0.02		
I_{CC} (max.)	mA		28^a				160		90		
I_{CC} (quiesc.)	mA			0.08^e	0.08					0.08	0.08
C_{pd} (typ.)	pF/channel			31^f	35^l					90	90
t_W (min.)	ns		20	20^g	25		7		5	4.5	4.5
t_{SU} (min.)	ns		20^b	15^h	15^m		5^q		2.5^r	4.5	5.5
t_H (min.)	ns		5	0^i	3^n		3		1	1	1
f_{MAX} (min.)	MHz		30^c	25^j	22^o		75		100	125	125
t_{pd} CP to Q (max.)	ns		27^d	40^k	48^p		15		10.5	11	11

[a] Signetics 28 mA with outputs HIGH and 35 mA with outputs LOW

[b] D to CP (Fairchild 10 ns); t_{SU} $\overline{CE}$ inactive (min.) is 10 ns, t_{SU} $\overline{CE}$ active (min.) 25 ns, except Signetics, which has t_{SU} $\overline{CE}$ HIGH or LOW (min.) of 20 ns, and Fairchild, which has t_{SU} $\overline{CE}$ HIGH (min.) of 10 ns and t_{SU} $\overline{CE}$ LOW (min.) of 20 ns

[c] Monolithic Memories 35 MHz

[d] Fairchild 25 ns

[e] SGS and Toshiba 0.04 mA

[f] RCA; Signetics 20 pF/channel, TI 30 pF/channel, SGS and Toshiba 34 pF/channel

[g] SGS 18 ns, Toshiba 19 ns, TI 25 ns

[h] D or $\overline{CE}$ to CP (RCA, Signetics); TI 25 ns. SGS has t_{SU} D to CP (min.) 18 ns but 30 ns for $\overline{CE}$ to CP, Toshiba 19 ns

for D but 32 ns for $\overline{CE}$

[i] D or $\overline{CE}$ (SGS, TI, Toshiba); Signetics has t_H D (min.) 3 ns and t_H $\overline{CE}$ 4 ns, while RCA has 3 ns for D but 5 ns for $\overline{CE}$

[j] TI 20 MHz, Signetics 24 MHzk, Toshiba 27 MHz

[k] SGS 39 ns, RCA 44 ns

[l] RCA; Signetics 20 pF/channel

[m] D or $\overline{CE}$ to CP (RCA); Signetics has t_{SU} D to CP (min.) 15 ns but $\overline{CE}$ to CP 28 ns

[n] D inputs (RCA); t_H $\overline{CE}$ (min.) 5 ns. Signetics has 2 ns at D, 3 ns at $\overline{CE}$

[o] Signetics; RCA 20 MHz

[p] RCA; Signetics 40 ns

[q] D to CP; t_{SU} (min.) $\overline{CE}$ to CP is 9 ns

[r] D to CP; t_{SU} (min.) $\overline{CE}$ to CP is 3 ns

8-BIT REGISTERS WITH 2-CYCLE LOAD

These single-sourced devices are especially useful for implementing storage files and hex/BCD serial-to-parallel converters. (See manufacturer's data for further details.)

396§   *8-bit registers with 2-cycle 4-bit parallel load.* These 16-pin LS
           registers have the usual eight output lines, but only four data
           inputs, through which one 4-bit word is loaded into the first four
           flip-flops of the register on each positive clock edge. On the next
           positive clock edge, these four bits are transferred to the second
           group of four flip-flops in the register while a second 4-bit word is
           loaded into the first group, and so on. All eight outputs can be
           held LOW by applying a HIGH level to a separate strobe input,
           which does not affect the read-and-transfer operation of the regis-
           ter itself. (Texas Instruments)

I_{OH} (max.)	-0.4	mA	t_W	(min.)	20 ns
I_{OL} (max.)	8	mA	t_{SU}	(min.)	20 ns
I_{IH} (max.)	0.04*	mA	t_H	(min.)	5 ns
I_{IL} (max.)	$-0.8^{\dagger}$	mA	f_{MAX}	(min.)	35 MHz
I_{CC} (max.)	40	mA	t_{pd} CP to Q	(max.)	30 ns
			t_{pd} $\overline{G}$ to Q	(max.)	30 ns

* CP; others 0.02 mA

† CP; others -0.4 mA

7

Bus Registers

A vital category of bus interface components are the edge-triggered bus registers. They are designed to read a data word from parallel input lines, store it, and then, when enabled by a separate output control, to place this data word on a set of parallel output lines. The 3-state or open-collector output sections of these registers are specifically designed to drive high-capacitance loads, such as microprocessor data buses or peripheral output buses, without requiring external drivers. (A group of edge-triggered registers with ordinary 2-state outputs is described in Chapter 6.)

Along with the bus latches (see Chapter 5), these registers are responsible for handling basic input and output tasks between word-oriented digital devices, especially among groups of such devices sharing a common bus. To prevent contention between devices sharing the same bus, the outputs are connected to the bus only when enabled by a separate output control.

Bus registers are distinguished from bus latches by the way they are enabled to read data. The latch outputs track or follow their inputs as long as the latch enable is active (HIGH), while the registers read their inputs only at the moment that the clock pulse undergoes a triggering voltage transition. As with the one-bit flip-flops of Chapter 2, triggering of the D flip-flops making up each register occurs when the clock signal reaches a certain voltage level and is not dependent on the transition rate of the clock pulse. All 7400-series registers are positive-edge-triggered, that is, triggered by the LOW-to-HIGH transition of the clock pulse, CP; the register inputs are disabled during all other phases of the clock cycle.

For most of these registers, the other basic control input is an active-LOW 3-state output enable, $\overline{\text{OE}}$. When $\overline{\text{OE}}$ is LOW, the register outputs are connected directly to the output bus. When $\overline{\text{OE}}$ is taken HIGH, the outputs are put into a high-impedance state that does not significantly load or drive the output lines, effectively disconnecting the register from the bus. Some of the more complex 3-state registers do not provide an external output control but automatically enable and disable the outputs in synchronization with the clock pulse. A few bus registers described in this chapter (the 383§, 605, and 607) have open-collector bus-driving outputs rather than 3-state outputs. In general, output enables have no effect on a register's read-and-store cycle, which is controlled by the clock input.

In addition to the CP and $\overline{\text{OE}}$ controls, some registers are provided with a preset (set) input, $\overline{\text{S}}$, a clear (reset) input, $\overline{\text{R}}$, or both. In almost all of these circuits (the single exception being the 4-bit 173), the preset or clear is active-LOW, and in most cases, the preset or clear is an asynchronous control that overrides the clock to directly load a HIGH or LOW level into all flip-flops within the register. Two types, the 878 and 879, provide a synchronous active-LOW clear instead; if held LOW, this $\overline{\text{R}}$ input does not clear the register directly but enables the internal flip-flops to be reset to LOW on arrival of the next high-going clock transition.

The switching characteristics of these bus registers are valid under the same test conditions used for bus latches. (See Chapter 5.) In particular, it should be noted that S components are specified with a load capacitance C_L of 15 pF instead of the 45 pF or 50 pF usually seen in 3-state timing specifications.

4-BIT BUS REGISTERS

There is only one type of 4-bit bus register in 7400-series logic, but this type is widely supported.

173   *4-bit 3-state registers with asynchronous clear, dual gated output enables, and dual gated clock enables.* The standard 4-bit bus register, supported by nearly all manufacturers and available in TTL, LS, HC, and HCT technologies. Since there are only four inputs and four noninverting outputs, the 16-pin DIP provides ample room for a full range of control inputs. The usual clock input, CP, is accompanied by two NOR-gated clock enables, $\overline{\text{CE1}}$ and $\overline{\text{CE2}}$. Either of these, when taken HIGH, will inhibit the clock, producing a "do nothing" or "hold" mode in which the register maintains its current state regardless of changes at the D or CP inputs. An active-HIGH direct clear (reset) input, R, will load a LOW level into the four flip-flops regardless of the other

inputs. A remaining pin is used to provide a second output enable, which is NOR-gated with the usual $\overline{OE}$ to provide dual control over the 3-state outputs. Either $\overline{OE}$ input, when taken HIGH, will produce a high-impedance condition at the register outputs that will effectively disconnect the register from its output bus regardless of the input at the other $\overline{OE}$.

FUNCTION TABLE
173

INPUTS							OUTPUTS	
$\overline{OE1}$	$\overline{OE2}$	R	$\overline{CE1}$	$\overline{CE2}$	CP	D	Q	
L	L	L	H	H	↑	L	L	Read data
L	L	L	H	H	↑	H	H	
L	L	L	L	.	↑	.	NO CHANGE	Clock inhibited
L	L	L	.	L	↑	.	NO CHANGE	
L	L	L	.	.	—	.	NO CHANGE	Clock inactive
L	L	H	.	.	.	.	L	Direct reset
H	.	.	.	.	.	.	Z	Outputs off
.	H	.	.	.	.	.	Z	

. = Either LOW or HIGH level
— = Clock inactive (L, H, or ↑)
Z = High impedance (disabled)

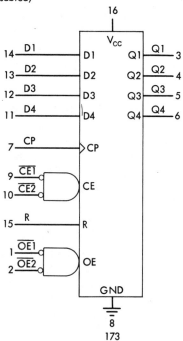

173

MANUFACTURERS	TTL	LS/A	HC	HCT	ALS	S	AS	F	AC	ACT
Fairchild	███	███								
Motorola			░░░							
National Semiconductor										
RCA			███	███						
SGS		░░░	░░░							
Signetics	███	███								
Texas Instruments	███	███								
Toshiba			███							

Missing change control symbol "A": Fairchild and Signetics 74LS173

KEY PARAMETERS		TTL	LS/A	HC	HCT	ALS	S	AS	F	AC	ACT
I_{OH} (max.)	mA	-5.2	-2.6^d	-6	-6						
I_{OL} (max.)	mA	16	24^e	6	6						
I_{IH} (max.)	mA	0.04	0.02								
I_{IL} (max.)	mA	-1.6	-0.4								
I_{CC} (max.)	mA	72	30^f								
I_{CC} (quiesc.)	mA			0.08^n	0.08						
C_{pd} (typ.)	pF/channel			30^o	32^x						
t_W (min.)	ns	20	25^g	20^p	25^y						
t_{SU} (min.)	ns	10^a	17^h	25^q	31^z						
t_H (min.)	ns	10^b	0^i	0^r	0^{aa}						
f_{MAX} (min.)	MHz	25	30	22^s	16^{bb}						
t_{pd} CP to Q (max.)	ns	43^c	30^j	44^t	54						
t_{pd} R to Q (max.)	ns	27	35^k	38^u	46^{cc}						
t_{en} (max.)	ns	30	27^l	38^v	44^{dd}						
t_{dis} (max.)	ns	20	17^m	38^w	44^{ee}						

[a] D to CP (and $\overline{R}$ inactive to CP, according to TI); t_{SU} $\overline{CE}$ (min.) is 17 ns

[b] D inputs; t_H $\overline{CE}$ (min.) is 2 ns

[c] National 28

[d] SGS -0.4 mA

[e] Fairchild 74LS173 8 mA

[f] TI 24 mA, Fairchild 28 mA

[g] Fairchild and National 17 ns, SGS and Signetics 20 ns

[h] D to CP (Fairchild and National 10 ns); t_{SU} $\overline{CE}$ (min.) is 35 ns (Fairchild and National 17 ns). TI further states that t_{SU} (min.) for $\overline{R}$ inactive is 10 ns

[i] TI has 0 ns for $\overline{CE}$ but requires t_H (min.) of 3 ns for D; Fairchild has t_H (min.) of 2 ns for $\overline{CE}$ and 10 ns for D

[j] Fairchild 40 ns at C_L = 15 pF (others are rated at 45 pF)

[k] National 30 ns; Fairchild 25 ns at C_L = 15 pF (others are rated at 45 pF)

[l] Fairchild 20 ns at C_L = 15 pF (others are rated at 45 pF)

[m] Fairchild 16 ns, National and TI 20 ns

[n] SGS and Toshiba 0.04 mA

[o] Toshiba; Signetics 25 pF/channel, RCA and TI 29 pF/channel, SGS 35 pF/channel, National 80 pF/channel. Motorola data not available

[p] CP or R (SGS 18 ns, Toshiba 19 ns). Signetics has t_W CP (min.) 23 ns and t_W R 20 ns

[q] D and $\overline{CE}$ (Motorola, National, and TI); RCA 15 ns. SGS has t_{SU} (min.) 18 ns for D and 24 ns for $\overline{CE}$, Toshiba 19 ns for D and 25 ns for $\overline{CE}$, Signetics 20 ns for D and 25 ns for $\overline{CE}$. TI further states t_{SU} (min.) = 23 ns for R inactive

[r] D and $\overline{CE}$ for all except RCA, which has t_H $\overline{CE}$ (min.) 0 ns but t_H D (min.) = 3 ns

[s] Signetics; Motorola and National 21 MHz, RCA and Toshiba 24 MHz, SGS and TI 25 MHz

[t] Motorola and National; TI 38 ns, SGS 40 ns, Toshiba 41 ns, RCA 50 ns, Signetics 56 ns

[u] SGS 40 ns, Toshiba 41 ns, RCA 44 ns, Signetics 46 ns

[v] SGS 26 ns, Toshiba 30 ns

[w] SGS 35 ns

[x] Signetics; RCA 34 pF/channel

[y] CP (Signetics); RCA 31 ns. Both have t_W R (min.) of 19 ns

[z] D to CP (Signetics); RCA 15 ns. From $\overline{CE}$, t_{SU} (min.) is 38 ns (Signetics) or 23 ns (RCA)

[aa] D and $\overline{CE}$

[bb] RCA; Signetics 20 MHz

[cc] RCA; Signetics 44 ns

[dd] Signetics; RCA 38 ns

[ee] RCA; Signetics 38 ns

4+4-BIT BUS REGISTERS

8-bit bus registers, like 8-bit bus latches, come in two basic configurations: a 4+4 configuration in which two 4-bit sections are triggered by independent clock inputs, and a more straightforward (but less flexible) arrangement in which all eight flip-flops are triggered by a single common clock input. The simpler 8-bit types, which are described in detail further on, satisfy the requirements of most byte-oriented I/O operations, but for certain applications the small group of components collected here provide somewhat more extensive control at the price of four additional input pins (24 instead of the 20 pins required by most common 8-bit designs).

Each of the 4+4-bit types listed here provides a separate clock input (CP), a separate preset ($\overline{S}$) or clear ($\overline{R}$) input, and a separate output enable ($\overline{OE}$) for each group of four flip-flops. The four types are differentiated by whether they provide inverting or noninverting logic and whether they have synchronous or asynchronous presets or clears.

874   *4+4-bit 3-state registers with asynchronous clears.* ALS and AS components that provide the basic 4+4-bit register functions plus two separate asynchronous active-LOW clears. When taken LOW, a clear input will override all other inputs and directly load a LOW level in each of the four flip-flops in the group to which it is assigned. Corresponding bus latch: 873.

876   *4+4-bit 3-state inverting registers with asynchronous presets.* Same as the 874, but with inverted outputs. The inputs that perform the direct clear function in the 874 behave exactly the same in the 876. In both cases, a LOW at one of these inputs (pin 1 or pin 13) will directly load a LOW level in the corresponding group of four flip-flops. Since the outputs of the 876 are inverted, however, these two inputs on the 876 are called "sets" or "presets" and labeled $\overline{S}$. Corresponding bus latch: 880.

878   *4+4-bit 3-state registers with synchronous clears.* Same as the noninverting 874, but with synchronous rather than asynchronous clears. When taken LOW, a given clear input does not override the clock to directly clear the four bits to which it is assigned. Instead, if held LOW, it enables the clock to clear the appropriate four flip-flops on the arrival of the next positive clock transition.

879   *4+4-bit 3-state inverting registers with synchronous clears.* Similar to the 878, but with inverting outputs and different clear (reset) logic. When held LOW, a given $\overline{R}$ input of the 879 will

cause the next high-going clock transition to synchronously load a HIGH level in the appropriate group of flip-flops rather than a LOW level as in the 878. Since the outputs of the 879 are inverted, however, this opposite effect is consistent with the overall logic of the circuit, and the inputs therefore continue to be named "clear" or "reset" rather than "set" or "preset."

FUNCTION TABLE
874

INPUTS				OUTPUTS
$\overline{OE}$	$\overline{R}$	CP	D	Q
L	H	↑	L	L
L	H	↑	H	H
L	H	–	.	NO CHANGE
L	L	.	.	L
H	.	.	.	Z

. = Either LOW or HIGH level
– = Clock inactive (L, H, or ↑)
Z = High impedance (disabled)

FUNCTION TABLE
876

INPUTS				OUTPUTS
$\overline{OE}$	$\overline{S}$	CP	D	$\overline{Q}$
L	H	↑	L	H
L	H	↑	H	L
L	H	–	.	NO CHANGE
L	L	.	.	L
H	.	.	.	Z

. = Either LOW or HIGH level
– = Clock inactive (L, H, or ↑)
Z = High impedance (disabled)

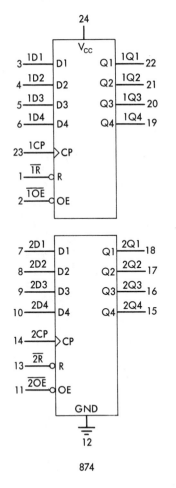

874

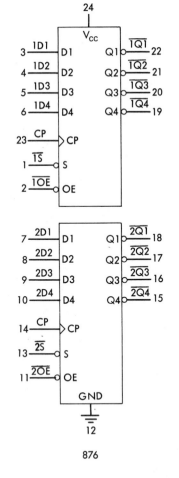

876

FUNCTION TABLE 878				
INPUTS				**OUTPUTS**
$\overline{OE}$	$\overline{R}$	CP	D	Q
L	H	↑	L	L
L	H	↑	H	H
L	H	−	.	NO CHANGE
L	L	↑	.	L
H	.	.	.	Z

. = Either LOW or HIGH level
− = Clock inactive (L, H, or ↓)
Z = High impedance (disabled)

FUNCTION TABLE 879				
INPUTS				**OUTPUTS**
$\overline{OE}$	$\overline{R}$	CP	D	$\overline{Q}$
L	H	↑	L	H
L	H	↑	H	L
L	H	−	.	NO CHANGE
L	L	↑	.	H
H	.	.	.	Z

. = Either LOW or HIGH level
− = Clock inactive (L, H, or ↓)
Z = High impedance (disabled)

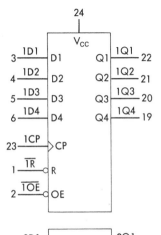

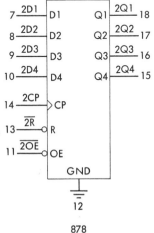

878

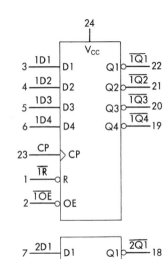

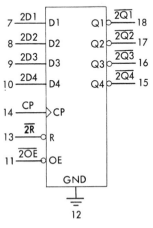

879

874 4+4-bit 3-state registers with asynchronous clears

MANUFACTURERS	TTL	LS	HC	HCT	ALS	S	AS	F	AC	ACT
National Semiconductor					■		■			
Texas Instruments					■		■			

Added change control symbol: TI 74ALS874B

KEY PARAMETERS		TTL	LS	HC	HCT	ALS	S	AS	F	AC	ACT
I_{OH} (max.)	mA					-2.6		-15			
I_{OL} (max.)	mA					24		48			
I_{IH} (max.)	mA					0.02		0.01^c			
I_{IL} (max.)	mA					-0.2		-0.5^d			
I_{CC} (max.)	mA					32		160			
t_W (min.)	ns					16.5^a		4^e			
t_{SU} (min.)	ns					15^b		2^f			
t_H (min.)	ns					0		1			
f_{MAX} (min.)	MHz					30		125			
t_{pd} CP to Q (max.)	ns					14		10.5			
t_{pd} $\overline{R}$ to Q (max.)	ns					17		9.5			
t_{en} (max.)	ns					18		10.5			
t_{dis} (max.)	ns					12		7.5			

[a] CP; t_W $\overline{R}$ (min.) is 10 ns

[b] D to CP; t_{SU} $\overline{R}$ inactive (min.) is 10 ns

[c] National 0.02 mA

[d] National; TI has I_{IH} (max.) of -0.5 mA at all inputs except the D inputs, where I_{IH} (max.) is -2 mA

[e] CP LOW; t_W CP HIGH (min.) is 3 ns and t_W $\overline{R}$ (min.) is 2 ns

[f] D to CP; t_{SU} $\overline{R}$ inactive (min.) is 4 ns

876   4+4-bit 3-state inverting registers with asynchronous presets

MANUFACTURERS	TTL	LS	HC	HCT	ALS	S	AS	F	AC	ACT
National Semiconductor					■		■			
Texas Instruments					■		■			

Added change control symbol: TI 74ALS876A

KEY PARAMETERS		TTL	LS	HC	HCT	ALS	S	AS	F	AC	ACT
I_{OH} (max.)	mA					-2.6		-15			
I_{OL} (max.)	mA					24		48			
I_{IH} (max.)	mA					0.02		0.01^e			
I_{IL} (max.)	mA					-0.2		-0.5^f			
I_{CC} (max.)	mA					32^a		160			
t_W (min.)	ns					16.5^b		4^g			
t_{SU} (min.)	ns					15^c		2^h			
t_H (min.)	ns					0		1			
f_{MAX} (min.)	MHz					30		125			
t_{pd} CP to $\overline{Q}$ (max.)	ns					14		10.5			
t_{pd} $\overline{S}$ to $\overline{Q}$ (max.)	ns					19		9.5			
t_{en} (max.)	ns					18		10.5			
t_{dis} (max.)	ns					12^d		6			

^{a} TI 31 mA
^{b} CP; t_W $\overline{S}$ (min.) is 10 ns
^{c} D to CP; t_{SU} $\overline{S}$ inactive (min.) is 10 ns
^{d} TI 13 ns
^{e} National 0.02 mA

^{f} National; TI has I_{IH} (max.) of -0.5 mA at all inputs except the D inputs, where I_{IH} (max.) is -2 mA
^{g} CP LOW; t_W CP HIGH (min.) is 3 ns and t_W $\overline{S}$ (min.) is 2 ns
^{h} D to CP; t_{SU} $\overline{S}$ inactive (min.) is 4 ns

878   4+4-bit 3-state registers with synchronous clears

MANUFACTURERS	TTL	LS	HC	HCT	ALS/A	S*	AS	F	AC	ACT
National Semiconductor							■			
Texas Instruments					■		■			

KEY PARAMETERS		TTL	LS	HC	HCT	ALS/A	S	AS	F	AC	ACT
I_{OH} (max.)	mA					-2.6		-15			
I_{OL} (max.)	mA					24		48			
I_{IH} (max.)	mA					0.02		0.02			
I_{IL} (max.)	mA					-0.2		-0.5^c			
I_{CC} (max.)	mA					33		160			
t_W (min.)	ns					16.5		4^d			
t_{SU} (min.)	ns					15^a		2^e			
t_H (min.)	ns					4^b		2^b			
f_{MAX} (min.)	MHz					30		125			
t_{pd} CP to Q (max.)	ns					16		10.5			
t_{en} (max.)	ns					20		10.5			
t_{dis} (max.)	ns					15		6			

[a] D to CP; t_{SU} $\overline{R}$ (min.) is 20 ns

[b] D inputs; t_H $\overline{R}$ (min.) is 0 ns

[c] National; TI has I_{IH} (max.) of -0.5 mA at all inputs except

the D inputs, where I_{IH} (max.) is -2 mA

[d] CP HIGH; t_W CP LOW (min.) is 2 ns

[e] D to CP; t_{SU} $\overline{R}$ (min.) is 5.5 ns

---

879   4+4-bit 3-state inverting registers with synchronous clears

MANUFACTURERS	TTL	LS	HC	HCT	ALS/A	S	AS	F	AC	ACT
National Semiconductor							■			
Texas Instruments					■		■			

KEY PARAMETERS		TTL	LS	HC	HCT	ALS/A	S	AS	F	AC	ACT
I_{OH} (max.)	mA					-2.6		-15			
I_{OL} (max.)	mA					24		48			
I_{IH} (max.)	mA					0.02		0.02			
I_{IL} (max.)	mA					-0.2		-0.5^c			
I_{CC} (max.)	mA					33		160			
t_W (min.)	ns					20		4^d			
t_{SU} (min.)	ns					15^a		2^e			
t_H (min.)	ns					4^b		2^b			
f_{MAX} (min.)	MHz					25		125			
t_{pd} CP to $\overline{Q}$ (max.)	ns					16		10.5			
t_{en} (max.)	ns					20		10.5			
t_{dis} (max.)	ns					15		6			

[a] D to CP; t_{SU} $\overline{R}$ (min.) is 20 ns

[b] D to CP; t_H $\overline{R}$ (min.) is 0 ns

[c] National; TI has I_{IH} (max.) of -0.5 mA at all inputs except

the D inputs, where I_{IH} (max.) is -2 mA

[d] CP HIGH; t_W CP LOW (min.) is 2 ns

[e] D to CP; t_{SU} $\overline{R}$ (min.) is 5.5 ns

8-BIT BUS REGISTERS

With their close relatives, the 8-bit bus latches, the simple 8-bit bus interface registers described here are the workhorses of digital I/O, appearing in nearly all byte-oriented processing or communications systems and especially at microprocessor input and output ports. Their function is to read an 8-bit data word from an input bus when triggered by an incoming high-going clock pulse, to store this byte, and to place the byte on a common output bus when signaled to do so by a LOW level at an output enable. The high drive capabilities of their 3-state output sections allow these interface registers to maintain the validity of the data they place on the output bus even if the bus is loaded by a number of other devices.

All of these designs are supplied in 20-pin DIPs, a minimum configuration that provides just a single edge-triggered clock and a single active-LOW output enable for the control of each register. As with the bus latches, two pinout arrangements are available: a "neighboring" pinout in which each data input pin is located adjacent to the corresponding output pin, and a "symmetrical" or "bus-structured" pinout in which each data input pin is located directly across from its corresponding output. The latter arrangement can simplify board layout in bus-oriented systems.

It should be remembered that the switching characteristics of the older S designs are given for $C_L = 15$ pF instead of the 50 pF normally used for 3-state output specifications.

Noninverting bus registers:

374   *8-bit 3-state bus registers.* The basic 8-bit register, supported by nearly every manufacturer and available in a wide variety of performance ranges. "Neighboring" data pinout. Corresponding bus latch: 373.

574   *8-bit 3-state bus registers.* Versions of the 374 with bus-structured pinout. Available in most technologies. Corresponding bus latch: 573.

532§   *8-bit 3-state bus registers.* S versions of the 374 ("neighboring" pinout) with I_{OL} (max.) increased to 32 mA from the usual 20 mA. Corresponding bus latch: 531§. (Monolithic Memories)

I_{OH} (max.)	−6.5	mA	t_W	(min.)	7.3	ns
I_{OL} (max.)	32	mA	t_{SU}	(min.)	5	ns
I_{IH} (max.)	0.05	mA	t_H	(min.)	2	ns
I_{IL} (max.)	−0.25	mA	f_{MAX}	(min.)	75	MHz
I_{CC} (max.)	140	mA	t_{pd} CP to Q	(max.)	17	ns
			t_{en}	(max.)	18	ns
			t_{dis}	(max.)	12	ns

364§   *8-bit 3-state bus registers with MOS driver outputs.* LS versions of the 374 ("neighboring" pinout) with higher-voltage outputs. The output HIGH level is about 1 V closer to V_{CC} than that of the normal 3-state buffer, over 3.5 V at minimum V_{CC}, making these devices ideal for driving MOS memories or microprocessors with thresholds of 2.4 V to 3.5 V. Corresponding bus latch: 363§. (Signetics)

I_{OH} (max.)	−2.6	mA	t_W	(min.)	15 ns
I_{OL} (max.)	24	mA	t_{SU}	(min.)	20 ns
I_{IH} (max.)	0.02	mA	t_H	(min.)	0 ns
I_{IL} (max.)	−0.4	mA	f_{MAX}	(min.)	35 MHz
I_{CC} (max.)	70	mA	t_{pd} CP to Q	(max.)	34 ns
			t_{en}	(max.)	36 ns
			t_{dis}	(max.)	24 ns

534   *8-bit 3-state inverting bus registers.* Same as the 374 ("neighboring" pinout), but inverting. Supported by most manufacturers and available in most technologies. Corresponding bus latch: 533.

564   *8-bit 3-state inverting bus registers.* Same as the 574 (bus-structured pinout), but inverting. Available in most technologies. Corresponding bus latch: 563.

576   *8-bit 3-state inverting bus registers.* ALS and AS versions of the 564 (bus-structured pinout). Corresponding bus latch: 580.

536§   *8-bit 3-state inverting bus registers.* S versions of the 534 ("neighboring" pinout) with I_{OL} (max.) increased to 32 mA from the usual 20 mA. Corresponding bus latch: 535§. (Monolithic Memories)

I_{OH} (max.)	−6.5	mA	t_W	(min.)	7.3 ns
I_{OL} (max.)	32	mA	t_{SU}	(min.)	5 ns
I_{IH} (max.)	0.05	mA	t_H	(min.)	5 ns
I_{IL} (max.)	−0.25	mA	f_{MAX}	(min.)	75 MHz
I_{CC} (max.)	140	mA	t_{pd} CP to Q	(max.)	20 ns
			t_{en}	(max.)	20 ns
			t_{dis}	(max.)	16 ns

FUNCTION TABLE
374, 574, 532§, 364§

INPUTS			OUTPUTS
$\overline{OE}$	CP	D	Q
L	↑	L	L
L	↑	H	H
L	–	.	NO CHANGE
H	.	.	Z

. = Either LOW or HIGH level
– = Clock inactive (L, H, or ↓)
Z = High impedance (disabled)

FUNCTION TABLE
534, 564, 576, 536§

INPUTS			OUTPUTS
$\overline{OE}$	CP	D	$\overline{Q}$
L	↑	L	H
L	↑	H	L
L	–	.	NO CHANGE
H	.	.	Z

. = Either LOW or HIGH level
– = Clock inactive (L, H, or ↓)
Z = High impedance (disabled)

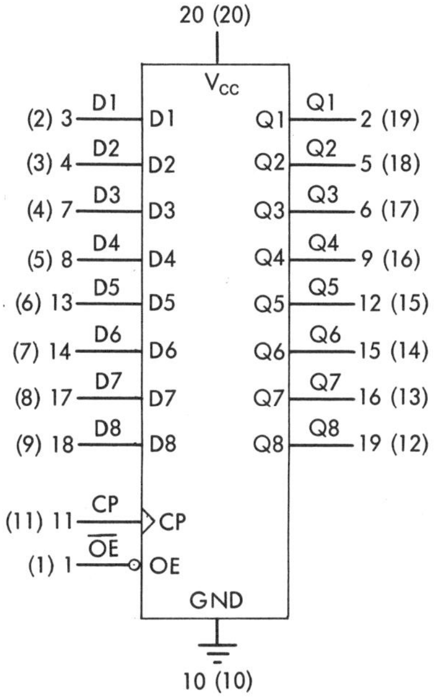

374, (574), 532§, 364§

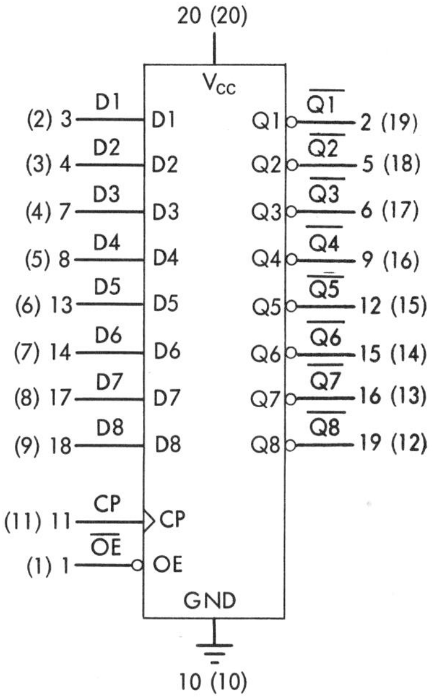

534, (564), (576), 536§

MANUFACTURERS	TTL	LS	HC	HCT	ALS	S	AS	F	AC	ACT
Fairchild		■						■		
Monolithic Memories						■				
Motorola			▨	D				■		
National Semiconductor				▨		■				
RCA			■						▨	▨
SGS			▨	D		■				
Signetics		■				■		■		
Texas Instruments				▨		■				
Toshiba									D	

KEY PARAMETERS		TTL	LS	HC	HCT	ALS	S	AS	F	AC	ACT
I_{OH} (max.)	mA		-2.6	-6	-6	-2.6	-6.5[w]	-15	-3	-24	-24
I_{OL} (max.)	mA		24	6	6	24	20	48	24	24	24
I_{IH} (max.)	mA		0.02			0.02	0.05	0.02	0.02		
I_{IL} (max.)	mA		-0.4			-0.2	-0.25	-0.5[aa]	-0.6		
I_{CC} (max.)	mA		40[a]			31	160[x]	128	86		
I_{CC} (quiesc.)	mA			0.08[e]	0.08[n]					0.08	0.08
C_{pd} (typ.)	pF/channel			50[f]	47[o]					80[dd]	80[dd]
t_W (min.)	ns		15[b]	20[g]	20[p]	14	7.3[y]	4[bb]	7[cc]	4.5[ee]	5[ii]
t_{SU} (min.)	ns		20	25[h]	25[q]	10	5	2	2	4.5[ff]	5.5[kk]
t_H (min.)	ns		0[c]	5[i]	5[r]	0	2	2	2	1.5[ff]	1.5[ll]
f_{MAX} (min.)	MHz		35	24[j]	24[s]	35	75[z]	125	70	100[gg]	90[mm]
t_{pd} CP to Q (max.)	ns		28	45[k]	45[l]	16	17	9	10	10.5[hh]	11.5[nn]
t_{en} (max.)	ns		28	38[l]	38[u]	18	18	10	12.5	9.5[ii]	10.5[oo]
t_{dis} (max.)	ns		25[d]	38[m]	37[v]	18	12	6	8	12.5[ii]	12.5[oo]

[a] Fairchild, National, and SGS 45 mA

[b] SGS 13 ns

[c] At frequencies below 10 MHz. According to TI, designs running faster than 10 MHz should use a minimum t_H of 5 ns

[d] TI 28 ns

[e] SGS and Toshiba 0.04 mA

[f] Motorola and National; Signetics 17 pF/channel, RCA 39 pF/channel, SGS and Toshiba 51 pF/channel, TI 100 pF/channel

[g] SGS 18 ns, Toshiba 19 ns

[h] Signetics 15 ns, SGS 24 ns

[i] SGS 6 ns

[j] SGS 20 MHz, RCA 25 MHz

[k] SGS 42 ns, Toshiba 44 ns

[l] SGS 35 ns, National 37 ns, RCA 44 ns

[m] National 37 ns, SGS 42 ns

[n] Toshiba 0.04 mA

[o] RCA; Signetics 17 pF/channel, Toshiba 60 pF/channel, TI 85 pF/channel. National data not available

[p] Signetics 24 ns, RCA 25 ns, Toshiba 32 ns

[q] Signetics 15 ns, Toshiba 19 ns

[r] Toshiba 0 ns

[s] National; Toshiba 20 MHz, Signetics 21 MHz, RCA and TI 25 MHz

[t] Signetics 40 ns, RCA 44 ns, Toshiba 50 ns

[u] National 37 ns, RCA 41 ns, Toshiba 53 ns

[v] National; RCA and Signetics 35 ns, TI 38 ns, Toshiba 40 ns

[w] TI data not available

[x] National 140 mA, Signetics 180 mA

[y] CP LOW; t_W CP HIGH (min.) is 6 ns

[z] National data not available

[aa] National; TI has I_{IL} (max.) of -0.5 mA for all inputs except the D inputs, where I_{IL} (max.) is -2 mA

[bb] CP HIGH; t_W CP LOW (min.) is 3 ns

[cc] CP HIGH; t_W CP LOW (min.) is 6 ns

[dd] RCA 60 pF/channel

[ee] RCA 5.9 ns

[ff] RCA 2 ns

[gg] RCA 85 MHz

[hh] RCA 10 ns

[ii] RCA 13 ns

[jj] RCA 6.7 ns

[kk] RCA 2 ns

[ll] RCA 3 ns

[mm] RCA 75 MHz

[nn] RCA 10.3 ns

[oo] RCA 13 ns

MANUFACTURERS	TTL	LS	HC	HCT	ALS	S	AS	F	AC	ACT
Fairchild		█						D	▒	█
Motorola			D							
National Semiconductor				▒						
RCA			█	█					▒	▒
SGS		▒	▒	▒						
Signetics			D	D				▒		
Texas Instruments			█	█				█		
Toshiba			█	█					D	

KEY PARAMETERS		TTL	LS	HC	HCT	ALS	S	AS	F	AC	ACT
I_{OH} (max.)	mA		-2.6^a	-6	-6	-2.6		-15	-3	-24	-24
I_{OL} (max.)	mA		24	6	6	24		48	24	24	24
I_{IH} (max.)	mA		0.02			0.02		0.02	0.02		
I_{IL} (max.)	mA		-0.4			-0.2		-0.5^u	-0.6		
I_{CC} (max.)	mA		45^b			28		134	86		
I_{CC} (quiesc.)	mA			0.08^d	0.08^d					0.08	0.08
C_{pd} (typ.)	pF/channel			51^e	57^m					40^w	40^y
t_W (min.)	ns		15^c	20^f	20^n	14		4^v	7	5.9^x	4^z
t_{SU} (min.)	ns		20^c	25^g	25^o	15		2	2	2^x	2.5^{aa}
t_H (min.)	ns		0^c	5^h	5^p	0		2	2	2^x	1^{aa}
f_{MAX} (min.)	MHz		35^c	24^i	24^q	35		125	70	85^x	85^{bb}
t_{pd} CP to Q (max.)	ns		28^c	38^j	45^r	14		9	10	11.3^x	12^{cc}
t_{en} (max.)	ns		28^c	38^k	44^s	18		10	12.5	13^x	10^{dd}
t_{dis} (max.)	ns		25^c	38^l	38^t	12		6	8	13^x	11.5^{dd}

[a] SGS -0.4 mA

[b] SGS 40 mA

[c] Fairchild (SGS data not available)

[d] SGS and Toshiba 0.04 mA

[e] SGS and Toshiba; RCA 39 pF/channel, National 50 pF/channel, TI 100 pF/channel. Signetics has 35 pF per package (not channel)

[f] SGS 18 ns, Toshiba 19 ns

[g] Signetics 15 ns, Toshiba 19 ns, SGS 24 ns

[h] National and Toshiba 0 ns, SGS 6 ns

[i] SGS 20 MHz, RCA 25 MHz

[j] Toshiba; National 29 ns, Signetics 35 ns, SGS 42 ns, RCA and TI 45 ns

[k] Signetics and TI; National, SGS, and Toshiba 35 ns, RCA 44 ns

[l] National 31 ns, SGS 42 ns

[m] SGS and Toshiba; RCA 47 pF/channel, TI 93 pF/channel. Signetics has 40 pF/package; National data not available

[n] Signetics and TI; Toshiba 19 ns, SGS 24 ns, RCA 25 ns. National data not available

[o] RCA and TI; Toshiba 13 ns, Signetics 15 ns, SGS 18 ns. National data not available

[p] SGS 6 ns. National data not available

[q] Signetics and TI; SGS 18 MHz, Toshiba 20 MHz, RCA 25 MHz. National data not available

[r] TI; RCA and Signetics 44 ns, SGS 50 ns, Toshiba 51 ns. National data not available

[s] TI 38 ns, RCA 41 ns. National data not available

[t] Signetics and TI; RCA 35 ns, SGS 40 ns, Toshiba 46 ns. National data not available

[u] National; TI has I_{IL} (max.) of -0.5 mA for all inputs except the D inputs, where I_{IL} (max.) is -2 mA

[v] CP HIGH; t_W CP LOW (min.) is 2 ns

[w] RCA 60 pF/channel

[x] RCA; comparable Fairchild data not available

[y] RCA 72 pF/channel

[z] RCA 6.7 ns

[aa] RCA 2 ns

[bb] RCA 75 MHz

[cc] RCA 11.4 ns

[dd] RCA 13 ns

MANUFACTURERS	TTL	LS	HC	HCT	ALS	S	AS	F	AC	ACT
Fairchild		■							▨	▨
Monolithic Memories						■				
Motorola			▨				■			
National Semiconductor			■	▨	■					
RCA									▨	▨
SGS		▨	■							
Signetics			■		■		■			
Texas Instruments			▨	D		■		■		
Toshiba			■	D					D	

KEY PARAMETERS		TTL	LS	HC	HCT	ALS	S	AS	F	AC	ACT
I_{OH} (max.)	mA		-2.6^a	-6	-6	-2.6	-6.5	-15	-3	-24	-24
I_{OL} (max.)	mA		24	6	6	24	20	48	24	24	24
I_{IH} (max.)	mA		0.02			0.02	0.05	0.02	0.02		
I_{IL} (max.)	mA		-0.4			-0.2^w	-0.25	-0.5^{aa}	-0.6		
I_{CC} (max.)	mA		45^b			31	180^x	128	86		
I_{CC} (quiesc.)	mA			0.08^f	0.08					0.08	0.08
C_{pd} (typ.)	pF/channel			51^g	o					60^{dd}	60^{dd}
t_W (min.)	ns		15^c	20^h	25^p	14	7.3^y	4^{bb}	7^{cc}	5.9^{dd}	6.7^{dd}
t_{SU} (min.)	ns		20^c	25^i	25^q	10	5	2	2	2^{dd}	2^{dd}
t_H (min.)	ns		0^c	5^j	5^r	0	2^z	2	2	2^{dd}	3^{dd}
f_{MAX} (min.)	MHz		35^c	24^k	20^s	35	75	125	70	85^{dd}	75^{dd}
t_{pd} CP to $\overline{Q}$ (max.)	ns		28^d	45^l	45^t	16	17	9	10	10.2^{dd}	10.6^{dd}
t_{en} (max.)	ns		28^d	38^m	38^u	18	18	10	12.5	13^{dd}	13^{dd}
t_{dis} (max.)	ns		25^e	38^n	38^v	14	12	6	8	13^{dd}	13^{dd}

[a] SGS -0.4 mA

[b] Fairchild; SGS 40 mA, Monolithic Memories 48 mA

[c] SGS data not available

[d] Monolithic Memories 30 ns; SGS data not available

[e] Monolithic Memories 29 ns; SGS data not available

[f] SGS and Toshiba 0.04 mA

[g] SGS and Toshiba; Signetics 19 pF/channel, RCA 32 pF/channel, Motorola and National 50 pF/channel, TI 100 pF/channel

[h] SGS 18 ns, Toshiba 19 ns

[i] RCA and Signetics 15 ns, SGS 24 ns

[j] Signetics 3 ns, SGS 6 ns

[k] SGS 20 MHz, RCA and TI 25 MHz

[l] RCA and Signetics 41 ns, SGS 42 ns, Toshiba 44 ns

[m] SGS 35 ns, National 37 ns

[n] National 37 ns, SGS 42 ns

[o] No representative value. Signetics 19 pF/channel, RCA 36 pF/channel, TI [93] pF/channel; National data not available

[p] RCA; National 20 ns, Signetics 29 ns

[q] Signetics 15 ns

[r] Signetics 3 ns

[s] RCA; Signetics 18 MHz, National 24 MHz

[t] National; Signetics 38 ns, RCA 44 ns

[u] Signetics; National 37 ns, RCA 44 ns

[v] National 37 ns

[w] National CP and $\overline{OE}$ inputs -0.1 mA

[x] Signetics; Monolithic Memories states 140 mA

[y] CP LOW; t_W CP HIGH (min.) is 6 ns

[z] Monolithic Memories 5 ns

[aa] National; TI has I_{IL} (max.) of -0.5 mA for all inputs except the D inputs, where I_{IL} (max.) is -2 mA

[bb] CP HIGH; t_W CP LOW (min.) is 3 ns

[cc] CP HIGH; t_W CP LOW (min.) is 6 ns

[dd] RCA; comparable Fairchild data not available

MANUFACTURERS	TTL	LS	HC	HCT	ALS	S	AS	F	AC	ACT
Fairchild		▓						D	▓	▓
Motorola			D							
National Semiconductor				▓	▓					
RCA			▓	▓					▓	▓
SGS			▓	▓						
Signetics			▓	▓				▓		
Texas Instruments			▓	▓	▓					
Toshiba									D	

Added change control symbol: TI 74ALS564A

KEY PARAMETERS		TTL	LS	HC	HCT	ALS	S	AS	F	AC	ACT
I_{OH} (max.)	mA		-2.6	-6	-6	-2.6			-3	-24	-24
I_{OL} (max.)	mA		24	6	6	24			24	24	24
I_{IH} (max.)	mA		0.02			0.02			0.02		
I_{IL} (max.)	mA		-0.4			-0.2			-0.6		
I_{CC} (max.)	mA		45			30^r			86		
I_{CC} (quiesc.)	mA			0.08^a	0.08^a					0.08	0.08
C_{pd} (typ.)	pF/channel			51^b	60^j					50^t	50^v
t_W (min.)	ns		15	20^c	23^k	14			7	5.9^u	3.5^w
t_{SU} (min.)	ns		20	24^d	25^l	15			2	2^u	3^x
t_H (min.)	ns		0	5^e	5^m	0			2	2^u	1^x
f_{MAX} (min.)	MHz		35	24^f	20^n	30			70	85^u	75
t_{pd} CP to $\overline{Q}$ (max.)	ns		28	41^g	45^o	14			10	10.2^u	11.5^y
t_{en} (max.)	ns		28	35^h	44^p	18			12.5	13^u	9.5^z
t_{dis} (max.)	ns		25	38^i	38^q	15^s			8	13^u	11.5^z

[a] SGS and Toshiba 0.04 mA

[b] SGS and Toshiba; Signetics 27 pF/channel, RCA 32 pF/channel, National 50 pF/channel, TI 100 pF/channel

[c] SGS 18 ns, Toshiba 19 ns

[d] SGS; RCA and Signetics 15 ns, Toshiba 19 ns, National and TI 25 ns

[e] National and Toshiba 0 ns, SGS 6 ns

[f] SGS 20 MHz, RCA and TI 25 MHz

[g] RCA and Signetics; National 29 ns, Toshiba 38 ns, SGS 42 ns, TI 45 ns

[h] RCA and TI 38 ns

[i] National 31 ns, RCA and Signetics 34 ns, SGS 42 ns

[j] SGS and Toshiba; Signetics 27 pF/channel, RCA 36 pF/channel, TI 93 pF/channel. National data not available

[k] Signetics; Toshiba 19 ns, TI 20 ns, SGS 24 ns, RCA 25 ns. National data not available

[l] Toshiba 13 ns, Signetics 15 ns, SGS 18 ns. National data not available

[m] TI and Toshiba; RCA and Signetics 3 ns, SGS 6 ns. National data not available

[n] RCA and Toshiba; SGS 18 MHz, Signetics 22 MHz, TI 25 MHz. National data not available

[o] TI; RCA and Signetics 44 ns, SGS 50 ns, Toshiba 51 ns. National data not available

[p] TI 38 ns. National data not available

[q] SGS 40 ns, Toshiba 46 ns. National data not available

[r] National 27 mA

[s] National 13 ns

[t] RCA 60 pF/channel

[u] RCA; comparable Fairchild data not available

[v] RCA 72 pF/channel

[w] RCA 6.7 ns

[x] RCA 2 ns

[y] RCA 10.6 ns

[z] RCA 13 ns

576   8-bit 3-state inverting registers

MANUFACTURERS	TTL	LS	HC	HCT	ALS/A	S	AS	F	AC	ACT
National Semiconductor					■		■			
Texas Instruments										

Missing change control symbol "A": National 74ALS576

KEY PARAMETERS		TTL	LS	HC	HCT	ALS	S	AS	F	AC	ACT
I_{OH} (max.)	mA					-2.6		-15			
I_{OL} (max.)	mA					24		48			
I_{IH} (max.)	mA					0.02		0.02			
I_{IL} (max.)	mA					-0.2		-0.5^c			
I_{CC} (max.)	mA					30^a		135			
t_W (min.)	ns					16.5		4^d			
t_{SU} (min.)	ns					15		2			
t_H (min.)	ns					0		2			
f_{MAX} (min.)	MHz					30		125			
t_{pd} CP to $\overline{Q}$ (max.)	ns					14		9			
t_{en} (max.)	ns					18		10			
t_{dis} (max.)	ns					15^b		6			

[a] National 27 ns

[b] National 13 ns

[c] National; TI has I_{IL} (max.) of -0.5 mA for all inputs except the D inputs, where I_{IL} (max.) is -2 mA

[d] CP HIGH; t_W CP LOW (min.) is 2 ns

8-BIT OPEN-COLLECTOR BUS REGISTERS WITH CLOCK ENABLE

This single-sourced component offers an open-collector alternative to 3-state bus registers like the 374 (see above).

383§   *8-bit bus registers with clock enable and open-collector outputs.* An open-collector S version of the 374 ("neighboring" pinout). A clock enable replaces the usual output enable at pin 1 of the 20-pin package. When taken HIGH, $\overline{CE}$ inhibits the clock pulse from triggering the register, leaving its contents undisturbed regardless of changes at the register's data or clock inputs. This type, the only available 8-bit TTL register with open-collector outputs, is especially suited to microprocessor-based systems in which a common bus serves a number of physically separated devices. See manufacturer's data on this component for a method of calculating R_L in wired-AND applications. (Monolithic Memories)

I_{OH} (max.)	0.25 mA	t_W	(min.)	7 ns
I_{OL} (max.)	24 mA	t_{SU}	(min.)	5* ns
I_{IH} (max.)	0.05 mA	t_H	(min.)	3† ns
I_{IL} (max.)	−0.25 mA	f_{MAX}	(min.)	75 MHz
I_{CC} (max.)	160 mA	t_{pd} CP to Q	(max.)	22 ns

* D to CP; t_{SU} $\overline{CE}$ to CP (min.) is 9 ns
† D and $\overline{CE}$ LOW; t_H $\overline{CE}$ HIGH (min.) is 0 ns

FUNCTION TABLE
383§

INPUTS			OUTPUTS
$\overline{CE}$	CP	D	Q
L	↑	L	L
L	↑	H	H
H	↑	.	NO CHANGE
.	−	.	NO CHANGE

. = Either LOW or HIGH level
− = Clock inactive (L, H, or ↑)

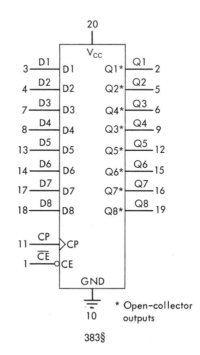

383§

8-BIT BUS REGISTERS WITH SYNCHRONOUS CLEAR

The two types listed here add a synchronous clear capability to the basic 8-bit bus register. Adding the clear input (pin 1) requires an expansion of the basic 20-pin package to the next larger standard size, a 24-pin DIP; the three extra pins (11, 13, 23) are left unused.

575   *8-bit 3-state registers with synchronous clear.* ALS and AS versions of the 574 (bus-structured pinout) with added synchronous clear capability. When held LOW, the $\overline{R}$ input enables the clock to load a LOW level into all eight flip-flops of the register on the next high-going clock transition, clearing the register in synchronization with the clock pulse.

577   *8-bit 3-state inverting registers with synchronous clear.* Similar to the 575, but with inverting outputs and different clear (reset) logic. When held LOW, the $\overline{R}$ input of the 577 will cause the next high-going clock transition to synchronously load a HIGH level into the eight flip-flops of the register rather than a LOW level as in the 575. Since the outputs of the 577 are inverted, this opposite effect is consistent with the overall logic of the circuit, and the input therefore continues to be named "clear" or "reset" rather than "set" or "preset."

INPUTS				OUTPUTS
$\overline{OE}$	$\overline{R}$	CP	D	Q
L	H	↑	L	L
L	H	↑	H	H
L	H	–	.	NO CHANGE
L	L	↑	.	L
H	H	.	.	Z

. = Either LOW or HIGH level
– = Clock inactive (L, H, or ↑)
Z = High impedance (disabled)

INPUTS				OUTPUTS
$\overline{OE}$	$\overline{R}$	CP	D	$\overline{Q}$
L	H	↑	L	H
L	H	↑	H	L
L	H	–	.	NO CHANGE
L	L	↑	.	H
H	H	.	.	Z

. = Either LOW or HIGH level
– = Clock inactive (L, H, or ↑)
Z = High impedance (disabled)

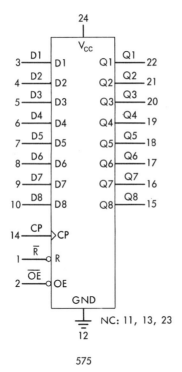

575

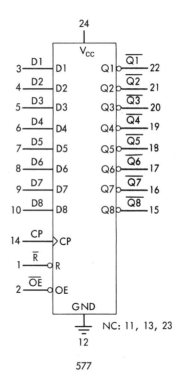

577

313

575   8-bit 3-state registers with synchronous clear

MANUFACTURERS	TTL	LS	HC	HCT	ALS/A	S	AS	F	AC	ACT
National Semiconductor							■			
Texas Instruments					■		■			

KEY PARAMETERS		TTL	LS	HC	HCT	ALS	S	AS	F	AC	ACT
I_{OH} (max.)	mA					-2.6		-15			
I_{OL} (max.)	mA					24		48			
I_{IH} (max.)	mA					0.02		0.02			
I_{IL} (max.)	mA					-0.2		-0.5[a]			
I_{CC} (max.)	mA					30		142			
t_W (min.)	ns					16.5		4[b]			
t_{SU} (min.)	ns					15		2[c]			
t_H (min.)	ns					0		2[d]			
f_{MAX} (min.)	MHz					30		125			
t_{pd} CP to Q (max.)	ns					14		9			
t_{en} (max.)	ns					18		10			
t_{dis} (max.)	ns					13		6			

[a] National; TI has I_{IL} (max.) of -0.5 mA for all inputs except the D inputs, where I_{IL} (max.) is -2 mA
[b] CP HIGH; t_W CP LOW (min.) is 2 ns
[c] D to CP; t_{SU} $\overline{R}$ (min.) is 5.5 ns
[d] D inputs; t_H $\overline{R}$ (min.) is 0 ns

577   8-bit 3-state inverting registers with synchronous clear

MANUFACTURERS	TTL	LS	HC	HCT	ALS/A	S	AS	F	AC	ACT
National Semiconductor							■			
Texas Instruments					■		■			

KEY PARAMETERS		TTL	LS	HC	HCT	ALS	S	AS	F	AC	ACT
I_{OH} (max.)	mA					-2.6		-15			
I_{OL} (max.)	mA					24		48			
I_{IH} (max.)	mA					0.02		0.02			
I_{IL} (max.)	mA					-0.2		-0.5[a]			
I_{CC} (max.)	mA					30		142			
t_W (min.)	ns					16.5		4[b]			
t_{SU} (min.)	ns					15		2[c]			
t_H (min.)	ns					0		2[d]			
f_{MAX} (min.)	MHz					30		125			
t_{pd} CP to $\overline{Q}$ (max.)	ns					14		9			
t_{en} (max.)	ns					18		10			
t_{dis} (max.)	ns					15		6			

[a] National; TI has I_{IL} (max.) of -0.5 mA for all inputs except the D inputs, where I_{IL} (max.) is -2 mA
[b] CP HIGH; t_W CP LOW (min.) is 2 ns
[c] D to CP; t_{SU} $\overline{R}$ (min.) is 5.5 ns
[d] D inputs; t_H $\overline{R}$ (min.) is 0 ns

8-BIT BUS REGISTERS WITH MULTIPLE CONTROLS

For applications in which a register must be controlled by more complex logic than that afforded by just a clock and single output enable, the two types described below offer a more complete set of control inputs at the price of four additional pins, which enlarge the 20-pin complement of the ordinary 8-bit bus register to 24.

Both of these types were originally designed by Advanced Micro Devices, which continues to supply them in its 29800 and 29800A series of bus interface components. While not identical to the usual 7400-series technology categories, these families are fully TTL-compatible and offer an important alternative to other sources. In the data pages that follow, the 29825 and 29826 (which have performance characteristics that fall roughly between those of S and AS devices) are listed in the columns usually reserved for S components, while the improved 29825A is described along with the 74AS825, which it strongly resembles.

825 *8-bit 3-state registers with clock enable, clear, and triple output enables.* Bus registers with bus-structured pinout and a full set of control inputs, consisting of clock pulse, CP; active-LOW clock enable, $\overline{\text{CE}}$; active-LOW reset or clear, $\overline{\text{R}}$; and three NOR-gated output enables, $\overline{\text{OE1}}$, $\overline{\text{OE2}}$, and $\overline{\text{OE3}}$. The 3-state outputs are in the high-impedance condition whenever any of the three output enables is taken HIGH, and enabled (put on the bus) only when all three output enables are held LOW. Thus, any one (or more) of up to three "users" can switch the register off the bus on its own initiative, without the permission of any of the other users. The clear (reset) control, when taken LOW, will override any other input condition and cause a LOW level to be stored in all eight flip-flops of the register. The clock enable, when taken HIGH, will inhibit the CP input, preserving the internal condition of the register regardless of changes at the clock or data inputs. TI's ALS and AS versions of the 825 feature "power-up high impedance state." Corresponding bus latch: 845.

826 *8-bit 3-state inverting registers with clock enable, clear, and triple output enables.* Same as 825, but inverting. NOTE: The outputs of the 826 are conventionally considered to be noninverting and the act of inversion is considered to take place at the inputs instead. This is logically the same as calling the data inputs normal and the outputs inverting, but it has the advantage of keeping the reset ($\overline{\text{R}}$) designation and functional description the same as that of the 825. In both cases, a LOW at $\overline{\text{R}}$ will asynchronously load a LOW level in all eight flip-flops of the register. Corresponding bus latch: 846.

FUNCTION TABLE
825

INPUTS							OUTPUTS	
$\overline{OE1}$	$\overline{OE2}$	$\overline{OE3}$	$\overline{R}$	$\overline{CE}$	CP	D	Q	
L	L	L	H	H	↑	L	L	Read data
L	L	L	H	H	↑	H	H	
L	L	L	H	L	↑	•	NO CHANGE	Clock inhibited
L	L	L	H	•	−	•	NO CHANGE	Clock inactive
L	L	L	L	•	•	•	L	Direct reset (clear)
H	•	•	•	•	•	•	Z	
•	H	•	•	•	•	•	Z	Outputs off
•	•	H	•	•	•	•	Z	

• = Either LOW or HIGH level
− = Clock inactive (L, H, or ↑)
Z = High impedance (disabled)

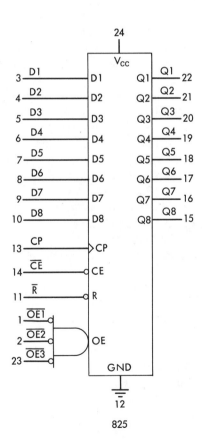

825

FUNCTION TABLE
826

INPUTS							OUTPUTS	
$\overline{OE1}$	$\overline{OE2}$	$\overline{OE3}$	$\overline{R}$	$\overline{CE}$	CP	$\overline{D}$	Q	
L	L	L	H	H	↑	H	L	Read data
L	L	L	H	H	↑	L	H	
L	L	L	H	L	↑	.	NO CHANGE	Clock inhibited
L	L	L	H	.	—	.	NO CHANGE	Clock inactive
L	L	L	L	.	.	.	L	Direct reset (clear)
H	.	.	.	.	.	.	Z	
.	H	.	.	.	.	.	Z	Outputs off
.	.	H	.	.	.	.	Z	

. = Either LOW or HIGH level
— = Clock inactive (L, H, or ↑)
Z = High impedance (disabled)

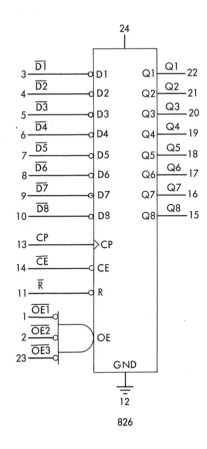

826

MANUFACTURERS	TTL	LS	HC	HCT	ALS	*	AS	F	AC	ACT
Advanced Micro Devices						▓▓				
Fairchild								D	▓▓	▓▓
Signetics							▓▓			
Texas Instruments							▓▓			

The * column refers to the Am29825 (comparable to S and AS)
The AS column includes the Am29825A, which is similar to the 74AS825

KEY PARAMETERS		TTL	LS	HC	HCT	ALS	*	AS	F	AC	ACT
I_{OH} (max.)	mA						-24	-24	-3	-24	-24
I_{OL} (max.)	mA						48	48	48	24	24
I_{IH} (max.)	mA						0.05	0.02^b	0.07^i		
I_{IL} (max.)	mA						-1^a	-0.5	-0.07^j		
I_{CC} (max.)	mA						140	95^c	86		
I_{CC} (quiesc.)	mA									0.08	0.08
C_{pd} (typ.)	pF/channel									n/a	n/a
t_W (min.)	ns						7	8^d	n/a	n/a	n/a
t_{SU} (min.)	ns						4	6^e	n/a	n/a	n/a
t_H (min.)	ns						2	0^f	n/a	n/a	n/a
f_{MAX} (min.)	MHz						n/a	n/a	n/a	n/a	n/a
t_{pd} CP to Q (max.)	ns						12	11^g	n/a	n/a	n/a
t_{pd} $\overline{R}$ to Q (max.)	ns						20	13^h	n/a	n/a	n/a
t_{en} (max.)	ns						14	12	n/a	n/a	n/a
t_{dis} (max.)	ns						16	8	n/a	n/a	n/a

[a] D and $\overline{R}$ inputs; I_{IL} CP, $\overline{CE}$, $\overline{OE}$ (max.) is -2 mA
[b] AMD 0.05 mA
[c] AMD 94 mA
[d] CP (AMD 7 ns); t_W $\overline{R}$ LOW is 4 ns (AMD 6 ns)
[e] D to CP (AMD 4 ns); t_{SU} $\overline{CE}$ to CP is 6 ns. TI further states that t_{SU} $\overline{R}$ HIGH (min.) is 8 ns

[f] AMD 2 ns
[g] AMD 10 ns
[h] AMD 14 ns
[i] Including 0.05 mA off-state output current
[j] Including -0.05 mA off-state output current

826   8-bit 3-state inverting registers with clock enable, clear, triple $\overline{OE}$

MANUFACTURERS	TTL	LS	HC	HCT	ALS	*	AS	F	AC	ACT
Advanced Micro Devices						■				
Fairchild								D	▓	▓
Signetics							▓			
Texas Instruments						■				

The * column refers to the Am29826 (comparable to S and AS)

KEY PARAMETERS		TTL	LS	HC	HCT	ALS	*	AS	F	AC	ACT
I_{OH} (max.)	mA						-24	-24	-3	-24	-24
I_{OL} (max.)	mA						48	48	48	24	24
I_{IH} (max.)	mA						0.05	0.02	0.07^d		
I_{IL} (max.)	mA						-1^a	-0.5	-0.07^e		
I_{CC} (max.)	mA						140	95	110		
I_{CC} (quiesc.)	mA									0.08	0.08
C_{pd} (typ.)	pF/channel									n/a	n/a
t_W (min.)	ns						7	8^b	n/a	n/a	n/a
t_{SU} (min.)	ns						4	6^c	n/a	n/a	n/a
t_H (min.)	ns						2	0	n/a	n/a	n/a
f_{MAX} (min.)	MHz						n/a	n/a	n/a	n/a	n/a
t_{pd} CP to Q (max.)	ns						12	11	n/a	n/a	n/a
t_{pd} $\overline{R}$ to Q (max.)	ns						20	13	n/a	n/a	n/a
t_{en} (max.)	ns						14	12	n/a	n/a	n/a
t_{dis} (max.)	ns						16	8	n/a	n/a	n/a

^{a} D and $\overline{R}$ inputs; I_{IL} CP, $\overline{CE}$, $\overline{OE}$ (max.) is -2 mA ^{d} Including 0.05 mA off-state output current
^{b} CP; t_W $\overline{R}$ LOW (min.) is 4 ns ^{e} Including -0.05 mA off-state output current
^{c} D or $\overline{CE}$ to CP; t_{SU} $\overline{R}$ HIGH (min.) is 8 ns

8-BIT READBACK AND DIAGNOSTIC REGISTERS

Two special-purpose versions of the standard 8-bit bus register are available for certain microprocessor-based system applications. See manufacturer's data for logic diagrams and further details on these single-sourced designs.

794§   *8-bit registers with readback.* LS and ACT components designed for I/O operations between the data bus of a microprocessor and one of its peripheral devices. The pinout is identical to a standard bus-structured 8-bit 3-state bus register like the 574, but a readback enable pin takes the place of the usual 3-state enable input. In ordinary operation the register reads 8-bit words from the data bus and provides these as output to the peripheral. When the readback enable is activated, the register's D inputs are enabled as bus outputs, placing the data word stored in the register back onto the microprocessor data bus from which it came. This allows the microprocessor to read the register as if it were an I/O port in order to verify or update the output data without storing a redundant copy of the data word in memory. Corresponding bus latch: 793§. (Monolithic Memories)

LS794

I_{OH} (max.)	−2.6	mA	t_W	(min.)	15 ns
I_{OL} (max.)	24	mA	t_{SU}	(min.)	15 ns
I_{IH} (max.)	0.04	mA	t_H	(min.)	0 ns
I_{IL} (max.)	−0.25	mA	f_{MAX}	(min.)	35 MHz
I_{CC} (max.)	120	mA	t_{pd} CP to Q	(max.)	20 ns
			t_{en}	(max.)	20 ns
			t_{dis}	(max.)	20 ns

ACT794

I_{OH} (max.)	−6	mA	t_W	(min.)	15 ns
I_{OL} (max.)	12	mA	t_{SU}	(min.)	25 ns
I_{CC} (quiesc.)	0.08	mA	t_H	(min.)	0 ns
			t_{pd} CP to Q	(max.)	40 ns
			t_{en}	(max.)	30 ns
			t_{dis}	(max.)	30 ns

818   *8-bit diagnostic registers.* These 24-pin components act like ordinary 8-bit registers but have four additional pins—diagnostic clock, mode, serial data input, and serial data output—to load, read, and control a separate 8-bit "shadow register" and internal multiplexer. The register can be made to perform a number of useful diagnostic functions while playing the role of a microprogram control storage register or microprocessor instruction, data, status, or address register. The device can also function as an interrupt mask register, pipeline register, general-purpose register, or parallel-serial/serial-parallel converter.

818   8-bit diagnostic registers

MANUFACTURERS	TTL	LS	HC	HCT	ALS	S	AS	F	AC	ACT
Fairchild									▓▓	▓▓
Monolithic Memories						▓				

KEY PARAMETERS		TTL	LS	HC	HCT	ALS	S	AS	F	AC	ACT
I_{OH} (max.)	mA						-6.5			-24	-24
I_{OL} (max.)	mA						32			24	24
I_{IH} (max.)	mA						0.05				
I_{IL} (max.)	mA						-0.25				
I_{CC} (max.)	mA						145				
I_{CC} (quiesc.)	mA									1	1
C_{pd} (typ.)	pF/channel									n/a	n/a
t_W CP (min.)	ns						13			n/a	n/a
t_{SU} D to CP (min.)	ns						14			n/a	n/a
t_H D (min.)	ns						0			n/a	n/a
f_{MAX} (min.)	MHz						40^a			n/a	n/a
t_{pd} CP to Q (max.)	ns						20^b			n/a	n/a
t_{en} (max.)	ns						20^b			n/a	n/a
t_{dis} (max.)	ns						25			n/a	n/a

[a] C_L = 50 pF. f_{MAX} (min.) of diagnostic clock is 20 MHz   [b] C_L = 50 pF
(cascaded) or 25 MHz (not cascaded)

8-BIT MULTIFUNCTIONAL REGISTERS WITH SYNCHRONOUS PRESET AND CLEAR

These single-sourced registers conveniently provide a number of general-purpose bus interface functions in a single 24-pin package.

380§   *8-bit 3-state multifunctional bus registers with synchronous true and complement load, preset, and clear.* These LS components use the extra four pins of a 24-pin DIP to provide synchronous $\overline{S}$ and $\overline{R}$ inputs as well as two parallel load controls, $\overline{LD}$ and POL. A LOW level held at $\overline{S}$ will cause the next high-going clock transition to load a HIGH level into each of the eight flip-flops, while the same signal applied at $\overline{R}$ will cause the positive clock edge to load all LOW levels into the register. If $\overline{LD}$ is held HIGH, the clock is disabled and the register is put into "hold" mode; if LOW, $\overline{LD}$ enables the clock to trigger the parallel load (read data) operation. The POL input determines the load logic; if POL is HIGH, the inputs are loaded unchanged into the register, and if POL is LOW, complements of each input are loaded

into the flip-flops, effectively transforming a non-inverting regis-
ter into an inverting one. $\overline{R}$ overrides $\overline{S}$ and $\overline{S}$ overrides $\overline{LD}$. See
manufacturer's data for function table and logic diagrams. (Mon-
olithic Memories)

I_{OH} (max.)	-3.2	mA	t_W	(min.)	40 ns
I_{OL} (max.)	24	mA	t_{SU}	(min.)	50 ns
I_{IH} (max.)	0.025	mA	t_H	(min.)	0 ns
I_{IL} (max.)	-0.25	mA	f_{MAX}	(min.)	12.5 MHz
I_{CC} (max.)	180	mA	t_{pd} CP to Q	(max.)	30 ns
			t_{en}	(max.)	45 ns
			t_{dis}	(max.)	45 ns

8/8-BIT BUS REGISTERS WITH MULTIPLEXED OUTPUTS

These 28-pin devices each contain two complete 8-bit registers, together with two
separate 8-line input sections that simultaneously load separate 8-bit words into
the two registers, a bus-driver output section, and a multiplexer that determines
which of the two data words simultaneously stored in the device will be placed on
its output bus. Multiplexed registers are especially useful where an output bus
must receive data selected from one of two input buses, or where a single 16-bit
word must be read in one clock cycle and then sequentially multiplexed out as two
8-bit data words—for example, where a single 16-bit microprocessor memory
address is applied as 8-bit row and column addresses to a 64K RAM circuit.

To keep the pin count to a minimum, no separate output enables are provided for these registers. Instead, output impedance is controlled by the clock
pulse. The driver outputs are in the high impedance or OFF state when the clock
is LOW and are enabled when the clock is HIGH.

The four types making up this category (604–607) can be classified into two
major subcategories according to output structure: the 604 and 606 have 3-state
outputs, and the 605 and 607 have open-collector outputs. Within each class these
very application-oriented devices provide a further pair of options: the 604 and
605 are optimized for high speed, while the 606 and 607 have been specially
designed to eliminate decoding voltage spikes.

604   *8/8-bit 3-state registers with multiplexed outputs.* LS, HC, and F
devices that simultaneously read and store two 8-bit words on the
positive edge of the clock and place one of these two words on an
output bus during the HIGH period of the clock cycle. Register
A or B is selected for output by the level at $A/\overline{B}$ (pin 2). Optimized for high speed.

606   *8/8-bit 3-state registers with multiplexed outputs.* LS version of the 604 that has been optimized for maximum suppression of decoding spikes rather than maximum speed.

605   *8/8-bit registers with multiplexed open-collector outputs.* Open-collector LS and F versions of the 604 (optimized for high speed.)

607   *8/8-bit registers with multiplexed open-collector outputs.* Open-collector LS versions of the 606 (optimized for maximum suppression of decoding spikes).

FUNCTION TABLE
604, 606

INPUTS				OUTPUTS
CP	A/$\overline{B}$	A	B	Q
⬆	L	data	data	A data
⬆	H	data	data	B data
H	L	.	.	stored A
H	H	.	.	stored B
L	.	.	.	Z

. = Either LOW or HIGH level
Z = High impedance (disabled)

FUNCTION TABLE
605, 607

INPUTS				OUTPUTS
CP	A/$\overline{B}$	A	B	Q
	L	data	data	A data
	H	data	data	B data
H	L	.	.	stored A
H	H	.	.	stored B
L	.	.	.	OFF

. = Either LOW or HIGH level
OFF = High impedance; H if pull-up
resistor is connected to
open-collector output

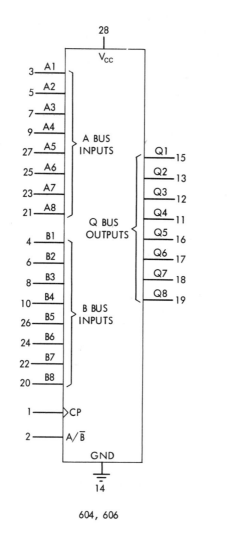

604, 606   605, 607

* Open-collector
outputs

604   8/8-bit 3-state registers with multiplexed outputs

MANUFACTURERS	TTL	LS	HC	HCT	ALS	S	AS	F	AC	ACT
Fairchild								D		
Motorola		■								
Signetics								■		
Texas Instruments		■								

KEY PARAMETERS		TTL	LS	HC	HCT	ALS	S	AS	F	AC	ACT
I_{OH} (max.)	mA		-2.6	-6					-3		
I_{OL} (max.)	mA		24	6					24		
I_{IH} (max.)	mA		0.02						0.07^b		
I_{IL} (max.)	mA		-0.4^a						-0.07^c		
I_{CC} (max.)	mA		70						100		
I_{CC} (quiesc.)	mA			0.08							
C_{pd} (typ.)	pF/channel			100							
t_W (min.)	ns		20	25					6		
t_{SU} (min.)	ns		20	19					3		
t_H (min.)	ns		0	5					1.5		
f_{MAX} (min.)	MHz		n/a	n/a					80		
t_{pd} A/$\overline{B}$ to Q (max.)	ns		45	43					11.5		
t_{en} from CP (max.)	ns		40	49					12		
t_{dis} from CP (max.)	ns		30	50					11		

[a] Data inputs; I_{IL} (max.) at CP and A/$\overline{B}$ inputs is -0.2 mA   [c] Including -0.05 mA off-state output current
[b] Including 0.05 mA off-state output current

606   8/8-bit 3-state registers with multiplexed outputs

MANUFACTURERS	TTL	LS	HC	HCT	ALS	S	AS	F	AC	ACT
Motorola		■								
Texas Instruments		■								

KEY PARAMETERS		TTL	LS	HC	HCT	ALS	S	AS	F	AC	ACT
I_{OH} (max.)	mA		-2.6								
I_{OL} (max.)	mA		24								
I_{IH} (max.)	mA		0.02								
I_{IL} (max.)	mA		-0.4^a								
I_{CC} (max.)	mA		70								
t_W (min.)	ns		20								
t_{SU} (min.)	ns		20								
t_H (min.)	ns		0								
f_{MAX} (min.)	MHz		n/a								
t_{pd} A/$\overline{B}$ to Q (max.)	ns		50								
t_{en} from CP (max.)	ns		50								
t_{dis} from CP (max.)	ns		30								

[a] Data inputs; I_{IL} (max.) at CP and A/$\overline{B}$ inputs is -0.2 mA

605   8/8-bit registers with multiplexed open-collector outputs

MANUFACTURERS	TTL	LS	HC	HCT	ALS	S	AS	F	AC	ACT
Motorola		■								
Signetics								■		
Texas Instruments		■								

KEY PARAMETERS		TTL	LS	HC	HCT	ALS	S	AS	F	AC	ACT
I_{OH} (max.)	mA		0.25						1		
I_{OL} (max.)	mA		24						24		
I_{IH} (max.)	mA		0.02						0.02		
I_{IL} (max.)	mA		-0.4[a]						-0.02		
I_{CC} (max.)	mA		60						105		
t_W (min.)	ns		20						6		
t_{SU} (min.)	ns		20						4		
t_H (min.)	ns		0						3		
f_{MAX} (min.)	MHz		n/a						80		
t_{pd} A/$\overline{B}$ to Q (max.)	ns		60						14.5		
t_{pd} CP (max.)	ns		40						14.5		

[a] Data inputs; I_{IL} (max.) at CP and A/$\overline{B}$ inputs is -0.2 mA

607   8/8-bit registers with multiplexed open-collector outputs

MANUFACTURERS	TTL	LS	HC	HCT	ALS	S	AS	F	AC	ACT
Motorola		■								
Texas Instruments		■								

KEY PARAMETERS		TTL	LS	HC	HCT	ALS	S	AS	F	AC	ACT
I_{OH} (max.)	mA		0.25								
I_{OL} (max.)	mA		24								
I_{IH} (max.)	mA		0.02								
I_{IL} (max.)	mA		-0.4[a]								
I_{CC} (max.)	mA		60								
t_W (min.)	ns		20								
t_{SU} (min.)	ns		20								
t_H (min.)	ns		0								
f_{MAX} (min.)	MHz		n/a								
t_{pd} A/$\overline{B}$ to Q (max.)	ns		70								
t_{pd} CP (max.)	ns		45								

[a] Data inputs; I_{IL} (max.) at CP and A/$\overline{B}$ inputs is -0.2 mA

8/8-BIT PIPELINED BUS REGISTERS

Like the 8/8-bit registers described above, these single-sourced components contain two separate 8-bit registers plus output multiplexing circuitry that allows the contents of either register to be placed on the output bus. Rather than twin sets of 8-line inputs, however, these registers have just a single set of eight input lines that are routed by input multiplexing circuitry to one of the two internal registers. In other words, both inputs and outputs are multiplexed to and from the internal registers. The elimination of eight input lines allows the pin count to be reduced from 28 to 24, while still leaving four extra pins for added control inputs that provide greatly increased flexibility of operation.

548§   *8/8-bit 3-state pipelined registers.* LS components that provide two 8-bit registers in the same 24-pin DIP. Internal multiplexing circuitry allows the registers to be configured for nose-to-tail or side-by-side operation. Applications include two-stage buffers for pipelined I/O, automatic backup storage for diagnostic purposes, video display character/attribute registers, etc. Corresponding bus latch: 549§. See manufacturer's data for further information. (Monolithic Memories)

I_{OH} (max.)	-2.6	mA	t_W	(min.)	11 ns
I_{OL} (max.)	32	mA	t_{SU}	(min.)	15 ns
I_{IH} (max.)	0.02	mA	t_H	(min.)	0 ns
I_{IL} (max.)	-0.25^*	mA	f_{MAX}	(min.)	45 MHz
I_{CC} (max.)	150	mA	t_{pd} CP to Q	(max.)	20 ns
			t_{en}	(max.)	20 ns
			t_{dis}	(max.)	17 ns

9-BIT BUS REGISTERS

Byte-oriented systems operate on data in 8-bit groups and are therefore usually well served by 8-bit components like the 8-bit registers listed above. In many applications, however, one or two extra bits per byte are needed for parity checking, expanded addressing, extra control lines, or other special requirements, resulting in buses whose lines are organized in groups of nine or ten. A minimum set of 9-bit and 10-bit registers is available for these applications. The 9-bit registers are listed below, and the 10-bit registers follow separately.

Both of the 9-bit designs described here were originally manufactured by Advanced Micro Devices, which continues to provide them in its 29800, 29800A, and 29C800 series of bus interface components. While not identical to the normal 7400-series technologies, the AMD families are fully TTL-compatible and provide an important source for these types. In the data pages that follow, the 29823 and 29824 (comparable to S or AS) are described in the columns usually reserved for S components; the improved 29823A is included with the 74AS823, which it strongly resembles; and the 29C823 (high-speed TTL-compatible CMOS) is included in the ACT column.

All of these 9-bit registers are 3-state devices supplied in 24-pin DIPs.

823   *9-bit 3-state registers with clock enable and clear.* Implementations of the basic 9-bit bus register. Data input (D) and output (Q) pins are assigned in a bus-structured configuration similar to that of the bus-structured 8-bit types such as the 574. Assignment of clock pulse (CP), active-LOW output enable ($\overline{OE}$), V_{CC}, and ground pins leaves two remaining control inputs, which are assigned the functions of active-LOW clock enable, $\overline{CE}$, and active-LOW direct clear (reset), $\overline{R}$. TI's ALS and AS versions of the 823 feature "power-up high impedance state." Corresponding bus latch: 843.

824   *9-bit 3-state inverting registers with clock enable and clear.* Same as 823, but inverting. NOTE: As in the case of the 8-bit inverting registers with clear, the 824 is considered to have noninverted outputs (Q), the inversion being considered to take place at the inputs ($\overline{D}$) instead. This convention has no effect on the basic data logic of the register, but it keeps the pin labeling and logical description of the 824's $\overline{R}$ inputs identical to that of the 823. Corresponding bus latch: 844.

FUNCTION TABLE
823

INPUTS					OUTPUTS	
$\overline{OE}$	$\overline{R}$	$\overline{CE}$	CP	D	Q	
L	H	L	↑	L	L	Read data
L	H	L	↑	H	H	
L	H	H	↑	•	NO CHANGE	Clock inhibited
L	H	•	–	•	NO CHANGE	Clock inactive
L	L	•	•	•	L	Direct clear (reset)
H	•	•	•	•	Z	Outputs off

• = Either LOW or HIGH level
– = Clock inactive (L, H, or ↓)
Z = High impedance (disabled)

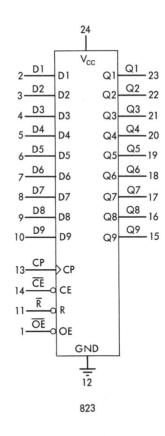

823

FUNCTION TABLE
824

INPUTS					OUTPUTS	
$\overline{OE}$	$\overline{R}$	$\overline{CE}$	CP	$\overline{D}$	Q	
L	H	L	↑	H	L	Read data
L	H	L	↑	L	H	
L	H	H	↑	.	NO CHANGE	Clock inhibited
L	H	.	−	.	NO CHANGE	Clock inactive
L	L	.	.	.	L	Direct clear (reset)
H	.	.	.	.	Z	Outputs off

. = Either LOW or HIGH level
− = Clock inactive (L, H, or ↓)
Z = High impedance (disabled)

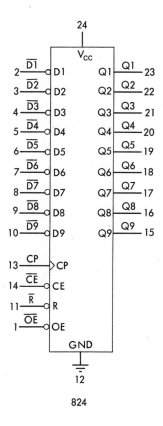

824

823   9-bit 3-state registers with clock enable and clear

MANUFACTURERS	TTL	LS	HC	HCT	ALS	*	AS	F	AC	ACT
Advanced Micro Devices						███				███
Fairchild								D		
Signetics								▒▒▒		
Texas Instruments						███				

The * column refers to the Am29823 (comparable to S and AS)
The AS column includes the Am29823A, which is similar to the 74AS823
The ACT column includes the Am29C823, which is comparable to the 74ACT823

KEY PARAMETERS		TTL	LS	HC	HCT	ALS	*	AS	F	AC	ACT
I_{OH} (max.)	mA						-24	-24	-3	-24	-24[i]
I_{OL} (max.)	mA						48	48	48	24	24
I_{IH} (max.)	mA						0.05	0.02[b]	0.07[m]		
I_{IL} (max.)	mA						-1[a]	-0.5	-0.07[n]		
I_{CC} (max.)	mA						140	103[c]	110		
I_{CC} (quiesc.)	mA									0.08	0.08[j]
C_{pd} (typ.)	pF/channel									n/a	k
t_W (min.)	ns						7	8[d]	n/a	n/a	7[l]
t_{SU} (min.)	ns						4	6[e]	n/a	n/a	4[l]
t_H (min.)	ns						2	0[f]	n/a	n/a	2[l]
f_{MAX} (min.)	MHz						n/a	n/a	n/a	n/a	n/a
t_{pd} CP to Q (max.)	ns						12	11[g]	n/a	n/a	12[l]
t_{pd} $\overline{R}$ to Q (max.)	ns						20	13[h]	n/a	n/a	13[l]
t_{en} (max.)	ns						14	12	n/a	n/a	12[l]
t_{dis} (max.)	ns						16	8	n/a	n/a	12[l]

[a] D and $\overline{R}$ inputs; I_{IL} CP, $\overline{CE}$, $\overline{OE}$ (max.) is -2 mA
[b] AMD 0.05 mA
[c] AMD 100 mA
[d] CP (AMD 7 ns); t_W $\overline{R}$ LOW is 4 ns (AMD 6 ns)
[e] D to CP (AMD 4 ns); t_{SU} $\overline{CE}$ to CP is 6 ns. TI further states that t_{SU} $\overline{R}$ HIGH (min.) is 8 ns
[f] AMD 2 ns
[g] AMD 10 ns
[h] AMD 14 ns
[i] AMD -15 mA
[j] AMD 0.12 mA
[k] AMD specifies dynamic supply current I_{CCD} (max.) of 0.275 mA/MHz/bit; Fairchild C_{pd} not given
[l] AMD; comparable Fairchild data not available
[m] Including 0.05 mA off-state output current
[n] Including -0.05 mA off-state output current

824   9-bit 3-state inverting registers with clock enable and clear

MANUFACTURERS	TTL	LS	HC	HCT	ALS	*	AS	F	AC	ACT
Advanced Micro Devices						■				
Fairchild								D	▨	▨
Signetics							▨			
Texas Instruments						■				

The * column refers to the Am29824 (comparable to S and AS)

KEY PARAMETERS		TTL	LS	HC	HCT	ALS	*	AS	F	AC	ACT
I_{OH} (max.)	mA						-24	-24	-3	-24	-24
I_{OL} (max.)	mA						48	48	48	24	24
I_{IH} (max.)	mA						0.05	0.02	0.07^d		
I_{IL} (max.)	mA						-1^a	-0.5	-0.07^e		
I_{CC} (max.)	mA						140	103	110		
I_{CC} (quiesc.)	mA									0.08	0.08
C_{pd} (typ.)	pF/channel									n/a	n/a
t_W (min.)	ns						7	8^b	n/a	n/a	n/a
t_{SU} (min.)	ns						4	6^c	n/a	n/a	n/a
t_H (min.)	ns						2	0	n/a	n/a	n/a
f_{MAX} (min.)	MHz						n/a	n/a	n/a	n/a	n/a
t_{pd} CP to Q (max.)	ns						12	11	n/a	n/a	n/a
t_{pd} $\overline{R}$ to Q (max.)	ns						20	13	n/a	n/a	n/a
t_{en} (max.)	ns						14	12	n/a	n/a	n/a
t_{dis} (max.)	ns						16	8	n/a	n/a	n/a

[a] D and $\overline{R}$ inputs; I_{IL} CP, $\overline{CE}$, $\overline{OE}$ (max.) is -2 mA
[b] CP; t_W $\overline{R}$ LOW (min.) is 4 ns
[c] D or $\overline{CE}$ to CP; t_{SU} $\overline{R}$ HIGH (min.) is 8 ns
[d] Including 0.05 mA off-state output current
[e] Including -0.05 mA off-state output current

10-BIT BUS REGISTERS

Like the 9-bit registers, these 10-bit types are all 3-state bus-driving devices supplied in 24-pin DIPs. They provide the widest data path possible in a non-multiplexed configuration limited to 24 pins. Since two additional pins are needed for the tenth input and output lines, the clock enable and clear provided in the 9-bit versions must be omitted to keep the 24-pin package, leaving the 10-bit types with just the basic 3-state register functions.

Advanced Micro Devices originated and remains an important source for these types in its 29800, 29800A, and 29C800 families of TTL-compatible bus interface circuits. As with the 9-bit registers, the 29821 and 29822 (comparable to S or AS devices) are listed in the columns usually reserved for S components, while the 29C821 (high-speed TTL-compatible CMOS) is included with the 74ACT821, and the 29821A is included with the 74AS821, which it strongly resembles.

821   *10-bit 3-state registers*. Components with a simple bus-struc-
           tured 10-line data pinout very similar to that of the 9-bit registers,
           plus the minimum clock (CP) and active-LOW output enable
           ($\overline{\text{OE}}$) controls. Logic is identical to that of the basic 8-bit types
           (374, 574, etc.). TI's ALS and AS versions of the 821 feature
           "power-up high impedance state." Corresponding bus latch:
           841.

822   *10-bit 3-state registers*. Same as 821, but inverting. Logic is
           identical to that of the basic 8-bit inverting types (534, 564, etc.).
           NOTE: The inverting function of the 822 is conventionally as-
           cribed to the inputs ($\overline{\text{D}}$) rather than the outputs (Q) to keep the
           style identical to that adopted for the 9-bit registers. Correspond-
           ing bus latch: 842.

FUNCTION TABLE
821

INPUTS			OUTPUTS
$\overline{OE}$	CP	D	Q
L	↟	L	L
L	↟	H	H
L	–	•	NO CHANGE
H	•	•	Z

. = Either LOW or HIGH level
– = Clock inactive (L, H, or ↟)
Z = High impedance (disabled)

FUNCTION TABLE
822

INPUTS			OUTPUTS
$\overline{OE}$	CP	$\overline{D}$	Q
L	↟	H	L
L	↟	L	H
L	–	•	NO CHANGE
H	•	•	Z

. = Either LOW or HIGH level
– = Clock inactive (L, H, or ↟)
Z = High impedance (disabled)

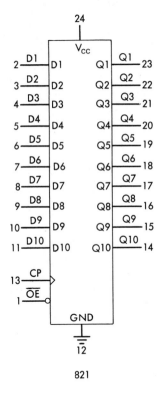

821

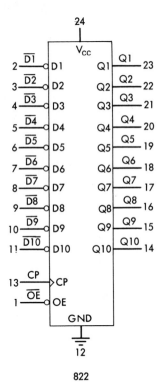

822

334

MANUFACTURERS	TTL	LS	HC	HCT	ALS	*	AS	F	AC	ACT
Advanced Micro Devices						■				■
Fairchild								D	▨	■
National Semiconductor							D			
Signetics								▨		
Texas Instruments						■				

The * column refers to the Am29821 (comparable to S and AS)
The AS column includes the Am29821A, which is similar to the 74AS821
The ACT column includes the Am29C821, which is comparable to the 74ACT821

KEY PARAMETERS		TTL	LS	HC	HCT	ALS	*	AS	F	AC	ACT
I_{OH} (max.)	mA						-24	-24	-3	-24	-24^h
I_{OL} (max.)	mA						48	48	48	24	24
I_{IH} (max.)	mA						0.05	0.02^b	0.07^p		
I_{IL} (max.)	mA						-1^a	-0.5	-0.07^q		
I_{CC} (max.)	mA						140	113^c	110		
I_{CC} (quiesc.)	mA									0.08	0.08^i
C_{pd} (typ.)	pF/channel									35	35^j
t_W (min.)	ns						7	8^d	n/a	n/a	5.5^k
t_{SU} (min.)	ns						4	6^e	n/a	n/a	2.5^l
t_H (min.)	ns						2	0^f	n/a	n/a	2.5^m
f_{MAX} (min.)	MHz						n/a	n/a	n/a	n/a	110^n
t_{pd} CP to Q (max.)	ns						12	10.5^g	n/a	n/a	10.5^o
t_{en} (max.)	ns						14	12	n/a	n/a	12
t_{dis} (max.)	ns						16	8	n/a	n/a	13°

[a] D inputs; I_{IL} CP and $\overline{OE}$ (max.) is -2 mA
[b] AMD 0.05 mA
[c] AMD 100 mA
[d] AMD 7 ns
[e] AMD 4 ns
[f] AMD 2 ns
[g] AMD 10 ns
[h] AMD -15 mA
[i] AMD 0.12 mA

[j] C_{pd} for AMD not given; instead, AMD specifies dynamic supply current I_{CCD} (max.) of 0.275 mA/MHz/bit
[k] AMD 7 ns
[l] AMD 4 ns
[m] AMD 2 ns
[n] AMD data not available
[o] AMD 12 ns
[p] Including 0.05 mA off-state output current
[q] Including -0.05 mA off-state output current

822   10-bit 3-state inverting registers

MANUFACTURERS	TTL	LS	HC	HCT	ALS	*	AS	F	AC	ACT
Advanced Micro Devices						■				
Fairchild								D	▨	■
Signetics								▨		
Texas Instruments						■				

The * column refers to the Am29822 (comparable to S and AS)
The ACT column includes the Am29C822, which is comparable to the 74ACT822

KEY PARAMETERS		TTL	LS	HC	HCT	ALS	*	AS	F	AC	ACT
I_{OH} (max.)	mA						-24	-24	-3	-24	-24
I_{OL} (max.)	mA						48	48	48	24	24
I_{IH} (max.)	mA						0.05	0.02	0.07^b		
I_{IL} (max.)	mA						-1^a	-0.5	-0.07^c		
I_{CC} (max.)	mA						140	113	110		
I_{CC} (quiesc.)	mA									0.08	0.08
C_{pd} (typ.)	pF/channel									35	35
t_W (min.)	ns						7	8	n/a	n/a	5.5
t_{SU} (min.)	ns						4	6	n/a	n/a	2.5
t_H (min.)	ns						2	0	n/a	n/a	2.5
f_{MAX} (min.)	MHz						n/a	n/a	n/a	n/a	110
t_{pd} CP to Q (max.)	ns						12	10.5	n/a	n/a	10.5
t_{en} (max.)	ns						14	12	n/a	n/a	12
t_{dis} (max.)	ns						16	8	n/a	n/a	13

[a] D inputs; I_{IL} CP and $\overline{CE}$ (max.) is -2 mA

[b] Including 0.05 mA off-state output current

[c] Including -0.05 mA off-state output current

8

Bus Transceivers

Devices specifically designed for asynchronous two-way communication between data buses are called *bus transceivers*. At their simplest, bus transceivers implement in one package the equivalent of two sets of bus drivers, one set for transmitting a parallel group of signals from bus A to bus B, and the other for transmitting a similar group of bits from bus B to bus A. Since these two sets of drivers share common input and output lines, a bus transceiver bears an external resemblance to an ordinary parallel bus driver like the ones described in Chapter 4. The main difference is that the direction of transmission is reversible depending on the state of the transceiver's control inputs. In one mode, the device transmits data from the A pins to the B pins, and in the other mode it transmits from the B pins to the A pins.

Beyond this resemblance to bus drivers, some transceivers can also function as a kind of storage device, and others include actual bus latches, bus registers, and even more sophisticated circuitry along with the two-way bus driver sections. For this reason, the transceivers are presented here at the end of Part II rather than being included with the simpler buffers/drivers of Chapter 4.

Because transceivers tend to be either quite simple or quite complex in their functioning, depending on the presence or absence of internal storage and multiplexing circuitry, somewhat different practices have been adopted in this chapter for the presentation of data. In particular, function tables have been omitted, since these are not necessary for understanding the operation of the simpler transceivers and are too complex to be given space here in the case of the more complex types, which are almost always single-sourced. For similar reasons, the usual block diagrams (in a somewhat condensed form) are retained for the simpler, widely supported transceivers but are omitted in the case of the more complex single-sourced types. For complete information on these latter types the interested reader is referred to the manufacturer's published data. Key parameters are given for all types, however, to enable the user of this book to form a quick judgment of whether a given type comes close enough to meeting the requirements of a particular application to make such further exploration of its characteristics worthwhile.

Since the simpler transceivers are basically just paired sets of bus drivers, the convention adopted for drivers and gates of putting the inverting versions before the generally slower noninverting types has been followed here as well. Likewise, it should be noted that the switching characteristics of these devices are specified under the same test conditions used for bus drivers. In all cases, including the few S versions, the load capacitance C_L is 50 pF (or 45 pF for some LS versions) for every parameter except the 3-state disable time, which is 5 pF in the case of LS and S versions and 50 pF for all others.

4-LINE BUS TRANSCEIVERS

The 4-line bus interfaces listed here are all supplied in 14-pin packages that provide two complementary output enables, $\overline{\text{EAB}}$ and EBA, to control the four pairs of I/O lines. (Two pins, 2 and 12, are left unconnected.) When $\overline{\text{EAB}}$ is active (LOW) and EBA is inactive (LOW), data flows from the A bus to the B bus; conversely, when EBA is active (HIGH) and $\overline{\text{EAB}}$ is inactive (HIGH), data flows from the B bus to the A bus. When both controls are inactive—that is, $\overline{\text{EAB}}$ is HIGH and EBA is LOW—both sets of I/O lines are put into the high-impedance state and the two buses are effectively disconnected from each other.

All of the 4-line transceivers listed below behave identically in the three cases just described, except that some invert the data being transmitted and some do not. In the fourth possible combination, however, where the two control inputs are both active ($\overline{\text{EAB}}$ is LOW and EBA is HIGH), different versions of the same type, or even the same version supplied by different manufacturers, can behave in one of two radically different ways. Since both sets of I/O lines are enabled to function simultaneously as inputs and outputs under these circumstances, the output levels at each bus will simultaneously serve as inputs to drive the other bus. Some transceivers are designed to make use of this self-reinforcing condition, and some are not.

In versions that can handle it, this condition makes the transceiver act as a "local bus latch"; that is, if all other data inputs to the two buses are at high impedance, both sets of bus lines will retain their states indefinitely as long as active levels are maintained at both transceiver enable inputs. But in most 4-line transceivers, the circular I/O condition will create excess currents that can damage the device, and in these cases the simultaneous enabling of both sets of outputs is forbidden. The versions that can safely function as local bus latches are identified in the individual descriptions.

Inverting transceivers:

242   *4-line 3-state inverting bus transceivers.* The basic inverting 4-line bus interface, supported by nearly all manufacturers and available in most technologies. LS and S versions provide a moderate amount of Schmitt-trigger hysteresis (min. 0.2 V, typ. 0.4 V) at the bus inputs to improve noise rejection. The LS version from Texas Instruments and the ALS and AS versions from Texas Instruments and National Semiconductor can function in the "local bus latch" mode.

1242   *4-line 3-state inverting bus transceivers.* Low-power versions of the ALS242 and F242.

2242§   *4-line 3-state inverting bus transceivers/MOS drivers.* ALS version of the 242 designed to drive MOS devices. I/O ports have 25 Ω series resistors so that no external resistors are required. Can function in the "local bus latch" mode. (Texas Instruments)

I_{OH} (max.)	-15	mA	t_{pd} (max.)	11 ns	
I_{OL} (max.)	30	mA	t_{en} (max.)	20 ns	
I_{IH} (max.)	0.02	mA	t_{dis} (max.)	14 ns	
I_{IL} (max.)	-0.1	mA			
I_{CC} (max.)	21	mA			

758§   *4-line inverting bus transceivers with open-collector outputs.* Open-collector ALS and AS versions of the 242. (Texas Instruments)

	ALS758		AS758				ALS758	AS758
I_{OH} (max.)	0.1	mA	0.1	mA	t_{pd} A or B (max.)		28 ns	19.5 ns
I_{OL} (max.)	24*	mA	64	mA	t_{pd} EAB (max.)		28 ns	21 ns
I_{IH} (max.)	0.02	mA	0.05†	mA	t_{pd} EBA (max.)		28 ns	19.5 ns
I_{IL} (max.)	-0.1	mA	-0.5	mA				
I_{CC} (max.)	16	mA	60	mA				

* -1 version has I_{OL} (max.) of 48 mA if I_{CC} is kept between 4.75 V and 5.25 V

† A and B inputs; control inputs 0.02 mA

Noninverting transceivers:

243   *4-line 3-state bus transceivers.* The basic noninverting 4-line bus interface, supported by nearly all manufacturers and available in most technologies. Same as the 242, but noninverting.

1243   *4-line 3-state bus transceivers.* Low-power versions of the ALS243 and F243.

759§   *4-line bus transceivers with open-collector outputs.* Open-collector AS version of the 243. (Texas Instruments)

I_{OH} (max.)	0.1	mA	t_{pd} A or B (max.)	20 ns
I_{OL} (max.)	64	mA	t_{pd} $\overline{EAB}$ (max.)	21 ns
I_{IH} (max.)	0.05*	mA	t_{pd} EBA (max.)	20 ns
I_{IL} (max.)	−1†	mA		
I_{CC} (max.)	74	mA		

* A and B inputs; control inputs 0.02 mA

† A and B inputs; control inputs −0.5 mA

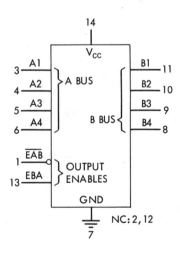

Inverting:
   3-state outputs: 242, 1242, 2242§
   Open-collector outputs: 758§
Noninverting:
   3-state outputs: 243, 1243
   Open-collector outputs: 759§

242 4-line 3-state inverting bus transceivers

MANUFACTURERS	TTL	LS	HC	HCT	ALS/A	S	AS	F	AC	ACT
Fairchild								D		
Motorola		■	▒						■	
National Semiconductor		■	■		■	■	■			
RCA			■							
SGS		▒	▒	■						
Signetics		■	■	■	■		■			
Texas Instruments		■	■	■	■					
Toshiba		■	■				■			

Variant change control symbol: TI 74ALS242B
ALS -1 versions available from National and TI

KEY PARAMETERS		TTL	LS	HC	HCT	ALS/A	S	AS	F	AC	ACT
I_{OH} (max.)	mA		-15	-6	-6	-15	-15	-15	-15		
I_{OL} (max.)	mA		24^a	6	6	24^l	64	64	64		
I_{IH} (max.)	mA		0.02^b			0.02	0.05	0.07^p	0.07		
I_{IL} (max.)	mA		-0.2			-0.1	-0.4^o	-0.5	-1.6^q		
I_{CC} (max.)	mA		50			21^m	150	60	75^r		
I_{CC} (quiesc.)	mA			0.08^c	0.08						
C_{pd} (typ.)	pF/channel			50^d	58^h						
t_{pd} (max.)	ns		18	25^e	38^i	11	7	6.5	7.5^s		
t_{en} (max.)	ns		30	38^f	50^j	20^n	15	8	10.5^t		
t_{dis} (max.)	ns		25	38^g	50^k	22	15	10.5	11^u		

a SGS 8 mA
b SGS 0.04
c SGS and Toshiba -0.04 mA
d TI 34 pF/channel, SGS and Toshiba 42 pF/channel, RCA 85 pF/channel
e RCA 23 ns, SGS 26 ns
f SGS and Toshiba 36 ns
g SGS 41 ns, Signetics 43 ns
h Signetics; TI 40 pF/channel, RCA 90 pF/channel
i Signetics 35 ns
j RCA 43 ns
k RCA 44 ns
l Versions with -1 suffix have I_{OL} (max.) = 48 mA if V_{CC} is kept between 4.75 V and 5.25 V

m National 22 mA
n National 21 ns
o Enable inputs -2 mA
p A and B inputs (including off-state output current); others 0.02 mA
q A and B inputs (Signetics specified at V_I = 0.4 V, Motorola at V_I = 0.5 V); enable inputs -1 mA at V_I = 0.5 V. Fairchild predicts I_{IL} any input (max.) = [-1] mA at V_I = 0.5 V
r Motorola; Signetics 55 mA, Fairchild [69] mA
s Signetics; Motorola 8 ns. Max. Fairchild data not available
t Signetics; Motorola 10 ns. Max. Fairchild data not available
u Signetics; Motorola 8 ns. Max. Fairchild data not available

1242 4-line 3-state inverting bus transceivers

MANUFACTURERS	TTL	LS	HC	HCT	ALS	S	AS	F	AC	ACT
National Semiconductor					■					
Signetics								■		
Texas Instruments					D					

ALS -1 version available from National

KEY PARAMETERS		TTL	LS	HC	HCT	ALS	S	AS	F	AC	ACT
I_{OH} (max.)	mA					-15			-15		
I_{OL} (max.)	mA					16^a			64		
I_{IH} (max.)	mA					0.02			0.07^b		
I_{IL} (max.)	mA					-0.1			-0.07^c		
I_{CC} (max.)	mA					14			72		
t_{pd} (max.)	ns					8			6.5		
t_{en} (max.)	ns					21			8		
t_{dis} (max.)	ns					16			9		

[a] -1 version has I_{OL} (max.) of 24 mA

[b] A and B inputs (including off-state output current); control inputs 0.02 mA

[c] A and B inputs (including off-state output current); control inputs -0.02 mA

MANUFACTURERS	TTL	LS	HC	HCT	ALS/A	S	AS	F	AC	ACT
Fairchild								■		
Motorola		■	▨							
National Semiconductor		■			■			■		
RCA		■	■							
SGS		▨	▨							
Signetics		■	■					■		
Texas Instruments		■	■		■		■			
Toshiba		■	■							

ALS -1 versions available from National and TI

KEY PARAMETERS		TTL	LS	HC	HCT	ALS/A	S	AS	F	AC	ACT
I_{OH} (max.)	mA		-15	-6	-6	-15	-15	-15	-15		
I_{OL} (max.)	mA		24^a	6	6	24^l	64	64	64		
I_{IH} (max.)	mA		0.02^b			0.02	0.05	0.07^n	0.07		
I_{IL} (max.)	mA		-0.2			-0.1	-0.4^m	-1^o	-1.6^p		
I_{CC} (max.)	mA		54			32	180	74	90		
I_{CC} (quiesc.)	mA			0.08^c	0.08						
C_{pd} (typ.)	pF/channel			50^d	54^h						
t_{pd} (max.)	ns		18	25^e	38^i	11	9	7.5	6.5		
t_{en} (max.)	ns		30	38^f	50^j	20	15	10.5	8.5		
t_{dis} (max.)	ns		25	38^g	50^k	22	15	11	7^q		

^{a} SGS 8 mA

^{b} SGS 0.04

^{c} SGS and Toshiba 0.04 mA

^{d} TI 34 pF/channel, Toshiba 36 pF/channel, SGS 42 pF/channel, RCA 80 pF/channel

^{e} RCA and Toshiba 23 ns, Signetics 28 ns, SGS 29 ns

^{f} SGS and Toshiba 36 ns, Signetics 44 ns

^{g} SGS 41 ns, Signetics 45 ns, RCA 50 ns

^{h} Signetics; TI 40 pF/channel, RCA 91 pF/channel

^{i} TI; RCA 28 ns, Signetics 34 ns

^{j} RCA 43 ns

^{k} RCA 44 ns

^{l} Versions with -1 suffix have I_{OL} (max.) = 48 mA if V_{CC} is kept between 4.75 V and 5.25 V

^{m} Enable inputs -2 mA

^{n} A and B inputs (including off-state output current); others 0.02 mA

^{o} A and B inputs (including off-state output current); others -0.5 mA

^{p} A and B inputs (Signetics specified at V_I = 0.4 V, Fairchild and Motorola specified at V_I = 0.5 V); enable inputs -1 mA at V_I = 0.5 V

^{q} Motorola 7.5 ns

1243   **4-line 3-state bus transceivers**

MANUFACTURERS	TTL	LS	HC	HCT	ALS	S	AS	F	AC	ACT
National Semiconductor					■					
Signetics								■		
Texas Instruments					D					

ALS -1 version available from National

KEY PARAMETERS		TTL	LS	HC	HCT	ALS	S	AS	F	AC	ACT
I_{OH} (max.)	mA					-15			-15		
I_{OL} (max.)	mA					16^a			64		
I_{IH} (max.)	mA					0.02			0.07^b		
I_{IL} (max.)	mA					-0.1			-0.07^c		
I_{CC} (max.)	mA					17			65		
t_{pd} (max.)	ns					12			7		
t_{en} (max.)	ns					22			8.5		
t_{dis} (max.)	ns					16			9		

[a] -1 version has I_{OL} (max.) of 24 mA

[b] A and B inputs (including off-state output current); control inputs 0.02 mA

[c] A and B inputs (including off-state output current); control inputs -0.02 mA

4-LINE BUS TRANSCEIVERS WITH INDIVIDUAL CHANNEL DIRECTION CONTROLS

These two single-sourced types, one inverting and one noninverting, are available for applications where individual lines of one bus must communicate with individual lines of the other. Each transceiver has four pairs of I/O lines and two active-LOW output enables, somewhat like the ordinary 4-line transceivers described above. In addition, however, each of these 16-pin devices has four direction controls, DIR1–DIR4, that enable or disable each channel individually.

Each line of one bus is isolated from the corresponding line of the other bus unless (a) the appropriate overall enable, $\overline{EAB}$ or $\overline{EBA}$, is at the LOW level that enables transmission in a particular direction; and (b) at the same time the DIR control for that pair of lines is at the appropriate level for transmission in the same direction, HIGH for transmission from A to B or LOW for transmission from B to A. Failure of a given DIR control to achieve the level that enables transmission in the same direction as that determined by the $\overline{EAB}$ and $\overline{EBA}$ controls will cause the corresponding pair of A and B lines to be isolated from one another. HIGH levels at both $\overline{EAB}$ and $\overline{EBA}$ will cause the two buses to be entirely isolated from each other regardless of the levels at the individual DIR controls.

446§   *4-line inverting bus transceivers with dual output enables and individual channel direction controls.* LS components that interface individual lines of two 4-line buses. A moderate amount of

Schmitt-trigger hysteresis (min. 0.2 V, typ. 0.4 V) at the bus inputs improves noise rejection. (Texas Instruments)

I_{OH} (max.)	−15	mA	t_{pd} (max.)	13	ns
I_{OL} (max.)	24	mA	t_{en} (max.)	40	ns
I_{IH} (max.)	0.02	mA	t_{dis} (max.)	25	ns
I_{IL} (max.)	−0.4	mA			
I_{CC} (max.)	68	mA			

449§ *4-line bus transceivers with dual output enables and individual channel direction controls.* Same as the 449, but noninverting. (Texas Instruments)

I_{OH} (max.)	−15	mA	t_{pd} (max.)	17	ns
I_{OL} (max.)	24	mA	t_{en} (max.)	35	ns
I_{IH} (max.)	0.02	mA	t_{dis} (max.)	25	ns
I_{IL} (max.)	−0.4	mA			
I_{CC} (max.)	80	mA			

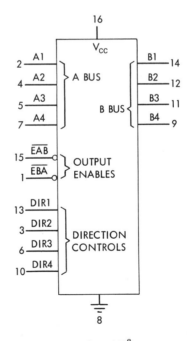

Inverting: 446§

Noninverting: 449§

4-LINE BUS TRANSCEIVERS WITH DUAL TRANSPARENT LATCHES

The single-sourced type below is a 4-line version of the more common 8-line latched bus transceivers described further on in this section. This 16-pin device combines a 4-line bus transceiver with a pair of 4-bit transparent latches, allowing the exchange of data between two 4-line buses in the equivalent of one clock cycle.

226§   *4-line 3-state latched bus transceivers.* S components that combine 4-line transceiver and latch functions in a single package. By selecting the appropriate combination of mode controls (S1, S2) and strobes (GAB, GBA) these devices can be made to directly pass data in either direction, like ordinary transceivers; to store data from both buses; or to read out stored data to either bus. In addition, two independent output enables allow the 3-state outputs to be put into the high-impedance state without affecting the latch functions in progress. A moderate amount of Schmitt-trigger hysteresis (min. 0.2 V, typ. 0.4 V) at the bus inputs improves noise rejection. See manufacturer's data for further information. (Texas Instruments)

I_{OH} (max.)	−10.3	mA	t_W	(min.)	20 ns
I_{OL} (max.)	15	mA	t_{SU}	(min.)	20 ns
I_{IH} (max.)	0.1	mA	t_H	(min.)	0 ns
I_{IL} (max.)	−1.6*	mA	t_{pd} A or B (max.)		30 ns
I_{CC} (max.)	185	mA	t_{pd} S or G (max.)		37 ns
			t_{en}	(max.)	20 ns
			t_{dis}	(max.)	15 ns

* GAB and GBA inputs −0.38 mA

4-LINE TRIDIRECTIONAL BUS TRANSCEIVERS

The single-sourced devices listed here can interface three independent 4-line buses rather than two. Because of this unique ability to provide three-way asynchronous communication, these components are called "tridirectional" bus transceivers, while all the others in this chapter are technically known as bidirectional bus transceivers.

Each of these tridirectional transceivers has two select inputs, S0 and S1, which determine the bus that data are to be transferred to, and three enable inputs, $\overline{GA}$, $\overline{GB}$, and $\overline{GC}$, which determine the bus or buses that the data are to be transferred from. In addition, a chip select ($\overline{CS}$) input can be used to put all the transceiver outputs in the high-impedance state and thus isolate all three buses from each other. A moderate amount of Schmitt-trigger hysteresis (min. 0.2 V, typ. 0.4 V) at the bus inputs improves noise rejection.

The six types offered in this small family of circuits provide a choice of inverting, noninverting, or combined inverting/noninverting logic and 3-state or open-collector outputs. All of them are LS devices in 20-pin packages. See manufacturer's data for function tables and diagrams. (Texas Instruments)

443§   *4-line tridirectional 3-state inverting bus transceivers.* The basic tridirectional bus transceiver.

I_{OH} (max.)	−15	mA	t_{pd} (max.)	14 ns	
I_{OL} (max.)	24	mA	t_{en} (max.)	42 ns	
I_{IH} (max.)	0.02	mA	t_{dis} (max.)	25 ns	
I_{IL} (max.)	−0.4	mA			
I_{CC} (max.)	95	mA			

441§   *4-line tridirectional inverting bus transceivers with open-collector outputs.* Open-collector version of the 443.

I_{OH} (max.)	0.1	mA	t_{pd} bus* (max.)	30 ns	
I_{OL} (max.)	24	mA	t_{PLH} ctrl† (max.)	40 ns	
I_{IH} (max.)	0.02	mA	t_{PHL} ctrl† (max.)	40 ns	
I_{IL} (max.)	−0.4	mA			
I_{CC} (max.)	95	mA			

* Any bus input
† Any control input

442§   *4-line tridirectional 3-state bus transceivers.* Same as the 443, but noninverting.

I_{OH} (max.)	−15	mA	t_{pd} (max.)	20 ns	
I_{OL} (max.)	24	mA	t_{en} (max.)	42 ns	
I_{IH} (max.)	0.02	mA	t_{dis} (max.)	25 ns	
I_{IL} (max.)	−0.4	mA			
I_{CC} (max.)	95	mA			

440§   *4-line tridirectional bus transceivers with open-collector outputs.* Open-collector version of the 442.

I_{OH} (max.)	0.1	mA	t_{pd} bus* (max.)	35 ns
I_{OL} (max.)	24	mA	t_{PLH} ctrl† (max.)	50 ns
I_{IH} (max.)	0.02	mA	t_{PHL} ctrl† (max.)	50 ns
I_{IL} (max.)	−0.4	mA		
I_{CC} (max.)	95	mA		

* Any bus input
† Any control input

444§   *4-line tridirectional 3-state inverting/non-inverting bus transceivers.* Same as the 443, but transactions with the A bus are inverting while those between the B bus and C bus are noninverting.

I_{OH} (max.)	−15	mA	t_{pd} (max.)	20 ns
I_{OL} (max.)	24	mA	t_{en} (max.)	42 ns
I_{IH} (max.)	0.02	mA	t_{dis} (max.)	25 ns
I_{IL} (max.)	−0.4	mA		
I_{CC} (max.)	95	mA		

448§   *4-line tridirectional inverting/non-inverting bus transceivers with open-collector outputs.* Open-collector version of the 444.

I_{OH} (max.)	0.1	mA	t_{pd} bus* (max.)	35 ns
I_{OL} (max.)	24	mA	t_{PLH} ctrl† (max.)	40 ns
I_{IH} (max.)	0.02	mA	t_{PHL} ctrl† (max.)	40 ns
I_{IL} (max.)	−0.4	mA		
I_{CC} (max.)	95	mA		

* Any bus input
† Any control input

8-LINE BUS TRANSCEIVERS

By far the most important and commonly encountered bus interfaces are the byte-oriented bidirectional transceivers described in the following pages. All of these devices are supplied in 20-pin packages, an allotment that provides only two pins for control inputs; and despite the bewildering variety of available types, there are really only two basic configurations, corresponding to the two basic ways these control inputs can be assigned.

In one configuration, the two controls are dual complementary output enables, EAB (pin 1) and $\overline{\text{EBA}}$ (pin 19), that operate much like the similarly named output controls of the 4-line transceivers described above. If EAB is active (HIGH) and $\overline{\text{EBA}}$ is inactive (HIGH), data are transmitted from the A pins to the B pins, and conversely, if $\overline{\text{EBA}}$ is active (LOW) and EAB is inactive (LOW), data are transmitted from the B pins to the A pins. If both controls are inactive (EAB LOW and $\overline{\text{EBA}}$ HIGH), all I/O pins are put into the high-impedance state, and the two buses are effectively isolated from each other. If both controls are active (EAB HIGH and $\overline{\text{EBA}}$ LOW), each output reinforces its input and the transceiver functions as a local bus latch, that is, both sets of bus lines will remain at their last states (assuming that all other data sources to the two sets of bus lines are at high impedance). This bus latch capability is provided by all the 8-line transceivers having the dual-enable configuration.

The other main category of 8-line transceivers consists of those in which one of the two control inputs is an output enable that simply enables or disables the entire transceiver, and the other is a direction control that determines the direction of data transmission, A to B or B to A. In the great majority of these devices, the direction control, DIR, is at pin 1, and the output enable, $\overline{\text{OE}}$, is at pin 19; two variant types, the 545 and the 588, put these controls at pins 11 and 9, respectively. Since the DIR control allows data transmission in one direction or the other, but not both, none of the transceivers with this configuration of control inputs can function as a local bus latch. This arrangement achieves maximum simplicity of control by sacrificing maximum flexibility.

Within each of these two major categories is the usual division into inverting and noninverting types, with the second configuration (those having a DIR control and an $\overline{\text{OE}}$ control) adding a group of inverting/noninverting transceivers that invert signals going from the A bus to the B bus but not those going the other way. Each of these minor divisions can be further divided into 3-state and open-collector types, with a few that provide 3-state outputs to one bus and open-collector outputs to the other. The wide range of options available to the designer is summed up in the following table.

Transceivers with dual complementary output enables and local latch capability:

Inverting types:

620   *8-line 3-state inverting bus transceivers.* The basic dual-enabled 8-line inverting bus transceiver, available from a number of manufacturers in a wide range of technologies. The dual-enable configuration permits a local bus-latch mode and allows maximum flexibility in timing. LS versions provide a moderate amount of Schmitt-trigger hysteresis (min. 0.2 V, typ. 0.4 V) at the bus inputs to improve noise rejection.

TABLE 8.1 8-LINE BUS TRANSCEIVERS

Output Controls	Polarity	A Outputs	B Outputs	Type
EAB, $\overline{\text{EBA}}$	Inverting	3-state	3-state	620 1620§
		3-state MOS driver	3-state MOS driver	2620
		open-collector	open-collector	622 1622§
	Noninverting	3-state	3-state	623 1623§
		3-state MOS driver	3-state MOS driver	2623
		open-collector	open-collector	621 1621§
DIR, $\overline{\text{OE}}$	Inverting	3-state	3-state	640 1640
		3-state MOS driver	3-state MOS driver	2640
		open-collector	open-collector	642 1642§
		open-collector	3-state	638 1638§
	Noninverting	3-state	3-state	245 1245§ 645 1645 545 588
		3-state MOS driver	3-state MOS driver	2645
		open-collector	open-collector	641 1641§
		open-collector	3-state	639 1639§
	Inverting and Noninverting	3-state	3-state	643 1643§
		open-collector	open-collector	644 1644§

1620§   *8-line 3-state inverting bus transceivers.* Low-power versions of the ALS620. (National Semiconductor)

I_{OH} (max.)	-15	mA	t_{pd} (max.)	n/a
I_{OL} (max.)	16*	mA	t_{en} (max.)	n/a
I_{IH} (max.)	0.02	mA	t_{dis} (max.)	n/a
I_{IL} (max.)	-0.1	mA		
I_{CC} (max.)	21	mA		

* -1 version has I_{OL} (max.) of 24 mA

2620   *8-line 3-state inverting bus transceivers/MOS drivers.* AS versions of the 620 specifically designed for driving MOS devices. The I/O ports are provided with 25 Ω series resistors so that no external resistors are required.

622   *8-line inverting bus transceivers with open-collector outputs.* Open-collector versions of the 620, widely supported and available in LS, ALS, AS, and F implementations. LS versions provide a moderate amount of Schmitt-trigger hysteresis (min. 0.2 V, typ. 0.4 V) at the bus inputs to improve noise rejection.

1622§   *8-line inverting bus transceivers with open-collector outputs.* Low-power versions of the ALS622. (National Semiconductor)

I_{OH} (max.)	0.1	mA	t_{pd} A or $\underline{B}$ (max.)	25 ns
I_{OL} (max.)	16*	mA	t_{pd} E or $\overline{E}$ (max.)	31 ns
I_{IH} (max.)	0.02	mA		
I_{IL} (max.)	-0.1	mA		
I_{CC} (max.)	18	mA		

* -1 version has I_{OL} (max.) of 24 mA

Noninverting types:

623   *8-line 3-state bus transceivers.* The basic dual-enabled noninverting 8-line bus transceiver, available from a number of manufacturers in a broad range of technologies. The dual-enable configuration permits a local bus-latch mode and allows maximum flexibility in timing. LS versions provide a moderate amount of Schmitt-trigger hysteresis (min. 0.2 V, typ. 0.4 V) at the bus inputs to improve noise rejection.

1623§   *8-line 3-state bus transceivers.* Low-power versions of the ALS623. (National Semiconductor)

I_{OH} (max.)	-15	mA	t_{pd} (max.)	n/a
I_{OL} (max.)	16*	mA	t_{en} (max.)	n/a
I_{IH} (max.)	0.02	mA	t_{dis} (max.)	n/a
I_{IL} (max.)	-0.1	mA		
I_{CC} (max.)	18	mA		

* -1 version has I_{OL} (max.) of 24 mA

2623   *8-line 3-state bus transceivers/MOS drivers.* AS versions of the 623 specifically designed for driving MOS devices. The I/O ports are provided with 25 Ω series resistors so that no external resistors are required.

621   *8-line bus transceivers with open-collector outputs.* Open-collector versions of the 623, widely supported and available in LS, ALS, AS, and F implementations. LS versions provide a moderate amount of Schmitt-trigger hysteresis (min. 0.2 V, typ. 0.4 V) at the bus inputs to improve noise rejection.

1621§   *8-line bus transceivers with open-collector outputs.* Low-power versions of the ALS621. (National Semiconductor)

I_{OH} (max.)	0.1	mA	t_{pd} A or $\overline{B}$ (max.)	22 ns
I_{OL} (max.)	16*	mA	t_{pd} E or $\overline{E}$ (max.)	33 ns
I_{IH} (max.)	0.02	mA		
I_{IL} (max.)	-0.1	mA		
I_{CC} (max.)	16	mA		

* -1 version has I_{OL} (max.) of 24 mA

Transceivers with single output-enable and direction-control inputs:

Inverting types:

640   *8-line 3-state inverting bus transceivers.* The basic inverting byte-oriented bus transceiver, widely supported and available in a broad range of technologies. LS versions provide a moderate amount of Schmitt-trigger hysteresis (min. 0.2 V, typ. 0.4 V) at the bus inputs to improve noise rejection.

1640   *8-line 3-state inverting bus transceivers.* Low-power versions of the ALS640.

2640   *8-line 3-state inverting bus transceivers/MOS drivers.* AS versions of the 640 specifically designed for driving MOS devices.

The I/O ports are provided with 25 Ω series resistors so that no external resistors are required.

642   *8-line inverting bus transceivers with open-collector outputs.* Open-collector versions of the 640, available in LS, ALS, AS, and F implementations. LS versions provide a moderate amount of Schmitt-trigger hysteresis (min. 0.2 V, typ. 0.4 V) at the bus inputs to improve noise rejection.

1642§   *8-line inverting bus transceivers with open-collector outputs.* Low-power versions of the ALS642. (National Semiconductor)

I_{OH} (max.)	0.1	mA	t_{pd} A or B (max.)	n/a
I_{OL} (max.)	16*	mA	t_{pd} OE to A (max.)	n/a
I_{IH} (max.)	0.02	mA		
I_{IL} (max.)	−0.1	mA		
I_{CC} (max.)	20	mA		

* −1 version has I_{OL} (max.) of 24 mA

638   *8-line 3-state/open-collector inverting bus transceivers.* LS, ALS, and AS versions of the 640 designed for asynchronous two-way communication between open-collector and 3-state buses (A outputs are open-collector, B outputs are 3-state).

1638§   *8-line 3-state/open-collector inverting bus transceivers.* Low-power ALS versions of the 638. (National Semiconductor)

I_{OH} (max.)	−15*	mA	t_{pd} A to B (max.)	n/a
I_{OL} (max.)	16†	mA	t_{pd} B to A (max.)	n/a
I_{IH} (max.)	0.02	mA	t_{pd} OE to A (max.)	n/a
I_{IL} (max.)	−0.1	mA	t_{en} B (max.)	n/a
I_{CC} (max.)	25	mA	t_{dis} B (max.)	n/a

* B outputs (3-state); A outputs (OC) have I_{OH} (max.) of 0.1 mA

† −1 version has I_{OL} (max.) of 24 mA

Noninverting types:

245   *8-line 3-state bus transceivers.* The most basic 8-line bus interface, universally supported and available in most technologies. LS versions provide a moderate amount of Schmitt-trigger hysteresis (min. 0.2 V, typ. 0.4 V) at the bus inputs to improve noise rejection.

1245§　　*8-line 3-state bus transceivers.* Alternative version of the 74F245 with NPN inputs for reduced loading. (Signetics)

I_{OH} (max.)	−15*	mA	t_{pd} (max.)	n/a
I_{OL} (max.)	64*	mA	t_{en} (max.)	n/a
I_{IH} (max.)	0.07†	mA	t_{dis} (max.)	n/a
I_{IL} (max.)	−0.6†	mA		
I_{CC} (max.)	21	mA		

* B outputs; A outputs have I_{OH} (max.) of −3 mA and I_{OL} of 24 mA

† Including off-state output current; alone, I_{IH} (max.) = 0.02 mA and I_{OH} (max.) = −0.02 mA

645　　*8-line 3-state bus transceivers.* A somewhat slower and less widely supported form of the 245. LS versions provide a moderate amount of Schmitt-trigger hysteresis (min. 0.2 V, typ. 0.4 V) at the bus inputs to improve noise rejection.

1645　　*8-line 3-state bus transceivers.* Low-power versions of the ALS645.

2645　　*8-line 3-state bus transceivers/MOS drivers.* AS versions of the 645 specifically designed for driving MOS devices. The I/O ports are provided with 25 Ω series resistors so that no external resistors are required.

545　　*8-line 3-state bus transceivers.* F versions of the 245/645 with a nonstandard pinout. The B outputs have a considerably greater drive capability than the A outputs, making the B side better suited to face the "outside" when this device is used as an I/O port. The DIR control is renamed T/$\overline{R}$ (transmit/receive) to reflect this asymmetrical drive capability.

588　　*8-line 3-state bus transceivers with IEEE-488 termination resistors.* An F component similar to the 545 except that the B outputs are equipped with termination resistors per IEEE-488 specifications.

641　　*8-line bus transceivers with open-collector outputs.* Open-collector LS, ALS, and AS versions of the 645. LS versions provide a moderate amount of Schmitt-trigger hysteresis (min. 0.2 V, typ. 0.4 V) at the bus inputs to improve noise rejection.

1641§　　*8-line bus transceivers with open-collector outputs.* Low-power versions of the ALS641. (National Semiconductor)

I_{OH} (max.)	0.1	mA	t_{pd} A or B (max.)	n/a
I_{OL} (max.)	16*	mA	t_{pd} OE to A (max.)	n/a
I_{IH} (max.)	0.02	mA		
I_{IL} (max.)	−0.1	mA		
I_{CC} (max.)	23	mA		

* −1 version has I_{OL} (max.) of 24 mA

639   *8-line 3-state/open-collector bus transceivers.* LS, ALS, and AS versions of the 641 designed for asynchronous two-way communication between open-collector and 3-state buses (A outputs are open-collector, B outputs are 3-state).

1639§   *8-line 3-state/open-collector inverting bus transceivers.* Low-power ALS versions of the 638. (National Semiconductor)

I_{OH} (max.)	−15*	mA	t_{pd} A to B (max.)	n/a
I_{OL} (max.)	16†	mA	t_{pd} B to A (max.)	n/a
I_{IH} (max.)	0.02	mA	t_{pd} OE to A (max.)	n/a
I_{IL} (max.)	−0.1	mA	t_{en} B (max.)	n/a
I_{CC} (max.)	25	mA	t_{dis} B (max.)	n/a

* B outputs (3-state); A outputs (OC) have I_{OH} (max.) of 0.1 mA

† −1 version has I_{OL} (max.) of 24 mA

Inverting/noninverting types:

643   *8-line 3-state inverting/noninverting bus transceivers.* Widely supported versions of the 640/645 that invert signals going from the A bus to the B bus but do not invert signals going from the B bus to the A bus. LS versions provide a moderate amount of Schmitt-trigger hysteresis (min. 0.2 V, typ. 0.4 V) at the bus inputs to improve noise rejection.

1643§   *8-line 3-state inverting/noninverting bus transceivers.* Low-power versions of the ALS643. (National Semiconductor)

I_{OH} (max.)	−15	mA	t_{pd} (max.)	n/a
I_{OL} (max.)	16*	mA	t_{en} (max.)	n/a
I_{IH} (max.)	0.02	mA	t_{dis} (max.)	n/a
I_{IL} (max.)	−0.1	mA		
I_{CC} (max.)	22	mA		

* −1 version has I_{OL} (max.) of 24 mA

644   *8-line inverting/noninverting bus transceivers with open-collector outputs.* Open-collector LS, ALS, and AS versions of the 643. LS versions provide a moderate amount of Schmitt-trigger hysteresis (min. 0.2 V, typ. 0.4 V) at the bus inputs to improve noise rejection.

1644§   *8-line inverting/noninverting bus transceivers with open-collector outputs.* Low-power versions of the ALS644. (National Semiconductor)

I_{OH} (max.)	0.1	mA	t_{pd} $\underline{A}$ or B (max.)	n/a
I_{OL} (max.)	16*	mA	t_{pd} $\overline{OE}$ to A (max.)	n/a
I_{IH} (max.)	0.02	mA		
I_{IL} (max.)	−0.1	mA		
I_{CC} (max.)	22	mA		

* −1 version has I_{OL} (max.) of 24 mA

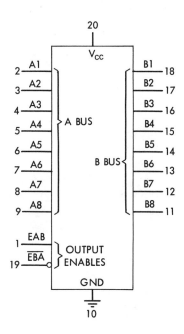

Inverting:
    3–state outputs: 620, 1620§, 2620
    Open–collector outputs: 622, 1622§
Noninverting:
    3–state outputs: 623, 1623§, 2623§
    Open–collector outputs: 621, 1621§

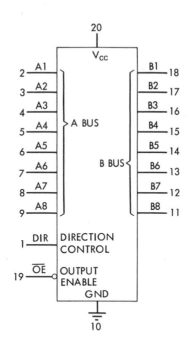

Inverting:
    3-state outputs: 640, 1640, 2640
    Open-collector outputs: 642, 1642§
    Open-collector A outputs and 3-state B outputs: 638, 1638§
Noninverting:
    3-state outputs: 245, 645, 1245§, 1645, 2645
    Open-collector outputs: 641, 1641§
    Open-collector A outputs and 3-state B outputs: 639, 1639§
Inverting/noninverting:
    3-state outputs: 643, 1643§
    Open-collector outputs: 644, 1644§

MANUFACTURERS	TTL	LS	HC	HCT	ALS/A	S	AS	F	AC	ACT
Fairchild								D		
Motorola		■								
National Semiconductor					■		■			
SGS			▒							
Signetics		▒						■		
Texas Instruments		■		▒	■					
Toshiba			■							

ALS -1 versions available from National and TI

KEY PARAMETERS		TTL	LS	HC	HCT	ALS/A	S	AS	F	AC	ACT
I_{OH} (max.)	mA		-15	-6	-6	-15		-15	-15^i		
I_{OL} (max.)	mA		24	6	6	24^f		64	64^j		
I_{IH} (max.)	mA		0.02			0.02		0.07^g	0.07^k		
I_{IL} (max.)	mA		-0.4			-0.1		-0.75^h	-0.07^l		
I_{CC} (max.)	mA		95			47		122	110		
I_{CC} (quiesc.)	mA			0.08^a	0.08						
C_{pd} (typ.)	pF/channel			40^b	40						
t_{pd} (max.)	ns		15	25^c	26	10		7	7.5		
t_{en} (max.)	ns		40	38^d	53	25		9	11.5		
t_{dis} (max.)	ns		25	44^e	38	18		13	10.5		

[a] SGS and Toshiba 0.04 mA

[b] SGS data not available

[c] Toshiba; SGS 24 ns, TI 26 ns

[d] Toshiba; SGS 36 ns, TI 53 ns

[e] SGS; TI 38 ns, Toshiba 45 ns

[f] Versions with -1 suffix have I_{OL} (max.) = 48 mA if V_{CC} is kept between 4.75 V and 5.25 V

[g] A and B inputs (including off-state output current); control inputs 0.02 mA

[h] A and B inputs (including off-state output current); control inputs -0.5 mA

[i] B outputs; A outputs -3 mA

[j] B outputs; A outputs 24 mA

[k] A and B inputs (including off-state output current); control inputs 0.02 mA

[l] A and B inputs (including off-state output current); control inputs -0.02 mA

2620   8-line 3-state inverting bus transceivers/MOS drivers

MANUFACTURERS	TTL	LS	HC	HCT	ALS	S	AS	F	AC	ACT
National Semiconductor							■			
Texas Instruments							■			

KEY PARAMETERS		TTL	LS	HC	HCT	ALS	S	AS	F	AC	ACT
I_{OH} (max.)	mA							-35			
I_{OL} (max.)	mA							35			
I_{IH} (max.)	mA							0.07[a]			
I_{IL} (max.)	mA							-0.75[b]			
I_{CC} (max.)	mA							121			
t_{pd} (max.)	ns							8			
t_{en} (max.)	ns							11			
t_{dis} (max.)	ns							12			

[a] A and B inputs (including off-state output current); control inputs 0.02 mA

[b] A and B inputs (including off-state output current); control inputs -0.5 mA

622   8-line inverting bus transceivers with OC outputs

MANUFACTURERS	TTL	LS	HC	HCT	ALS/A	S	AS	F	AC	ACT
Fairchild								D		
Motorola		■								
National Semiconductor					■		■			
Signetics		▒						■		
Texas Instruments		■			■		■			

ALS -1 versions available from National and TI

KEY PARAMETERS		TTL	LS	HC	HCT	ALS/A	S	AS	F	AC	ACT
I_{OH} (max.)	mA		0.1			0.1		0.1	-15[d]		
I_{OL} (max.)	mA		24			24[a]		64	64[e]		
I_{IH} (max.)	mA		0.02			0.02		0.07[b]	0.02		
I_{IL} (max.)	mA		-0.4			-0.1		-0.75[c]	-0.02		
I_{CC} (max.)	mA		90			28		103	90		
t_{pd} A or B (max.)	ns		25			35		25	13.5		
t_{pd} E or $\overline{E}$ (max.)	ns		60			38		23	15.5		

[a] Versions with -1 suffix have I_{OL} (max.) = 48 mA if V_{CC} is kept between 4.75 V and 5.25 V

[b] A and B inputs (including off-state output current); control inputs 0.02 mA

[c] A and B inputs (including off-state output current); control inputs -0.5 mA

[d] B outputs; A outputs -3 mA

[e] B outputs; A outputs 24 mA

MANUFACTURERS	TTL	LS	HC	HCT	ALS/A	S	AS	F	AC	ACT
Fairchild								D		
Motorola		■								
National Semiconductor					■		■			
RCA									▨	▨
SGS			▨							
Signetics		■						■		
Texas Instruments			■	▨	■		■			
Toshiba		■	■							

ALS -1 versions available from National and TI

KEY PARAMETERS		TTL	LS	HC	HCT	ALS/A	S	AS	F	AC	ACT
I_{OH} (max.)	mA		-15	-6	-6	-15		-15	-15[i]	-24	-24
I_{OL} (max.)	mA		24	6	6	24[f]		64	64[j]	24	24
I_{IH} (max.)	mA		0.02			0.02		0.07[g]	0.07[k]		
I_{IL} (max.)	mA		-0.4			-0.1		-0.75[h]	-0.07[l]		
I_{CC} (max.)	mA		95			55		189	140		
I_{CC} (quiesc.)	mA			0.08[a]	0.08					0.08	0.08
C_{pd} (typ.)	pF/channel			40[b]	40					66	79
t_{pd} (max.)	ns		15	21[c]	26	13		9	7.5	8.6	9.6
t_{en} (max.)	ns		40	38[d]	53	22		11.5	12	12	13
t_{dis} (max.)	ns		25	44[e]	38	19		11.5	10	12	13

[a] SGS and Toshiba 0.04 mA

[b] TI; Toshiba 35 pF/channel. SGS data not available

[c] TI 26 ns

[d] Toshiba; SGS 36 ns, TI 53 ns

[e] SGS; TI 38 ns, Toshiba 45 ns

[f] Versions with -1 suffix have I_{OL} (max.) = 48 mA if V_{CC} is
kept between 4.75 V and 5.25 V

[g] A and B inputs (including off-state output current); control inputs 0.02 mA

[h] A and B inputs (including off-state output current); control inputs -0.5 mA

[i] B outputs; A outputs -3 mA

[j] B outputs; A outputs 24 mA

[k] A and B inputs (including off-state output current); control inputs 0.02 mA

[l] A and B inputs (including off-state output current); control inputs -0.02 mA

2623   8-line 3-state bus transceivers/MOS drivers

MANUFACTURERS	TTL	LS	HC	HCT	ALS	S	AS	F	AC	ACT
National Semiconductor							■			
Texas Instruments										

KEY PARAMETERS		TTL	LS	HC	HCT	ALS	S	AS	F	AC	ACT
I_{OH} (max.)	mA							-35			
I_{OL} (max.)	mA							35			
I_{IH} (max.)	mA							0.07[a]			
I_{IL} (max.)	mA							-0.75[b]			
I_{CC} (max.)	mA							189			
t_{pd} (max.)	ns							9			
t_{en} (max.)	ns							12			
t_{dis} (max.)	ns							12.5			

[a] A and B inputs (including off-state output current); control inputs 0.02 mA

[b] A and B inputs (including off-state output current); control inputs -0.5 mA

621   8-line bus transceivers with OC outputs

MANUFACTURERS	TTL	LS	HC	HCT	ALS/A	S	AS	F	AC	ACT
Fairchild								D		
Motorola		■								
National Semiconductor					■		■			
Signetics		▓						■		
Texas Instruments		■			■		■			

ALS -1 versions available from National and TI

KEY PARAMETERS		TTL	LS	HC	HCT	ALS/A	S	AS	F	AC	ACT
I_{OH} (max.)	mA		0.1			0.1		0.1	-15[d]		
I_{OL} (max.)	mA		24			24[a]		64	64[e]		
I_{IH} (max.)	mA		0.02			0.02		0.07[b]	0.02		
I_{IL} (max.)	mA		-0.4			-0.1		-0.75[c]	-0.02		
I_{CC} (max.)	mA		90			48		189	140		
t_{pd} A or B (max.)	ns		25			33		24	13		
t_{pd} E or $\overline{E}$ (max.)	ns		50			39		22	17		

[a] Versions with -1 suffix have I_{OL} (max.) = 48 mA if V_{CC} is kept between 4.75 V and 5.25 V

[b] A and B inputs (including off-state output current); control inputs 0.02 mA

[c] A and B inputs (including off-state output current); control inputs -0.5 mA

[d] B outputs; A outputs -3 mA

[e] B outputs; A outputs 24 mA

640   8-line 3-state inverting bus transceivers

MANUFACTURERS	TTL	LS	HC	HCT	ALS/A	S	AS	F	AC	ACT
Fairchild									■	■
Motorola		■	D	D						
National Semiconductor			■				■			
RCA										
SGS		■	■	D						
Signetics		■						■		
Texas Instruments		■	■	■				■		
Toshiba				■					D	

LS -1 versions available from Signetics and TI
ALS -1 version available from TI

KEY PARAMETERS		TTL	LS	HC	HCT	ALS	S	AS	F	AC	ACT
I_{OH} (max.)	mA		-15	-6	-6	-15		-15	-15^n	-24	-24
I_{OL} (max.)	mA		24^a	6	6	24^a		64	64^o	24	24
I_{IH} (max.)	mA		0.02			0.02		0.07^l	0.07^p		
I_{IL} (max.)	mA		-0.4			-0.1		-0.75^m	-0.07^q		
I_{CC} (max.)	mA		95			43		123	120		
I_{CC} (quiesc.)	mA			0.08^b	0.08^g					0.08	0.08
C_{pd} (typ.)	pF/channel			40^c	41^h					n/a	n/a
t_{pd} (max.)	ns		15	25^d	28^i	11		7	8	n/a	n/a
t_{en} (max.)	ns		40	43^e	53^j	24		10	13	n/a	n/a
t_{dis} (max.)	ns		25	38^f	38^k	15		13	9	n/a	n/a

[a] Versions with -1 suffix have I_{OL} (max.) = 48 mA if V_{CC} is kept between 4.75 V and 5.25 V

[b] SGS and Toshiba 0.04 mA

[c] Signetics 35 pF/channel, RCA 38 pF/channel, National 120 pF/channel

[d] RCA; National 22 ns, Signetics 23 ns, TI 26 ns, SGS 27 ns, Toshiba 28 ns

[e] RCA; Signetics 38 ns, SGS and Toshiba 40 ns, National 56 ns, TI 58 ns

[f] Signetics and TI; RCA 43 ns, SGS 46 ns, Toshiba 48 ns, National 52 ns

[g] Toshiba 0.04 mA

[h] RCA; Signetics 35 pF/channel, TI 40 pF/channel, Toshiba 44 pF/channel, National 100 pF/channel

[i] RCA and Signetics; TI 26 ns, National 29 ns, Toshiba 35 ns

[j] National and Toshiba; Signetics 38 ns, RCA 43 ns, TI 44 ns

[k] RCA 43 ns, Toshiba 50 ns

[l] A and B inputs (including off-state output current); control inputs 0.02 mA

[m] A and B inputs (including off-state output current); control inputs -0.5 mA

[n] B outputs; A outputs -3 mA

[o] B outputs; A outputs 24 mA

[p] A and B inputs (including off-state output current); control inputs 0.04 mA

[q] A and B inputs (including off-state output current); control inputs -0.04 mA

1640 8-line 3-state inverting bus transceivers

MANUFACTURERS	TTL	LS	HC	HCT	ALS	S	AS	F	AC	ACT
National Semiconductor					■					
Texas Instruments					■					

Added change control symbol: TI 74ALS1640A
ALS -1 versions available from National and TI

KEY PARAMETERS		TTL	LS	HC	HCT	ALS	S	AS	F	AC	ACT
I_{OH} (max.)	mA					-15					
I_{OL} (max.)	mA					16^a					
I_{IH} (max.)	mA					0.02					
I_{IL} (max.)	mA					-0.1					
I_{CC} (max.)	mA					32					
t_{pd} (max.)	ns					15					
t_{en} (max.)	ns					22					
t_{dis} (max.)	ns					13					

[a] Versions with -1 suffix have I_{OL} (max.) = 24 mA if V_{CC} is
kept between 4.75 V and 5.25 V

2640 8-line 3-state inverting bus transceivers/MOS drivers

MANUFACTURERS	TTL	LS	HC	HCT	ALS	S	AS	F	AC	ACT
National Semiconductor							■			
Texas Instruments							■			

KEY PARAMETERS		TTL	LS	HC	HCT	ALS	S	AS	F	AC	ACT
I_{OH} (max.)	mA							-35^a			
I_{OL} (max.)	mA							35^a			
I_{IH} (max.)	mA							0.07^b			
I_{IL} (max.)	mA							-0.75^c			
I_{CC} (max.)	mA							123			
t_{pd} (max.)	ns							7.5			
t_{en} (max.)	ns							10			
t_{dis} (max.)	ns							13			

[a] TI data (National data not available)
[b] A and B inputs (including off-state output current); control inputs 0.02 mA
[c] A and B inputs (including off-state output current); control inputs -0.5 mA

642   8-line inverting bus transceivers with OC outputs

MANUFACTURERS	TTL	LS	HC	HCT	ALS/A	S	AS	F	AC	ACT
Motorola		■					■			
National Semiconductor							■			
Signetics		■						■		
Texas Instruments		■			■		■			

LS -1 versions available from Signetics and TI
ALS -1 version available from TI

KEY PARAMETERS		TTL	LS	HC	HCT	ALS/A	S	AS	F	AC	ACT
I_{OH} (max.)	mA		0.1			0.1		0.1	0.25		
I_{OL} (max.)	mA		24^a			24^a		64	64		
I_{IH} (max.)	mA		0.02			0.02		0.07^c	0.02		
I_{IL} (max.)	mA		-0.4			-0.1		-0.75^d	-0.02		
I_{CC} (max.)	mA		95			28		104	98		
t_{pd} A or B (max.)	ns		25			30		24	14.5		
t_{pd} $\overline{OE}$ or DIR (max.)	ns		60^b			38		23.5	14		

^{a} Versions with -1 suffix have I_{OL} (max.) = 48 mA if V_{CC} is kept between 4.75 V and 5.25 V
^{b} Signetics 50 ns
^{c} A and B inputs (including off-state output current); control inputs 0.02 mA
^{d} A and B inputs (including off-state output current); control inputs -0.5 mA

638   8-line 3-state/OC inverting bus transceivers

MANUFACTURERS	TTL	LS	HC	HCT	ALS/A	S	AS	F	AC	ACT
National Semiconductor							■			
Texas Instruments		■			■		■			

ALS -1 version available from TI

KEY PARAMETERS		TTL	LS	HC	HCT	ALS/A	S	AS	F	AC	ACT
I_{OH} (max.)	mA		-15^a			-15^a		-15^a			
I_{OL} (max.)	mA		24			24^b		64			
I_{IH} (max.)	mA		0.02			0.02		0.07^c			
I_{IL} (max.)	mA		-0.4			-0.1		-0.75^d			
I_{CC} (max.)	mA		95			41		122			
t_{pd} A to B (max.)	ns		15			12		7			
t_{pd} B to A (max.)	ns		25			30		20			
t_{pd} $\overline{OE}$ to A (max.)	ns		60			45		19			
t_{en} B (max.)	ns		40			22		10			
t_{dis} B (max.)	ns		25			15		10			

^{a} B outputs (3-state); A outputs (OC) have I_{OH} (max.) of 0.1 mA
^{b} Version with -1 suffix has I_{OL} (max.) = 48 mA if V_{CC} is kept between 4.75 V and 5.25 V
^{c} A and B inputs (including off-state output current); control inputs 0.02 mA
^{d} A and B inputs (including off-state output current); control inputs -0.5 mA

MANUFACTURERS	TTL	LS	HC	HCT	ALS/A	S	AS	F	AC	ACT
Fairchild										
Monolithic Memories										
Motorola			D	D						
National Semiconductor										
RCA										
SGS				D						
Signetics										
Texas Instruments										
Toshiba									D	

Added change control symbol: National 74AS245A
ALS -1 versions available from National and TI

KEY PARAMETERS		TTL	LS	HC	HCT	ALS/A	S	AS	F	AC	ACT
I_{OH} (max.)	mA		-15^a	-6	-6	-15		-15	-15^t	-24	-24
I_{OL} (max.)	mA		24	6	6	24^o		64^p	64^u	24	24
I_{IH} (max.)	mA		0.02			0.02		0.07^q	0.07^v		
I_{IL} (max.)	mA		-0.2			-0.1		-0.75^r	-1^w		
I_{CC} (max.)	mA		95			58		143^s	143^x		
I_{CC} (quiesc.)	mA			0.08^e	0.08^j					0.08	0.08
C_{pd} (typ.)	pF/channel			40^f	46^k					45^z	45^{cc}
t_{pd} (max.)	ns		12^b	23^g	33^i	10		7.5^s	7	7^{aa}	9^{dd}
t_{en} (max.)	ns		40^c	40^h	53^m	20		9^s	9^y	9.5^{bb}	12^{ee}
t_{dis} (max.)	ns		25^d	38^i	45^n	15		9.5^s	7.5	10^{bb}	11^{ee}

[a] Fairchild -3 mA

[b] Fairchild 18 ns

[c] Fairchild 30 ns

[d] TI 28 ns

[e] SGS and Toshiba 0.04 mA

[f] TI; Signetics 30 pF/channel, Toshiba 33 pF/channel, RCA 53 pF/channel, National 100 pF/channel

[g] Signetics and Toshiba; National 22 ns, TI 26 ns, SGS 27 ns, RCA 28 ns

[h] SGS and Toshiba; RCA and Signetics 38 ns, National 56 ns, TI 58 ns

[i] RCA and Signetics; SGS 46 ns, Toshiba 48 ns, TI 50 ns, National 52 ns

[j] Toshiba 0.04 mA

[k] Toshiba; Signetics 30 pF/channel, TI 40 pF/channel, RCA 55 pF/channel, National 100 pF/channel

[l] RCA; TI 28 ns, National 29 ns, Signetics and Toshiba 35 ns

[m] National and Toshiba; RCA 38 ns, Signetics 45 ns, TI 58 ns

[n] Signetics; National 38 ns, RCA 44 ns, Toshiba and TI 50 ns

[o] Versions with -1 suffix have I_{OL} (max.) = 48 mA if V_{CC} is kept between 4.75 V and 5.25 V

[p] National 48 mA

[q] A and B inputs, including off-state output current (National 0.05 mA); control inputs 0.02 mA

[r] A and B inputs, including off-state output current; control inputs -0.5 mA (National -0.1 mA)

[s] TI; max. values for National not available

[t] B outputs; -3 mA at A outputs (Fairchild -3 mA at both A and B outputs)

[u] B outputs; 20 mA at A outputs (Signetics 24 mA at A outputs)

[v] A and B inputs; 0.02 mA at control inputs (Fairchild, Motorola). Signetics 0.07 mA at A and B inputs, 0.04 mA at control inputs

[w] A and B inputs; -0.8 mA at T/$\overline{R}$ (DIR), -1.6 at $\overline{OE}$ (Motorola). Fairchild has -0.65 mA at A and B inputs, -1.2 mA at control inputs; Signetics has -0.6 mA at A and B inputs, -1.2 mA at control inputs

[x] Fairchild 120 mA

[y] Signetics 11 ns

[z] RCA 66 pF/channel

[aa] RCA 8.6 ns

[bb] RCA 12 ns

[cc] RCA 79 pF/channel

[dd] RCA 9.6 ns

[ee] RCA 13 ns

645 8-line 3-state bus transceivers

MANUFACTURERS	TTL	LS	HC	HCT	ALS/A	S	AS	F	AC	ACT
Monolithic Memories		■								
Motorola		■								
National Semiconductor		■					■			
SGS		▨								
Signetics		■								
Texas Instruments		■	▨	▨	■		■			

LS -1 versions available from Monolithic Memories, Signetics, and TI
ALS -1 version available from TI

KEY PARAMETERS		TTL	LS	HC	HCT	ALS/A	S	AS	F	AC	ACT
I_{OH} (max.)	mA		-15	-6	-6	-15		-15			
I_{OL} (max.)	mA		24^{a}	6	6	24^{a}		64			
I_{IH} (max.)	mA		0.02			0.02		0.07^{b}			
I_{IL} (max.)	mA		-0.4			-0.1		-0.75^{c}			
I_{CC} (max.)	mA		95			58		149			
I_{CC} (quiesc.)	mA			0.08	0.08						
C_{pd} (typ.)	pF/channel			40	40						
t_{pd} (max.)	ns		15	26	28	10		9.5			
t_{en} (max.)	ns		40	58	58	20		11			
t_{dis} (max.)	ns		25	50	50	15		12			

[a] Versions with -1 suffix have I_{OL} (max.) = 48 mA if V_{CC} is kept between 4.75 V and 5.25 V
[b] A and B inputs (including off-state output current); control inputs 0.02 mA
[c] A and B inputs (including off-state output current); control inputs -0.5 mA

1645 8-line 3-state bus transceivers

MANUFACTURERS	TTL	LS	HC	HCT	ALS	S	AS	F	AC	ACT
National Semiconductor					■					
Texas Instruments					■					

Added change control symbol: TI 74ALS1645A
ALS -1 versions available from National and TI
NOTE: The 74ALS1645A is also known as the 74ALS1245A

KEY PARAMETERS		TTL	LS	HC	HCT	ALS	S	AS	F	AC	ACT
I_{OH} (max.)	mA					-15					
I_{OL} (max.)	mA					16^{a}					
I_{IH} (max.)	mA					0.02					
I_{IL} (max.)	mA					-0.1					
I_{CC} (max.)	mA					36					
t_{pd} (max.)	ns					13					
t_{en} (max.)	ns					25					
t_{dis} (max.)	ns					18					

[a] Versions with -1 suffix have I_{OL} (max.) = 24 mA if V_{CC} is kept between 4.75 V and 5.25 V

2645 8-line 3-state bus transceivers/MOS drivers

MANUFACTURERS	TTL	LS	HC	HCT	ALS	S	AS	F	AC	ACT
National Semiconductor							■			
Texas Instruments							■			

KEY PARAMETERS		TTL	LS	HC	HCT	ALS	S	AS	F	AC	ACT
I_{OH} (max.)	mA							-35^a			
I_{OL} (max.)	mA							35^a			
I_{IH} (max.)	mA							0.07^b			
I_{IL} (max.)	mA							-0.75^c			
I_{CC} (max.)	mA							155			
t_{pd} (max.)	ns							10			
t_{en} (max.)	ns							11.5			
t_{dis} (max.)	ns							12			

[a] TI data (National data not available)
[b] A and B inputs (including off-state output current); control inputs 0.02 mA
[c] A and B inputs (including off-state output current); control inputs -0.5 mA

545 8-line 3-state bus transceivers

MANUFACTURERS	TTL	LS	HC	HCT	ALS	S	AS	F	AC	ACT
Fairchild								■		
Signetics								■		

KEY PARAMETERS		TTL	LS	HC	HCT	ALS	S	AS	F	AC	ACT
I_{OH} (max.)	mA								-15^a		
I_{OL} (max.)	mA								64^b		
I_{IH} (max.)	mA								0.07^c		
I_{IL} (max.)	mA								-0.07^d		
I_{CC} (max.)	mA								120		
t_{pd} (max.)	ns								7		
t_{en} (max.)	ns								11^e		
t_{dis} (max.)	ns								8^f		

[a] B outputs; -3 mA at A outputs
[b] B outputs; 24 mA at A outputs
[c] Bus inputs; control inputs 0.04 mA (Fairchild 0.02 mA)
[d] Bus inputs (Fairchild -0.65 mA); control inputs -0.04 mA
(Signetics) or -1.2 mA (Fairchild)
[e] Fairchild 9 ns
[f] Fairchild 7.5 ns

588   8-line 3-state bus transceivers with IEEE-488 termination resistors

MANUFACTURERS	TTL	LS	HC	HCT	ALS	S	AS	F	AC	ACT
Fairchild								■		
Signetics										

KEY PARAMETERS		TTL	LS	HC	HCT	ALS	S	AS	F	AC	ACT
I_{OH} (max.)	mA								-15^a		
I_{OL} (max.)	mA								64^b		
I_{IH} (max.)	mA								2.5^c		
I_{IL} (max.)	mA								-3.2^d		
I_{CC} (max.)	mA								135		
t_{pd} (max.)	ns								7.5		
t_{en} (max.)	ns								11^e		
t_{dis} (max.)	ns								8		

[a] B outputs; -3 mA at A outputs (Fairchild -3 mA at both A and B outputs)

[b] B outputs; 24 mA at A outputs (Fairchild 24 mA at B outputs, 20 mA at A outputs)

[c] B inputs (including off-state current); A inputs 0.07 mA (including off-state current), control inputs 0.04 mA (Fairchild 0.02 mA)

[d] B inputs (including off-state current); A inputs -0.07 mA (Fairchild -0.65 mA) including off-state current, control inputs -0.04 mA (Fairchild -1.2 mA)

[e] Fairchild 10 ns

641   8-line bus transceivers with OC outputs

MANUFACTURERS	TTL	LS	HC	HCT	ALS/A	S	AS	F	AC	ACT
Motorola		■								
National Semiconductor							■			
SGS		▓								
Signetics		■						■		
Texas Instruments		■			■		■			

LS -1 versions available from Signetics and TI
ALS -1 version available from TI

KEY PARAMETERS		TTL	LS	HC	HCT	ALS/A	S	AS	F	AC	ACT
I_{OH} (max.)	mA		0.1			0.1		0.1	0.25		
I_{OL} (max.)	mA		24^a			24^a		64	64^d		
I_{IH} (max.)	mA		0.02			0.02		0.07^b	0.02^e		
I_{IL} (max.)	mA		-0.4			-0.1		-0.75^c	-0.02^f		
I_{CC} (max.)	mA		95			47		136	120		
t_{pd} A or B (max.)	ns		25			25		21	13		
t_{pd} $\overline{OE}$ or DIR (max.)	ns		50			32		22	13.5		

[a] Versions with -1 suffix have I_{OL} (max.) = 48 mA if V_{CC} is kept between 4.75 V and 5.25 V

[b] A and B inputs (including off-state output current); control inputs 0.02 mA

[c] A and B inputs (including off-state output current); control inputs -0.5 mA

[d] B outputs; A outputs 20 mA

[e] A and B inputs; control inputs 0.04 mA

[f] A and B inputs; control inputs -0.04 mA

639   8-line 3-state/OC bus transceivers

MANUFACTURERS	TTL	LS	HC	HCT	ALS/A	S	AS	F	AC	ACT
National Semiconductor							■			
Texas Instruments		■			■		■			

ALS -1 version available from TI

KEY PARAMETERS		TTL	LS	HC	HCT	ALS/A	S	AS	F	AC	ACT
I_{OH} (max.)	mA		-15^a			-15^a		-15^a			
I_{OL} (max.)	mA		24			24^b		64			
I_{IH} (max.)	mA		0.02			0.02		0.07^c			
I_{IL} (max.)	mA		-0.4			-0.1		-0.75^d			
I_{CC} (max.)	mA		95			54		154			
t_{pd} A to B (max.)	ns		15			12		9.5			
t_{pd} B to A (max.)	ns		25			30		22			
t_{pd} $\overline{OE}$ to A (max.)	ns		50			35		21.5			
t_{en} B (max.)	ns		40			25		10.5			
t_{dis} B (max.)	ns		25			16		10.5			

[a] B outputs (3-state); A outputs (OC) have I_{OH} (max.) of 0.1 mA

[b] Version with -1 suffix has I_{OL} (max.) = 48 mA if V_{CC} is kept between 4.75 V and 5.25 V

[c] A and B inputs (including off-state output current); control inputs 0.02 mA

[d] A and B inputs (including off-state output current); control inputs -0.5 mA

643   8-line 3-state inverting/non-inverting bus transceivers

MANUFACTURERS	TTL	LS	HC	HCT	ALS/A	S	AS	F	AC	ACT
Fairchild									▓	▓
Motorola		■	D	D						
National Semiconductor			■	■			■			
RCA								⋅		
SGS			▓	D						
Signetics										
Texas Instruments		■	▓	▓	■		■			
Toshiba			■	■						

LS and ALS -1 versions available from TI

KEY PARAMETERS		TTL	LS	HC	HCT	ALS	S	AS	F	AC	ACT
I_{OH} (max.)	mA		-15	-6	-6	-15		-15		-24	-24
I_{OL} (max.)	mA		24^a	6	6	24^a		64		24	24
I_{IH} (max.)	mA		0.02			0.02		0.07^l			
I_{IL} (max.)	mA		-0.4			-0.1		-0.75^m			
I_{CC} (max.)	mA		95			48		123			
I_{CC} (quiesc.)	mA			0.08^b	0.08^g					0.08	0.08
C_{pd} (typ.)	pF/channel			40^c	44^h					n/a	n/a
t_{pd} (max.)	ns		15	28^d	29^i	13		10		n/a	n/a
t_{en} (max.)	ns		45	40^e	53^j	25		11		n/a	n/a
t_{dis} (max.)	ns		25	38^f	38^k	17		10.5		n/a	n/a

[a] Versions with -1 suffix have I_{OL} (max.) = 48 mA if V_{CC} is kept between 4.75 V and 5.25 V

[b] SGS and Toshiba 0.04 mA

[c] Signetics 42 pF/channel, RCA 45 pF/channel, National 120 pF/channel

[d] National 22 ns, Signetics 23 ns, SGS 27 ns

[e] SGS and Toshiba; RCA and Signetics 38 ns, National 56 ns, TI 58 ns

[f] Signetics and TI; RCA 43 ns, SGS 46 ns, Toshiba 48 ns, National 52 ns

[g] Toshiba 0.04 mA

[h] Signetics and Toshiba; TI 40 pF/channel, RCA 55 pF/channel, National 100 pF/channel

[i] Signetics and National; TI 26 ns, RCA 33 ns, Toshiba 35 ns

[j] National and Toshiba; Signetics 38 ns, TI 44 ns, RCA 45 ns

[k] RCA 45 ns, Toshiba 50 ns

[l] A and B inputs (including off-state output current); control inputs 0.02 mA

[m] A and B inputs (including off-state output current); control inputs -0.5 mA

| **644** | **8-line inverting/non-inverting bus transceivers with OC outputs** |

MANUFACTURERS	TTL	LS	HC	HCT	ALS/A	S	AS	F	AC	ACT
Motorola		■								
National Semiconductor							■			
Texas Instruments		■			■		■			

LS and ALS -1 versions available from TI

KEY PARAMETERS		TTL	LS	HC	HCT	ALS/A	S	AS	F	AC	ACT
I_{OH} (max.)	mA		0.1			0.1		0.1			
I_{OL} (max.)	mA		24^a			24^a		64			
I_{IH} (max.)	mA		0.02			0.02		0.07^b			
I_{IL} (max.)	mA		-0.4			-0.1		-0.75^c			
I_{CC} (max.)	mA		95			40		124			
t_{pd} A or B (max.)	ns		25			30		24			
t_{pd} $\overline{OE}$ or DIR (max.)	ns		60			35		22			

^{a} Versions with -1 suffix have I_{OL} (max.) = 48 mA if V_{CC} is kept between 4.75 V and 5.25 V

^{b} A and B inputs (including off-state output current); con-

trol inputs 0.02 mA

^{c} A and B inputs (including off-state output current); control inputs -0.5 mA

8-LINE BUS TRANSCEIVERS WITH PARITY

The small group of single-sourced devices listed below have the ability to generate a parity bit while transmitting a byte of data between two 8-line buses. This parity bit can be used to check the validity of the data when received by some other device or when stored in and then retrieved from a memory array. (The 657§ can also check the parity of transmitted data.) All of these circuits are housed in 24-pin packages. Like the ordinary 8-line transceivers, there are two basic configurations—one in which the device is controlled by dual output enables, and one in which it is controlled by a direction input and a single output enable. (See manufacturer's data for function tables and diagrams.)

658§   *8-line 3-state inverting bus transceivers with parity generator and dual complementary output enables.* Inverting HC and HCT bus transceivers with parity generation capability. These 24-pin devices are very similar to ordinary 20-pin 8-line inverting bus transceivers, but two additional output pins, APO and BPO, provide levels that indicate the parity of data on the A bus and B bus in conjunction with the levels at two additional input pins, API and BPI. To determine the level at APO, for example, the circuit adds up the HIGH levels among the eight lines of the A bus and then adds one if the level at API is also HIGH. If the sum is even (0, 2, 4, 6, 8) then the level at APO is HIGH; if it is odd (1, 3, 5, 7, 9) then the level at APO is LOW. The level at BPO is similarly a function of the HIGH levels at the B bus and the BPI inputs. These transceivers have active pull-ups and pull-downs that prevent the inputs from floating when the corresponding outputs are disabled, eliminating the need for external pull-up or pull-down resistors. See manufacturer's data for application notes. (Texas Instruments)

I_{OH} (max.)		−6	mA	t_{pd} A/B or B/A (max.)	38 ns
I_{OL} (max.)		6	mA	t_{pd} A/APO or B/BPO (max.)	58 ns
C_{pd} (typ.)	HC	56	pF	t_{pd} API/APO or BPI/BPO (max.)	39 ns
	HCT	62	pF	t_{en} (max.)	59 ns
I_{CC} (quiesc.)		0.08	mA	t_{dis} (max.)	59 ns

659§   *8-line 3-state bus transceivers with parity generator and dual complementary output enables.* Same as the 658§, but noninverting. (Texas Instruments)

I_{OH} (max.)		−6	mA	t_{pd} A/B or B/A (max.)	35 ns
I_{OL} (max.)		6	mA	t_{pd} A/APO or B/BPO (max.)	58 ns
C_{pd} (typ.)	HC	56	pF	t_{pd} API/APO or BPI/BPO (max.)	39 ns
	HCT	62	pF	t_{en} (max.)	59 ns
I_{CC} (quiesc.)		0.08	mA	t_{dis} (max.)	59 ns

664§   *8-line 3-state inverting bus transceivers with parity generator and single direction control and output enable.* Same as the 658§, but with alternative control configuration. (Texas Instruments)

I_{OH} (max.)		−6	mA	t_{pd} A/B or B/A (max.)	38 ns
I_{OL} (max.)		6	mA	t_{pd} A/APO or B/BPO (max.)	58 ns
C_{pd} (typ.)	HC	56	pF	t_{pd} API/APO or BPI/BPO (max.)	39 ns
	HCT	62	pF	t_{en} DIR or $\overline{OE}$ (max.)	64 ns
I_{CC} (quiesc.)		0.08	mA	t_{dis} DIR or $\overline{OE}$ (max.)	64 ns

665§   *8-line 3-state bus transceivers with parity generator and single direction control and output enable.* Same as the 664§, but noninverting. (Texas Instruments)

I_{OH} (max.)		−6	mA	t_{pd} A/B or B/A (max.)	HC	35 ns
					HCT	50 ns
I_{OL} (max.)		6	mA	t_{pd} A/APO or B/BPO (max.)		58 ns
C_{pd} (typ.)	HC	56	pF	t_{pd} API/APO or BPI/BPO (max.)		39 ns
	HCT	62	pF	t_{en} DIR or $\overline{OE}$ (max.)		64 ns
I_{CC} (quiesc.)		0.08	mA	t_{dis} DIR or $\overline{OE}$ (max.)		64 ns

657§   *8-line 3-state bus transceivers with parity generator/checker, direction control, and dual complementary output enables.* Noninverting F transceivers that include both parity generation and parity checking functions together with a transmit/receive control and two complementary 3-state output enables. Although these are bidirectional devices, the B outputs have greater drive capability and are therefore considered the ''sending'' outputs. The internal parity generator in each transceiver detects whether an even or odd number of bits are HIGH at the A bus, depending on the condition of the EVEN/$\overline{\text{ODD}}$ input; the result of this determination is made available at the PARITY output. The parity of data received at the B inputs is likewise compared with the setting at the EVEN/$\overline{\text{ODD}}$ input, and the result is made available at the $\overline{\text{ERROR}}$ output. (Signetics; under development by Fairchild)

I_{OH} (max.)	$-15*$	mA	t_{pd} A/B or B/A	(max.)	8	ns
I_{OL} (max.)	64†	mA	t_{pd} A to PARITY	(max.)	16	ns
I_{IH} (max.)	0.07‡	mA	t_{pd} EVEN/ODD to PARITY/ERROR			
I_{IL} (max.)	-0.07‡	mA		(max.)	12.5	ns
I_{CC} (max.)	150	mA	t_{pd} B to $\overline{\text{ERROR}}$	(max.)	22.5	ns
			t_{pd} PARITY to $\overline{\text{ERROR}}$	(max.)	17	ns
			t_{en}	(max.)	11	ns
			t_{dis}	(max.)	8	ns

* B outputs; A outputs have I_{OH} (max.) of -3 mA

† B outputs; A outputs have I_{OL} (max.) of 24 mA

‡ B inputs, including off-state output current. A inputs take 0.1 and -0.1 mA (max.), $\overline{\text{ERROR}}$ input takes 0.5 and -0.5 mA (max.)

8-LINE BUS TRANSCEIVERS WITH TRANSPARENT LATCHES

As explained in Chapter 5, many data interface applications require latching capabilities at either end of a parallel data transfer. The devices listed in this section can minimize component count in these applications by providing both latches and transceivers in the same package.

Two sets of choices are available in this category, the inverting 544 and the noninverting 543 from Fairchild and Signetics, and the inverting 567§ and the noninverting 547§ from Monolithic Memories. Both kinds are 24-pin devices with eight A bus lines, eight B bus lines, an A-to-B output enable, a B-to-A output enable, and, of course, power supply and ground pins. However, the two basic types differ in their pinout and, more important, in the functioning of the remaining four inputs that control the operation of the two internal latches.

In the 547§ and 567§ transceivers from Monolithic Memories, each of the two latches is controlled by a pair of complementary latch enables that are simply ORed to provide the enable input, so that an active level at either input will put the given latch in the transparent mode, with output control exercised separately through the 3-state output enable. In the 543 and 544 from Fairchild and Signetics, an overall active-LOW A-to-B or B-to-A enable control for each latch is NORed, not just with the latch enable for that latch, but also with the 3-state output enable for that latch, making it possible to control the overall data flow in a given direction by means of a single input.

Consult the published data on these devices for logic diagrams and further information.

544   *8-line 3-state latched inverting bus transceivers.* F components with a latch enable and an A-to-B or B-to-A enable for each internal latch. The B outputs have considerably higher drive capability than the A outputs.

567§   *8-line 3-state latched inverting bus transceivers.* LS and ACT components with complementary latch enables for each internal latch. A small amount of hysteresis (0.3 V typ.) at the latch-enable inputs aids noise rejection in the LS version. (Monolithic Memories)

LS567§

I_{OH} (max.)	−2.6	mA	t_W	(min.)	16† ns
I_{OL} (max.)	32	mA	t_{SU}	(min.)	22† ns
I_{IH} (max.)	0.02	mA	t_H	(min.)	5† ns
I_{IL} (max.)	−0.25*	mA	t_{pd} A or B (max.)		23 ns
I_{CC} (max.)	180	mA	t_{pd} LE	(max.)	24 ns
			t_{en}	(max.)	21 ns
			t_{dis}	(max.)	19 ns

* A or B inputs; all others −0.4 mA

† Active-LOW latch enable; active-HIGH requires t_W (min.) 8 ns, t_{SU} (min.) 13 ns, t_H (min.) 11 ns

ACT567§

I_{OH} (max.)	−6	mA	t_W	(min.)	12 ns
I_{OL} (max.)	12	mA	t_{SU}	(min.)	8 ns
I_{CC} (quiesc.)	0.08	mA	t_H	(min.)	8 ns
			t_{pd} A, B, LE (max.)		32 ns
			t_{en}	(max.)	34 ns
			t_{dis}	(max.)	34 ns

543   *8-line 3-state latched bus transceivers.* Same as the 544, but noninverting.

547§   *8-line 3-state latched bus transceivers.* Same as the 567§, but noninverting. (Monolithic Memories) NUMBERING CONFLICT:

Note that the 74F547 is an 8-line latched decoder/demultiplexer manufactured by Fairchild and Signetics, unrelated to the 74LS547 and 74ACT547 latched bus transceiver described here.

LS547§

I_{OH} (max.)	−2.6	mA	t_W	(min.)	16†	ns
I_{OL} (max.)	32	mA	t_{SU}	(min.)	15†	ns
I_{IH} (max.)	0.02	mA	t_H	(min.)	5†	ns
I_{IL} (max.)	−0.25*	mA	t_{pd} A or B	(max.)	18	ns
I_{CC} (max.)	180	mA	t_{pd} LE	(max.)	24	ns
			t_{en}	(max.)	21	ns
			t_{dis}	(max.)	19	ns

* A or B inputs; all others −0.4 mA

† Active-LOW latch enable; active-HIGH requires t_W (min.) 8 ns, t_{SU} (min.) 5 ns, t_H (min.) 13 ns

ACT547§

I_{OH} (max.)	−6	mA	t_W	(min.)	12 ns
I_{OL} (max.)	12	mA	t_{SU}	(min.)	8 ns
I_{CC} (quiesc.)	0.08	mA	t_H	(min.)	8 ns
			t_{pd} A, B, LE	(max.)	32 ns
			t_{en}	(max.)	34 ns
			t_{dis}	(max.)	34 ns

544 8-line 3-state latched inverting bus transceivers

MANUFACTURERS	TTL	LS	HC	HCT	ALS	S	AS	F	AC	ACT
Fairchild								■		
Signetics								▨		

KEY PARAMETERS		TTL	LS	HC	HCT	ALS	S	AS	F	AC	ACT
I_{OH} (max.)	mA								-15^a		
I_{OL} (max.)	mA								64^b		
I_{IH} (max.)	mA								0.07^c		
I_{IL} (max.)	mA								-0.6^d		
I_{CC} (max.)	mA								130		
t_W (min.)	ns								7.5^e		
t_{SU} (min.)	ns								3		
t_H (min.)	ns								3		
t_{pd} (max.)	ns								10.5^f		
t_{en} (max.)	ns								12		
t_{dis} (max.)	ns								9		

[a] B outputs; -1 mA at A outputs (Fairchild -3 mA at both A and B outputs)

[b] B outputs; 24 mA at A outputs

[c] Bus inputs; control inputs 0.02 mA

[d] Bus inputs (Fairchild -0.65 mA); latch enables and 3-state output enables -0.6 mA, A-to-B and B-to-A enables -1.2 mA

[e] Signetics data not available

[f] Bus to bus; max. delay for enable to bus is 14.5 ns

543   8-line 3-state latched bus transceivers

MANUFACTURERS	TTL	LS	HC	HCT	ALS	S	AS	F	AC	ACT
Fairchild								██		
Signetics								▒▒		

KEY PARAMETERS		TTL	LS	HC	HCT	ALS	S	AS	F	AC	ACT
I_{OH} (max.)	mA								-15^a		
I_{OL} (max.)	mA								64^b		
I_{IH} (max.)	mA								0.07^c		
I_{IL} (max.)	mA								-0.6^d		
I_{CC} (max.)	mA								125		
t_W (min.)	ns								9^e		
t_{SU} (min.)	ns								3^f		
t_H (min.)	ns								3^f		
t_{pd} (max.)	ns								8.5^g		
t_{en} (max.)	ns								12		
t_{dis} (max.)	ns								9		

[a] B outputs; -1 mA at A outputs (Fairchild -3 mA at both A and B outputs)
[b] B outputs; 24 mA at A outputs
[c] Bus inputs; control inputs 0.02 mA
[d] Bus inputs (Fairchild -0.65 mA); latch enables and 3-state output enables -0.6 mA, A-to-B and B-to-A enables -1.2 mA
[e] Signetics data not available
[f] Fairchild 3.5 ns
[g] Bus to bus; max. delay for enable to bus is 12.5 ns

8-LINE BUS TRANSCEIVERS WITH EDGE-TRIGGERED REGISTERS

Like the latched transceivers discussed just above, the pair of devices described here can reduce component count in many bus-to-bus data transfer applications by providing an 8-line bus transceiver and two edge-triggered 8-bit registers in a single package. Both of these single-sourced designs are supplied in 24-pin packages. The pin complement allows for the usual 8-line data ports with a 3-state output enable for each bus, plus a positive-edge-triggered clock input and an active-LOW clock enable input for each of the two internal registers. Consult the published data on these devices for logic diagrams and further information.

566§   *8-line 3-state registered inverting bus transceivers.* LS components that combine transceiver and register functions in the same package. A small amount of hysteresis (0.3 V typ.) at the clock and clock-enable inputs aids in noise rejection. (Monolithic Memories)

I_{OH} (max.)	−2.6	mA	t_W	(min.)	15	ns
I_{OL} (max.)	32	mA	t_{SU}	(min.)	11†	ns
I_{IH} (max.)	0.02	mA	t_H	(min.)	0†	ns
I_{IL} (max.)	−0.25*	mA	f_{MAX}	(min.)	43	MHz
I_{CC} (max.)	180	mA	t_{pd} CP	(max.)	21	ns
			t_{en}	(max.)	21	ns
			t_{dis}	(max.)	19	ns

* A or B inputs; all others −0.4 mA

† Clock inputs; clock enable inputs require t_{SU} (min.) of 11 ns and t_H (min.) of 4 ns

546§   *8-line 3-state registered bus transceivers.* Same as the 566§, but noninverting. Key parameters are identical to those of the 566§. (Monolithic Memories)

8-LINE BUS TRANSCEIVERS WITH EDGE-TRIGGERED REGISTERS AND STATUS FLAGS

These complex single-sourced devices combine an 8-line 3-state bus transceiver with two edge-triggered 8-bit registers and two status flip-flops that are set automatically when the associated register is loaded. For each register, there are eight I/O lines, an active-LOW output enable, a positive-edge-triggered clock input, an active-LOW clock enable, a positive-edge-triggered flag clear input, and an active-HIGH flag output. Consult the published data on these devices for logic diagrams and further information.

551§   *8-line 3-state registered inverting bus transceivers with status flags.* F components specifically designed for demand-response data transfer. (Fairchild)

I_{OH} (max.)	−3	mA	t_W	(min.)	3.5 ns
I_{OL} (max.)	64*	mA	t_{SU}	(min.)	4.5 ns
I_{IH} (max.)	0.07†	mA	t_H	(min.)	2.5 ns
I_{IL} (max.)	−0.65†	mA	f_{MAX}	(min.)	n/a
I_{CC} (max.)	190	mA	t_{pd} CP	(max.)	12.5 ns
			t_{en}	(max.)	10.5 ns
			t_{dis}	(max.)	10 ns

* B outputs; A outputs have I_{OL} (max.) of 24 mA

† Bus inputs; see manufacturer's data for other inputs

550§   *8-line 3-state registered bus transceivers with status flags.* Same as 551§, but noninverting. Key parameters are identical to those of the 551§. (Fairchild)

8-LINE BUS TRANSCEIVERS WITH EDGE-TRIGGERED REGISTERS, STATUS FLAGS, AND PARITY

Designed for demand-response data transfer with parity, this single-sourced type adds parity generation and checking to the capabilities of the registered transceivers with status flags described just above. The 28-pin package is retained by replacing the flag clear inputs of the 550§ and 551§ with parity and error outputs; instead of clearing in response to separate external control, each status flip-flop is cleared automatically when the corresponding output enable returns to HIGH after reading the output port.

Consult the published data on these devices for logic diagrams and further information.

552§   *8-line 3-state registered bus transceivers with status flags and parity generator/checker.* Noninverting F components specifically designed for demand-response data transfer with parity checking. For each internal 8-bit register there are eight I/O lines; an active-LOW output enable that doubles as a positive-edge-triggered flag clear and, in the case of the B port, as an active-LOW parity output enable; a positive-edge-triggered clock input; an active-LOW clock enable; and an active-HIGH flag output. In addition, there is a parity output that generates a parity bit when data are transferred from A to B and an error output that checks the parity of data transferred from B to A. (Fairchild)

I_{OH} (max.)	-3	mA	t_W CP	(min.)	7	ns
I_{OL} (max.)	64*	mA	t_{SU} A or B to CP	(min.)	8.5	ns
I_{IH} (max.)	0.07†	mA	t_H	(min.)	0	ns
I_{IL} (max.)	-0.65†	mA	f_{MAX}	(min.)	n/a	
I_{CC} (max.)	165	mA	t_{pd} CP to A or B	(max.)	10.5	ns
			t_{en} $\overline{OE}$ to A or B	(max.)	10.5	ns
			t_{dis} $\overline{OE}$ to A or B	(max.)	9.5	ns

* B outputs; A outputs have I_{OL} (max.) of 24 mA

† Bus inputs; see manufacturer's data for other inputs

8-LINE BUS TRANSCEIVERS WITH EDGE-TRIGGERED REGISTERS AND MULTIPLEXED REAL-TIME OR STORED-DATA TRANSMISSION

A choice between registered and unregistered bus interface modes is offered by this widely supported family of transceivers, which can operate as straight-through bus transceivers or as edge-triggered registered transceivers. Depending on the condition of various control inputs, these devices can transfer real-time data from one bus to the other, store data from either bus into the register of the opposing bus, or read out data from either register into its own bus.

The eight types listed below are all supplied in 24-pin packages having eight I/O lines for each of two buses, A and B; two positive-edge-triggered clock inputs, CAB and CBA; two inputs that select between real-time and stored-data transmission, SAB and SBA; and two output controls. The functions assigned to this last pair of inputs determine which of two broad categories a given type falls into, as in the case of the ordinary 4-line and 8-line bus transceivers described earlier.

In one configuration, the two output controls are dual active-LOW output enables, $\overline{EAB}$ and $\overline{EBA}$, which can be activated either individually or together, providing maximum flexibility in the timing and control of transceiver operations. Depending on the state of its various control inputs, a device in this category can perform one of five basic kinds of bus management functions:

1. Isolation of bus A from bus B
2. Real-time transfer of data from bus A to bus B
3. Real-time transfer of data from bus B to bus A
4. Storage of data from bus A into register B, or from bus B into register A, or both of these operations at the same time
5. Readout of data from register A to bus A, or from register B to bus B, or both of these operations at the same time

When in the real-time transfer mode, devices having this dual-enable configuration can latch data on both buses without using the internal registers when both enables are activated simultaneously. If all other data sources to the two sets of bus lines are in the high-impedance state, each set of bus lines will remain at its last state as long as the transceiver controls are maintained in this condition.

In the other basic configuration, the two output controls are a direction control, DIR, that determines the direction of data transmission (A to B or B to A), and an overall output enable, $\overline{OE}$, that enables or disables the outputs to both buses. This arrangement provides maximum simplicity of transceiver control at the cost of a certain amount of operational flexibility. Devices in this category can perform the following bus management functions:

1. Isolation of bus A from bus B
2. Real-time transfer of data from bus A to bus B
3. Real-time transfer of data from bus B to bus A
4. Storage of data from bus A into register B, or from bus B into register A, but NOT both of these operations at the same time
5. Readout of data from register A to bus A, or from register B to bus B, but NOT both of these operations at the same time

Also, devices in this category cannot enable both sets of outputs at the same time and therefore cannot function in the "local bus latch" mode.

Within each of these two major categories is the usual division into inverting and noninverting types. Each of these subdivisions is further divided into types with 3-state outputs to each bus and types in which one or both sets of outputs are open-collector. The array of options available to the designer is summed up in Table 8.2.

TABLE 8.2 8-LINE REAL-TIME/STORED DATA BUS TRANSCEIVERS

Output Controls	Polarity	A Outputs	B Outputs	Type
EAB, EBA	Inverting	3-state	3-state	651
		open-collector	3-state	653
	Non inverting	3-state	3-state	652
		open-collector	3-state	654
DIR, OE	Inverting	3-state	3-state	648
		open-collector	open-collector	649
	Non inverting	3-state	3-state	646
		open-collector	open-collector	647

Transceivers with dual complementary output enables and local latch capability:

651 *8-line 3-state inverting registered bus transceivers.* **Dual-enable real-time/stored-data transceivers available in a variety of implementations.**

653 *8-line 3-state/open-collector inverting registered bus transceivers.* **LS, ALS, and F versions of the 651 with open-collector A outputs.**

652 *8-line 3-state registered bus transceivers.* **Same as the 651, but noninverting.**

654 *8-line 3-state/open-collector registered bus transceivers.* **Same as the 653, but noninverting.**

Transceivers with single output-enable and direction-control inputs:

648 *8-line 3-state inverting registered bus transceivers.* **The basic inverting real-time/stored-data transceiver, available from a number of manufacturers in a wide range of implementations. The LS ver-**

sions feature a moderate amount of Schmitt-trigger hysteresis (min. 0.2 V, typ. 0.4 V) at the bus inputs for improved noise rejection.

649   *8-line inverting registered bus transceivers with open-collector outputs.* Open-collector LS, ALS, and F versions of the 648.

646   *8-line 3-state registered bus transceivers.* Same as the 648, but noninverting.

647   *8-line registered bus transceivers with open-collector outputs.* Same as the 649, but noninverting.

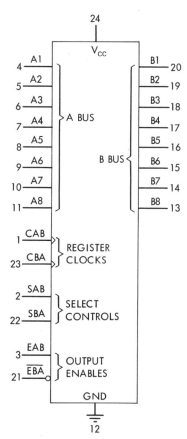

Inverting:
    3-state A and B outputs: 651
    Open-collector A outputs and 3-state B outputs: 653
Noninverting:
    3-state A and B outputs: 652
    Open-collector A outputs and 3-state B outputs: 654

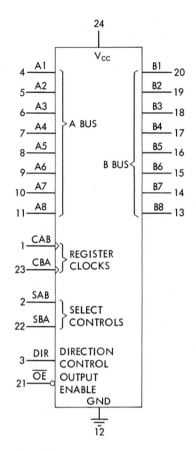

Inverting:
    3-state outputs: 648
     Open-collector outputs: 649
Noninverting:
    3-state outputs: 646
     Open-collector outputs: 647

651   8-line 3-state inverting registered bus transceivers

MANUFACTURERS	TTL	LS	HC	HCT	ALS	S	AS	F	AC	ACT
Monolithic Memories		■								■
National Semiconductor							■			
RCA									D	D
SGS			D	D						
Signetics								▨		
Texas Instruments		■	■	■			■			
Toshiba		■	■	■						

ALS -1 version available from TI

KEY PARAMETERS		TTL	LS	HC	HCT	ALS	S	AS	F	AC	ACT
I_{OH} (max.)	mA		-15	-6	-6	-15		-15	-15^y	[-24]	-6
I_{OL} (max.)	mA		24	6	6	24^v		48	64^z	[24]	12
I_{IH} (max.)	mA		0.02			0.02		0.07^w	0.07^{aa}		
I_{IL} (max.)	mA		-0.4			-0.2		-0.75^x	-0.07^{bb}		
I_{CC} (max.)	mA		165			82		195	58		
I_{CC} (quiesc.)	mA			0.08^f	0.08^f					[0.08]	0.04
C_{pd} (typ.)	pF/channel			50^g	50^o					n/a	n/a
t_W (min.)	ns		30^a	19	25	12.5		6	6	n/a	20
t_{SU} (min.)	ns		15^a	25^h	19^p	10		6	6	n/a	25
t_H (min.)	ns		0	5^i	5	0		0	0	n/a	0
f_{MAX} (min.)	MHz		n/a	22^j	17^q	40		90	90	n/a	n/a
t_{pd} A, B (max.)	ns		30^b	41^k	41^r	18		8	8	[12.7]	48
t_{pd} CAB, CBA (max.)	ns		35^c	54^l	54^s	32		9	9	[13.9]	44
t_{pd} SAB, SBA (max.)	ns		47^d	57^m	57^l	38		11	11	[13.9]	44
t_{pd} EAB, $\overline{EBA}$ (max.)	ns		60^e	74^n	74^u	22		16	16	[13.9]	38

[a] Monolithic Memories 20 ns
[b] Monolithic Memories 17 ns
[c] Monolithic Memories 30 ns
[d] Monolithic Memories 35 ns
[e] Monolithic Memories 40 ns
[f] TI; Toshiba 0.04 mA
[g] TI; Toshiba 46 pF/channel
[h] TI; Toshiba 13 ns
[i] TI; Toshiba 6 ns
[j] TI; Toshiba data not available
[k] TI; Toshiba 45 ns
[l] TI; Toshiba 60 ns
[m] TI; Toshiba 55 ns
[n] TI; Toshiba 45 ns
[o] TI; Toshiba 52 ns
[p] TI; Toshiba 13 ns
[q] TI; Toshiba data not available

[r] TI; Toshiba 39 ns
[s] TI; Toshiba 59 ns
[t] TI; Toshiba 60 ns
[u] TI; Toshiba 48 ns
[v] Version with -1 suffix has I_{OL} (max.) = 48 mA if V_{CC} is kept between 4.75 V and 5.25 V
[w] A and B inputs (including off-state output current); control inputs 0.02 mA
[x] A and B inputs (including off-state output current); control inputs -0.5 mA
[y] B outputs; A outputs -3 mA
[z] B outputs; A outputs 24 mA
[aa] A and B inputs (including off-state output current); control inputs 0.02 mA
[bb] A and B inputs (including off-state output current); control inputs -0.02 mA

MANUFACTURERS	TTL	LS	HC	HCT	ALS	S	AS	F	AC	ACT
Monolithic Memories		■								
RCA									D	D
Signetics								▓▓		
Texas Instruments		■			■					

ALS -1 version available from TI

KEY PARAMETERS		TTL	LS	HC	HCT	ALS	S	AS	F	AC	ACT
I_{OH} (max.)	mA		-15^a			-15^a			-15^l	[-24]	[-24]
I_{OL} (max.)	mA		24			24^g			64^m	[24]	[24]
I_{IH} (max.)	mA		0.02			0.02			0.07^n		
I_{IL} (max.)	mA		-0.4			-0.2			-0.07^o		
I_{CC} (max.)	mA		165			88			100		
I_{CC} (quiesc.)	mA									[0.08]	[0.08]
C_{pd} (typ.)	pF/channel									n/a	n/a
t_W (min.)	ns		30^b			14.5			6	n/a	n/a
t_{SU} (min.)	ns		15^b			10			6	n/a	n/a
t_H (min.)	ns		0			0			0	n/a	n/a
f_{MAX} (min.)	MHz		n/a			35			90	n/a	n/a
t_{pd} A, B (max.)	ns		32^c			56^h			9	[15.7]	[15.7]
t_{pd} CAB, CBA (max.)	ns		39^d			64^i			10	[16.9]	[16.9]
t_{pd} SAB, SBA (max.)	ns		57^e			62^j			12	[16.9]	[16.9]
t_{pd} EAB, $\overline{EBA}$ (max.)	ns		55^f			30^k			16	[13.9]	[13.9]

[a] B outputs (3-state); A outputs (OC) have I_{OH} (max.) of 0.1 mA

[b] Monolithic Memories 20 ns

[c] Monolithic Memories 25 ns

[d] Monolithic Memories 30 ns

[e] Monolithic Memories 45 ns

[f] Monolithic Memories 35 ns

[g] Version with -1 suffix has I_{OL} (max.) = 48 mA if V_{CC} is kept between 4.75 V and 5.25 V

[h] B inputs to A outputs (OC); A to B (3-state) 18 ns

[i] To A outputs (OC); 30 ns to B outputs (3-state)

[j] To A outputs (OC); 35 ns to B outputs (3-state)

[k] To A outputs (OC); 22 ns to B outputs (3-state)

[l] B outputs (3-state); A outputs (OC) have I_{OH} (max.) = 0.25 mA

[m] B outputs; A outputs 24 mA

[n] B inputs (including off-state output current); A inputs and control inputs 0.02 mA

[o] B inputs (including off-state output current); A inputs and control inputs -0.02 mA

652 8-line 3-state registered bus transceivers

MANUFACTURERS	TTL	LS	HC	HCT	ALS	S	AS	F	AC	ACT
Monolithic Memories		■								■
National Semiconductor							■			
RCA									D	D
SGS			D	D						
Signetics								▨		
Texas Instruments		■	■	■			■			
Toshiba			■	■						

ALS -1 version available from TI

KEY PARAMETERS		TTL	LS	HC	HCT	ALS	S	AS	F	AC	ACT
I_{OH} (max.)	mA		-15	-6	-6	-15		-15	-15^y	[-24]	-6
I_{OL} (max.)	mA		24	6	6	24^v		48	64^z	[24]	12
I_{IH} (max.)	mA		0.02			0.02		0.07^w	0.07^{aa}		
I_{IL} (max.)	mA		-0.4			-0.2		-0.75^x	-0.07^{bb}		
I_{CC} (max.)	mA		180^a			88		211	68		
I_{CC} (quiesc.)	mA			0.08^g	0.08^g					[0.08]	0.04
C_{pd} (typ.)	pF/channel			50^h	50^p					n/a	n/a
t_W (min.)	ns		30^a	19	25	12.5		6	6	n/a	25
t_{SU} (min.)	ns		15^b	25^i	19^q	10		6	6	n/a	20
t_H (min.)	ns		0	5^j	5	0		0	0	n/a	0
f_{MAX} (min.)	MHz		n/a	27^k	20^k	40		90	90	n/a	n/a
t_{pd} A, B (max.)	ns		20^c	34^l	34^r	18		9	8	[12.7]	48
t_{pd} CAB, CBA (max.)	ns		36^d	45^m	45^s	30		9	9	[13.9]	44
t_{pd} SAB, SBA (max.)	ns		50^e	48^n	48^t	35		11	11	[13.9]	44
t_{pd} EAB, $\overline{EBA}$ (max.)	ns		54^f	61^o	61^u	22		16	16	[13.9]	38

[a] Monolithic Memories 165 mA
[b] Monolithic Memories 20 ns
[c] Monolithic Memories 22 ns
[d] Monolithic Memories 30 ns
[e] Monolithic Memories 35 ns
[f] Monolithic Memories 40 ns
[g] Toshiba 0.04 mA
[h] TI; Toshiba 46 pF/channel
[i] TI; Toshiba 13 ns
[j] TI; Toshiba 6 ns
[k] TI; Toshiba data not available
[l] TI; Toshiba 45 ns
[m] TI; Toshiba 60 ns
[n] TI; Toshiba 55 ns
[o] TI; Toshiba 45 ns
[p] TI; Toshiba 52 pF/channel
[q] TI; Toshiba 13 ns

[r] TI; Toshiba 39 ns
[s] TI; Toshiba 59 ns
[t] TI; Toshiba 60 ns
[u] TI; Toshiba 48 ns
[v] Version with -1 suffix has I_{OL} (max.) = 48 mA if V_{CC} is kept between 4.75 V and 5.25 V
[w] A and B inputs (including off-state output current); control inputs 0.02 mA
[x] A and B inputs (including off-state output current); control inputs -0.5 mA
[y] B outputs; A outputs -3 mA
[z] B outputs; A outputs 24 mA
[aa] A and B inputs (including off-state output current); control inputs 0.02 mA
[bb] A and B inputs (including off-state output current); control inputs -0.02 mA

385

MANUFACTURERS	TTL	LS	HC	HCT	ALS	S	AS	F	AC	ACT
Monolithic Memories		■								
RCA									D	D
Signetics								▨		
Texas Instruments		■			■					

ALS -1 version available from TI

KEY PARAMETERS		TTL	LS	HC	HCT	ALS	S	AS	F	AC	ACT
I_{OH} (max.)	mA		-15[a]			-15[a]			-15[m]		
I_{OL} (max.)	mA		24			24[h]			64n		
I_{IH} (max.)	mA		0.02			0.02			0.07[o]		
I_{IL} (max.)	mA		-0.4			-0.2			-0.07[p]		
I_{CC} (max.)	mA		180[b]			88			100		
I_{CC} (quiesc.)	mA									[0.08]	[0.08]
C_{pd} (typ.)	pF/channel									n/a	n/a
t_W (min.)	ns		30[c]			14.5			6	n/a	n/a
t_{SU} (min.)	ns		15[c]			10			6	n/a	n/a
t_H (min.)	ns		0			0			0	n/a	n/a
f_{MAX} (min.)	MHz		n/a			35			90	n/a	n/a
t_{pd} A, B (max.)	ns		30[d]			56[i]			9	[15.7]	[15.7]
t_{pd} CAB, CBA (max.)	ns		39[e]			64[j]			10	[16.9]	[16.9]
t_{pd} SAB, SBA (max.)	ns		54[f]			62[k]			12	[16.9]	[16.9]
t_{pd} EAB, $\overline{EBA}$ (max.)	ns		53[g]			30[l]			16	[13.9]	[13.9]

[a] B outputs (3-state); A outputs (OC) have I_{OH} (max.) of 0.1 mA

[b] Monolithic Memories 165 mA

[c] Monolithic Memories 20 ns

[d] Monolithic Memories 25 ns

[e] Monolithic Memories 30 ns

[f] Monolithic Memories 45 ns

[g] Monolithic Memories 35 ns

[h] Version with -1 suffix has I_{OL} (max.) = 48 mA if V_{CC} is kept between 4.75 V and 5.25 V

[i] B inputs to A outputs (OC); A to B (3-state) 18 ns

[j] To A outputs (OC); 30 ns to B outputs (3-state)

[k] To A outputs (OC); 35 ns to B outputs (3-state)

[l] To A outputs (OC); 22 ns to B outputs (3-state)

[m] B outputs (3-state); A outputs (OC) have I_{OH} (max.) = 0.25 mA

[n] B outputs; A outputs 24 mA

[o] B inputs (including off-state output current); A inputs and control inputs 0.02 mA

[p] B inputs (including off-state output current); A inputs and control inputs -0.02 mA

MANUFACTURERS	TTL	LS	HC	HCT	ALS	S	AS	F	AC	ACT
Fairchild								D	■	
Monolithic Memories		■								■
Motorola			▨							
National Semiconductor			■				▨			
RCA			▨	▨					D	D
SGS			D	D						
Signetics			D	D			▨			
Texas Instruments	■	■	■	■	■		■			
Toshiba		■	■	■						

ALS -1 version available from TI

KEY PARAMETERS		TTL	LS	HC	HCT	ALS	S	AS	F	AC	ACT
I_{OH} (max.)	mA		-15	-6	-6	-15		-15	-3	-24	-6
I_{OL} (max.)	mA		24	6	6	24^u		48	24	24	12
I_{IH} (max.)	mA		0.02			0.02		0.07^v	0.07^y		
I_{IL} (max.)	mA		-0.4			-0.2		-0.75^w	-0.07^z		
I_{CC} (max.)	mA		180^a			88		195	100		
I_{CC} (quiesc.)	mA			0.08^c	0.08^c					0.08	0.04
C_{pd} (typ.)	pF/channel			52^d	52^m					65	n/a
t_W (min.)	ns		30^b	20^e	25^n	12.5		6	4^{aa}	3	20
t_{SU} (min.)	ns		15^b	25^f	25^o	10		6	3^{aa}	2	25
t_H (min.)	ns		0	5^g	5	0		0	1^{aa}	1	0
f_{MAX} (min.)	MHz		n/a	21^h	16^p	40		90	n/a	n/a	n/a
t_{pd} A, B (max.)	ns		25	38^i	39^q	17		8	11^{aa}	8	45
t_{pd} CAB, CBA (max.)	ns		40	55^j	58^r	33		9	13^{aa}	12	38
t_{pd} SAB, SBA (max.)	ns		55	55^k	60^s	39		11	13^{aa}	10.5	35
t_{pd} DIR, $\overline{OE}$ (max.)	ns		55	44^l	56^t	27		18^x	13.5^{aa}	11	45

[a] Monolithic Memories 165 mA

[b] Monolithic Memories 20 ns

[c] Toshiba 0.04 mA

[d] RCA; Toshiba 46 pF/channel, TI 50 pF/channel, Motorola 60 pF/channel. National data not available

[e] Motorola and National; TI and Toshiba 19 ns, RCA 23 ns

[f] Toshiba 13 ns, RCA 23 ns

[g] Motorola and National 0 ns, Toshiba 6 ns

[h] Motorola and National; RCA 22 MHz, TI 27 MHz. Toshiba data not available

[i] RCA; TI 34 ns, National 37 ns, Motorola 43 ns, Toshiba 45 ns

[j] TI 45 ns, RCA and Toshiba 60 ns

[k] RCA and Toshiba; Motorola 43 ns, TI 48 ns, National 72 ns

[l] Toshiba 50 ns, TI 61 ns

[m] TI 50 pF/channel

[n] Toshiba; TI 19 ns, RCA 31 ns

[o] TI; Toshiba 13 ns, RCA 31 ns

[p] RCA; TI 27 MHz. Toshiba data not available

[q] Toshiba; TI 34 ns, RCA 46 ns

[r] Toshiba; TI 45 ns, RCA 68 ns

[s] Toshiba; TI 48 ns, RCA 64 ns

[t] RCA; TI 61 ns, Toshiba 48 ns from $\overline{OE}$, 50 ns from DIR

[u] Versions with -1 suffix have I_{OL} (max.) = 48 mA if V_{CC} is kept between 4.75 V and 5.25 V

[v] A and B inputs (including off-state output current); control inputs 0.02 mA

[w] A and B inputs (including off-state output current); control inputs -0.5 mA

[x] National 20 ns

[y] Including 0.05 mA off-state output current

[z] Including -0.05 mA off-state output current

[aa] Preliminary data

MANUFACTURERS	TTL	LS	HC	HCT	ALS	S	AS	F	AC	ACT
Monolithic Memories		■								
RCA									D	D
Signetics								▒		
Texas Instruments		■			■					

ALS -1 version available from TI

KEY PARAMETERS		TTL	LS	HC	HCT	ALS	S	AS	F	AC	ACT
I_{OH} (max.)	mA		0.1			0.1			0.25		
I_{OL} (max.)	mA		24			24^b			24		
I_{IH} (max.)	mA		0.02			0.02			0.02		
I_{IL} (max.)	mA		-0.4			-0.2			-0.02		
I_{CC} (max.)	mA		150			70			100		
I_{CC} (quiesc.)	mA									[0.08]	[0.08]
C_{pd} (typ.)	pF/channel									n/a	n/a
t_W (min.)	ns		30			16.5			4	n/a	n/a
t_{SU} (min.)	ns		15			10			3	n/a	n/a
t_H (min.)	ns		0			0			1	n/a	n/a
f_{MAX} (min.)	MHz		n/a			30			n/a	n/a	n/a
t_{pd} A, B (max.)	ns		30^a			50			11	[15.7]	[15.7]
t_{pd} CAB, CBA (max.)	ns		45			62			13	[16.9]	[16.9]
t_{pd} SAB, SBA (max.)	ns		55			55			13	[16.9]	[16.9]
t_{pd} DIR, $\overline{OE}$ (max.)	ns		50			25			13	[13.9]	[13.9]

^{a} Monolithic Memories 35 ns

^{b} Version with -1 suffix has I_{OL} (max.) = 48 mA if V_{CC} is kept between 4.75 V and 5.25 V

MANUFACTURERS	TTL	LS	HC	HCT	ALS	S	AS	F	AC	ACT
Fairchild								D	■	
Monolithic Memories		■								■
Motorola			■							
National Semiconductor			■				■			
RCA			■	■					D	D
SGS			D	D						
Signetics			D	D				■		
Texas Instruments		■	■	■				■		
Toshiba		■	■	■						

ALS -1 version available from TI

KEY PARAMETERS		TTL	LS	HC	HCT	ALS	S	AS	F	AC	ACT
I_{OH} (max.)	mA		-15	-6	-6	-15		-15	-3	-24	-6
I_{OL} (max.)	mA		24	6	6	24^v		48	24	24	12
I_{IH} (max.)	mA		0.02			0.02		0.07^w	0.07^z		
I_{IL} (max.)	mA		-0.4			-0.2		-0.75^x	-0.07^{aa}		
I_{CC} (max.)	mA		165			88		211	100		
I_{CC} (quiesc.)	mA			0.08^c	0.08^m					0.08	0.04
C_{pd} (typ.)	pF/channel			52^d	52^n					60	n/a
t_W (min.)	ns		30^a	20^e	25^o	12.5		6	4^{bb}	3.5	20
t_{SU} (min.)	ns		15^a	25^f	25^p	10		6	3^{bb}	4.5	25
t_H (min.)	ns		0	5^g	5	0		0	1^{bb}	1	0
f_{MAX} (min.)	MHz		n/a	21^h	16^q	40		90	n/a	n/a	n/a
t_{pd} A, B (max.)	ns		20^b	38^i	39^r	20		9	11^{bb}	13	45
t_{pd} CAB, CBA (max.)	ns		35	55^j	55^s	30		9	13^{bb}	9.5	38
t_{pd} SAB, SBA (max.)	ns		50	55^k	60^t	35		11	13^{bb}	11	35
t_{pd} DIR, $\overline{OE}$ (max.)	ns		65	44^l	56^u	30		18^y	13.5^{bb}	11	45

[a] Monolithic Memories 20 ns
[b] Monolithic Memories 25 ns
[c] Toshiba 0.04 mA
[d] RCA; Toshiba 46 pF/channel, TI 50 pF/channel, Motorola 60 pF/channel. National data not available
[e] Motorola and National; TI and Toshiba 19 ns, RCA 23 ns
[f] Toshiba 13 ns, RCA 23 ns
[g] Motorola and National 0 ns, Toshiba 6 ns
[h] Motorola and National; RCA 22 MHz, TI 27 MHz. Toshiba data not available
[i] RCA; TI 34 ns, National 37 ns, Motorola 43 ns, Toshiba 45 ns
[j] TI 45 ns, Toshiba 60 ns
[k] RCA and Toshiba; Motorola 43 ns, TI 48 ns, National 72 ns
[l] Toshiba 50 ns, TI 61 ns
[m] Toshiba 0.04 mA
[n] RCA; TI 50 pF/channel, Toshiba 55 pF/channel

[o] Toshiba; TI 19 ns, RCA 31 ns
[p] TI; Toshiba 13 ns, RCA 31 ns
[q] RCA; TI 27 MHz. Toshiba data not available
[r] Toshiba; TI 34 ns, RCA 46 ns
[s] RCA; TI 45 ns, Toshiba 58 ns
[t] Toshiba; TI 48 ns, RCA 64 ns
[u] RCA; TI 61 ns, Toshiba 48 ns from $\overline{OE}$, 50 ns from DIR
[v] Versions with -1 suffix have I_{OL} (max.) = 48 mA if V_{CC} is kept between 4.75 V and 5.25 V
[w] A and B inputs (including off-state output current); control inputs 0.02 mA
[x] A and B inputs (including off-state output current); control inputs -0.5 mA
[y] National 20 ns
[z] Including 0.05 mA off-state output current
[aa] Including -0.05 mA off-state output current
[bb] Preliminary data

647   8-line registered bus transceivers with open-collector outputs

MANUFACTURERS	TTL	LS	HC	HCT	ALS	S	AS	F	AC	ACT
Monolithic Memories		■								
RCA									D	D
Signetics								▨		
Texas Instruments		■			■					

ALS -1 version available from TI

KEY PARAMETERS		TTL	LS	HC	HCT	ALS	S	AS	F	AC	ACT
I_{OH} (max.)	mA		0.1			0.1			0.25	[-24]	[-24]
I_{OL} (max.)	mA		24			24^b			24	[24]	[24]
I_{IH} (max.)	mA		0.02			0.02			0.02		
I_{IL} (max.)	mA		-0.4			-0.2			-0.02		
I_{CC} (max.)	mA		150			65			100		
I_{CC} (quiesc.)	mA									[0.08]	[0.08]
C_{pd} (typ.)	pF/channel									n/a	n/a
t_W (min.)	ns		30			16.5			4	n/a	n/a
t_{SU} (min.)	ns		15			10			3	n/a	n/a
t_H (min.)	ns		0			0			1	n/a	n/a
f_{MAX} (min.)	MHz		n/a			30			n/a	n/a	n/a
t_{pd} A, B (max.)	ns		27^a			54			11	[15.7]	[15.7]
t_{pd} CAB, CBA (max.)	ns		45			58			13	[16.9]	[16.9]
t_{pd} SAB, SBA (max.)	ns		60			60			13	[16.9]	[16.9]
t_{pd} DIR, $\overline{OE}$ (max.)	ns		50			31			13	[13.9]	[13.9]

[a] Monolithic Memories 32 ns

[b] Version with -1 suffix has I_{OL} (max.) = 48 mA if V_{CC} is kept between 4.75 V and 5.25 V

8-LINE UNIVERSAL TRANSCEIVER PORT CONTROLLERS

The ultimate in bus management is provided by a small group of single-sourced 24-pin AS devices that combine a bus transceiver with an internal edge-triggered parallel/serial-in, parallel/serial-out shift-right register. These components begin to approach microprocessors in complexity, and their detailed workings are far beyond the scope of this book. The following brief descriptions are paraphrased from manufacturer's data (*TI ALS/AS Logic Data Book,* 1986); see published materials for complete descriptions and application notes. (Texas Instruments)

852§   *8-line universal transceiver port controllers.* Eight selectable transceiver/port functions: A to B; B to A; register to A or B; shifted to A from B; shifted to B from A; off-line shifts (A and B ports transceiving or in high-impedance state). Serial register provides parallel storage of either A or B input data and serial transmission of data from either A or B port. Particularly suitable for use in diagnostics circuitry.

I_{OH} (max.)	-15^*	mA	t_W CP	(min.)	10	ns
I_{OL} (max.)	48^*	mA	t_{SU}	(min.)	5.5	ns
I_{IH} (max.)	$0.07\dagger$	mA	t_H	(min.)	0	ns
I_{IL} (max.)	$-0.5\dagger$	mA	f_{MAX}	(min.)	50	ns
I_{CC} (max.)	220	mA	t_{pd} CP to A or B (max.)		12.5	ns
			t_{en}	(max.)	10.5	ns
			t_{dis}	(max.)	10.5	ns

* Parallel A and B outputs; serial Q8 output has I_{OH} (max.) of -2 mA and I_{OL} (max.) of 20 mA

† Bus inputs; see manufacturer's data for other inputs

856§   *8-line universal transceiver port controllers.* Eight selectable transceiver/port functions: B to A; register to A; register to B; register to A and B; shifted to A; shifted to B; shifted to A and B; off-line shifts (A and B ports in high-impedance state). Serial register provides parallel storage of either A or B input data, serial transmission of data from either A or B port, and readback mode (B to A). Particularly suitable for use in diagnostics analysis circuitry.

I_{OH} (max.)	-15^*	mA	t_W CP	(min.)	10	ns
I_{OL} (max.)	48^*	mA	t_{SU}	(min.)	5.5	ns
I_{IH} (max.)	$0.07\dagger$	mA	t_H	(min.)	0	ns
I_{IL} (max.)	$-0.5\dagger$	mA	f_{MAX}	(min.)	50	ns
I_{CC} (max.)	200	mA	t_{pd} CP to A or B (max.)		11	ns
			t_{en}	(max.)	10	ns
			t_{dis}	(max.)	9.5	ns

* Parallel A and B outputs; serial Q8 output has I_{OH} (max.) of -2 mA and I_{OL} (max.) of 20 mA

† Bus inputs; see manufacturer's data for other inputs

877§   *8-line universal transceiver port controllers with clear.* Eight selectable transceiver/port functions: A to B; B to A; register to A; register to B; shifted to A; shifted to B; off-line shifts (A and B ports in high-impedance state); register clear. Serial register provides parallel storage of either A or B input data and serial transmission of data from either A or B port. Particularly suitable for use in signature analysis circuitry.

I_{OH} (max.)	-15^*	mA	t_W CP	(min.)	10	ns
I_{OL} (max.)	48^*	mA	t_{SU}	(min.)	5.5	ns
I_{IH} (max.)	$0.07\dagger$	mA	t_H	(min.)	0	ns
I_{IL} (max.)	$-0.5\dagger$	mA	f_{MAX}	(min.)	50	ns
I_{CC} (max.)	220	mA	t_{pd} CP to A or B	(max.)	11.5	ns
			t_{en}	(max.)	9.5	ns
			t_{dis}	(max.)	10.5	ns

* Parallel A and B outputs; serial Q8 output has I_{OH} (max.) of -2 mA and I_{OL} (max.) of 20 mA

† Bus inputs; see manufacturer's data for other inputs